California Vegetation

V. L. Holland and David J. Keil

KENDALL/HUNT PUBLISHING COMPANY
4050 Westmark Drive Dubuque, Iowa 52002

This volume is dedicated to our wives,
Janie Holland and Kathy Keil.
Without their encouragement and support this
book would not have been possible.

ISBN 0-7872-0733-0

Printed in the United States of America

10 9 8 7 6 5 4 3 2 1

Contents

Acknowledgments

Many people have contributed to our knowledge and interest in California's vegetation over the years. We wish we could acknowledge all of them because all were important in the publication of *California Vegetation*. Most significantly, we wish to thank the hundreds of students who have taken our field botany course over the years. Their enthusiasm and curiosity about California's flora and vegetation was a constant stimulus to us in our profession, and their probing questions made us dig a little deeper for more information. It is really our students who are primarily responsible for us writing this book. Many of these students have gone on to successful careers in biology and are contributing to the continuing information about California's botanical resources.

We must acknowledge Kimberly Shirley, Michelle Crites, and Tonya Willard who helped prepare the various maps in the text. Carmel Ruiz assisted with many of the illustrations. We gratefully acknowledge those individuals who gave us permission to use their photographs in the book. A very special thanks goes to Lynne Dee Oyler who helped with many aspects of the final preparation of the book including editing and assisting with maps and illustrations. More importantly Lynne Dee was a regular source of enthusiasm and encouragement. Dr. Tom Rice, Chair of the Soil Science Department at Cal Poly assisted with information about California soils. Special thanks go to Dr. Michael Barbour of U. C. Davis who provided valuable input to the content of California Vegetation. Many professors and colleagues have encouraged us and taught us about California's plants over the years—far too many to recognize here. However, we would like to acknowledge Drs. Herbert Baker, Robert Ornduff, Lincoln Constance, Howard Latimer, John Weiler and George Morgan among the many. Finally we wish to give a very special thanks to our families and friends who have been a source of love and inspiration even when they wondered if the book would ever be finished.

Preface

California's vegetation is wonderfully diverse and complex. Even many long-time residents of the state are unaware of the great variation of the state's plant life. Many students have seen only a limited portion of the state and its varied vegetation. People who fail to closely examine the vegetation or to travel off the main highways miss much of the fascinating diversity of the state's landscape and plant life.

When we started teaching at Cal Poly, we found that, as at many other universities and colleges, there was not a course in the curriculum that introduced students to the interesting and diverse flora and vegetation of California. After instituting our Field Botany course, we found its popularity was so great that it was difficult to accommodate all of the students. At the same time, we discovered that there was not a suitable text for the class. Although numerous scientific studies had been published that examine aspects of California's vegetation, only a few attempted to cover the vegetation of the state as a whole. At the time that we began our field course there were only two potential texts available: Robert Ornduff's *Introduction to California Plant Life* (1974) and *Terrestrial Vegetation of California*, a multiauthored reference edited by Michael G. Barbour and Jack Major (1977). Ornduff's book is comparatively brief and we considered its coverage of individual topics to be too limited for our course. *Terrestrial Vegetation of California is* an invaluable reference to any professional or serious student of the vegetation of the state. However, its high price and the detail of some of the contributions limit its usefulness as a general textbook. The lack of a suitable text, along with encouragement from our field botany students and colleagues, led us to write *California Vegetation*.

We have found that there is both an interest and a need for a an up-to-date book that discusses the vegetation of California at a level that is useful to both professionals and amateurs. Early versions of *California Vegetation* were written for local use and published through our campus bookstore. We were somewhat surprised when we began to receive inquiries about the book from instructors at other colleges and universities. Colleagues at several other institutions adopted the text for use in their classes. Copies were also purchased by botanical consultants and other professionals. Various individuals urged us to prepare the book for wider use. The potential audience for California Vegetation includes biologists, foresters, horticulturists, landscape architects, environmental planners, range managers, soil scientists and agriculturists as well as those in the general public interested in knowing more about the state's botanical resources. Thus, we have aimed the book for an intermediate level in hopes it could meet this need. We have attempted to make the book readable to general audiences. It is more comprehensive in coverage of topics than *Introduction to California Plant Life* and less detailed than *Terrestrial Vegetation of California*.

Since we prepared the early versions of *California Vegetation* several other books have been published that also consider the diversity of the state's plant life. *California's Changing Landscapes* (1993) by Michael Barbour, Bruce Pavlik, Frank Drysdale and Susan Lindstrom, discusses California's landscapes and plant resources in relationship to how they have changed over the years, what has replaced them, and what the future holds for them. In *An Island Called California* Elna Bakker (1984) describes California's natural communities and provides an excellent basic discussion of California's major biotic realms for general audiences. Allan Schoenherr's *A Natural History of California* (1992) is a more comprehensive book that discusses basic ecology, climate, rocks, soils, plants, and animals in each region of the state. Our text differs from each of these books in some aspects of vegetational classification and in various other features.

California Vegetation should also be useful as a complementary text with the new CNPS classification of plant communities. We have cross-referenced our book with the new CNPS descriptions and have adopted the plant nomenclature used in *The Jepson Manual*, the new flora of California. We hope this will make *California Vegetation* useful to those conducting botanical studies and assessments as well as to those involved in managing wildlands in the state. *California Vegetation* will be a required textbook in classes at Cal Poly that study the state's vegetation, and we will continue to solicit input and suggestions from students and colleagues in making it better. For those colleges without a course that deals with California's vegetation, we encourage you to consider adding such a course. It has become a very popular and important course for students at Cal Poly who plan to work with California's vegetation and wildlands.

Numerous sources of information were used in preparing the text. These are included among the bibliographic entries. Those wishing to learn additional information about particular topics discussed in the text are encouraged to consult the references for individual chapters. These bibliographies are selective in coverage and are not intended to be exhaustive searches of the literature. Many of the references (through their literature cited sections) can lead in turn to other studies. In most cases we have not cited specific sources of information in the text, but we acknowledge here the many authorities upon whose studies we have relied. Responsibility for any errors must be ours alone.

V. L. Holland

David Keil

Chapter 1

Introduction

California has the most remarkable and diverse assemblage of native plants and natural habitats in all of temperate North America. Few, if any, temperate regions in the world approach California's biodiversity and variety of natural settings. California has a native flora rich in species, many of which occur nowhere else. The state's vegetation ranges from the tall coniferous forests of the moist northwest coast to the arid scrublands of the southeastern deserts. It is a state with many botanical treasures, from the giant redwoods of the coast and Sierra to the ancient bristlecone pines of the White Mountains.

The remarkable biological diversity found in California is due in part to the location and size of the state (Fig. 1). California is roughly 1287 km (800 miles) long and 322 km (200 miles) wide and has an area of almost 411,000 square km (158,297 square miles) making it the third largest state in the Union. Size alone, however, does not account for the tremendous diversity found within the state. If that were the case, one would expect Alaska to be equally or more diverse, which it is not. Besides being large, California is a land of remarkable habitat diversity. It has a latitudinal range from about 33° to 42° N, greater than that of any other state except Alaska. Its range of topographic, climatic and geologic features is unmatched elsewhere in the temperate regions. With its elevational range from 85 m (276 ft) below sea level in Death Valley to 4,420 m (14,495 ft) on top of Mt. Whitney, California has a most extraordinary range of plant environments. In fact, these two areas, which are only a short distance apart, are the lowest and the highest elevations in the continental United States. Only in Alaska can one find taller mountains. The more than 1600 km (1000 miles) of the California coastline provides a wide range of marine and coastal habitats.

Flora and Vegetation

California's diversity of plant life can be examined both in terms of its flora and its vegetation. The **flora** of an area consists

Fig. 1-1. Map illustrating the size and topographic variability of California. The latitudinal extent of California and its topographic diversity contribute greatly to the remarkable variety of plant communities that occur in the state.

of all the different kinds of plants that occur in a particular location and can be described by listing all the kinds of plants (species, genera, families, etc.) that grow there. There are numerous books and scientific papers that deal with the flora of parts or all of California (e.g., *Marin Flora* by Howell; *Vascular Plants of San Luis Obispo County* by Hoover; *A Flora of the Santa Barbara Region* by Smith; *Flora of Kern County* by Twissleman) and others that deal with the flora of the whole state (e.g., *The Jepson Manual* by Hickman et al., and *A California Flora* by Munz).

The flora of California comprises both vascular and non-vascular plants. Vascular plants are those with specialized tissues (vascular tissue) called xylem and phloem that function to transport water, minerals, organic compounds and other substances throughout the plant body. Vascular plants include the club mosses and horsetails (sometimes referred to as fern allies), ferns, conifers (gymnosperms) and flowering plants (angiosperms). These are the plants, especially flowering plants, that are the dominant and most conspicuous portion of the flora and vegetation of California. Non-vascular plants (algae, liverworts, mosses) and fungi are also common and important components of California's varied flora. However, because the primary focus of this book is vascular plants and most non-vascular plants and fungi are comparatively small and inconspicuous components of their communities, they receive only incidental mention in the discussions that follow.

Vegetation refers to the life-form or general aspect (physiognomy) and species composition of the plant life in a particular site or region. The term **plant community** is a general term for an assemblage of plant species growing together in a particular area (habitat or environment). Within a community some species are more common than others, some kinds of plants are tall, some are short, some are woody and others are herbaceous, etc. Plants grow together in various combinations.. The vegetation of an area in the Sierran foothills might consist of foothill woodland, chaparral, or grassland communities. Each of these vegetation types has its own species composition and physical structure. California has many different vegetation types including various kinds of forests and woodlands (where trees are the dominant life-form), shrublands (where shrubs are dominant), grasslands (where grasses and other herbaceous plants are dominant), and marshlands (where aquatic and semi-aquatic herbaceous plants dominate).

California's floristic and vegetational diversity have developed in response to the interaction of plants with the state's many different environments. For example, California's vegetation types or plant communities include densely vegetated

Table 1-1. Percent cover of major vegetation types in California (based on maps by Küchler, 1964).

Vegetation type	Percent of California's land area
Forests	27
Woodlands(excluding deserts)	10
Shrublands (excluding deserts)	12
Grasslands	13
Deserts Shrublands	30
Desert Woodlands	4
Marshlands	2
Others (Alpine, Sand Dunes, etc.)	2

temperate rain forests in the wet, mild coastal regions of northwestern part of the state (where annual average precipitation is over 240 cm (100 in); rock-garden-like alpine fellfields above timberline on high mountains where winter temperatures are very cold, precipitation is below 76 cm (30 in) annually, and the growing season is just a few weeks; and sparsely vegetated deserts in the hot, dry interior regions of southeastern California where annual average precipitation is less than 10 cm (4 in) and elevations are near or below sea level. Among these extremes, California has many intermediate environments that support many other vegetation types including a diversity of forests, woodlands, chaparral, coastal and desert shrublands, grasslands, and marshes.

An appreciation for the overall composition of California's vegetation, can be gained by examining the approximate land area covered or capable of being covered by the various types of vegetation (Table 1-1). For example, California's forests, mostly dominated by tall conifers, occupy most of the mountainous and northern coastal regions of the state. These forests cover over one quarter of California's land area. Woodlands of lower stature, mostly dominated by various species of oaks, dominate the drier foothill regions of the state, and cover about 10% of the state. In foothill and coastal regions, where topographic and soil features result in environmental conditions too dry to support trees, various types of shrublands (coastal scrub and chaparral) dominate the landscape, especially in southern California. These shrublands cover about 12% of the state. Grasslands are the most common vegetation type of the interior valleys and some of the drier coastal valleys; in addition, they occur in some of the drier foothill regions. They cover about 13% of the state. Desert communities, most of which occur east of the major interior mountains, comprise over one-third (34%) of California's vegetation. Freshwater and saltwater marshlands, specialized habitats that occur where the presence of standing water is the

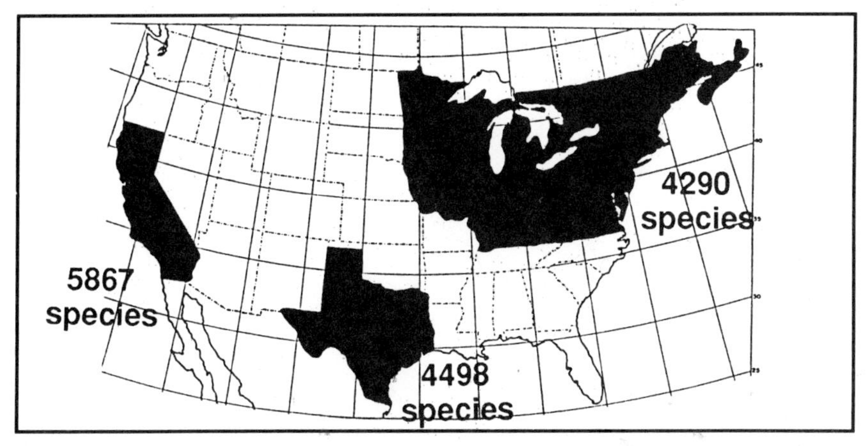

Fig. 1-2. Comparison of California's area and flora with those of Texas and the northeastern United States and adjacent Canada.

controlling factor, are found in locations scattered across the state. Together they cover about 2% of the state. There are also several other vegetation types that are ecologically significant but cover only a small portion of the state.

Floral and vegetation diversity are directly related to the variety of habitats available for plant growth and the harshness of the environment making up these plant habitats. The most diverse floras in the world are found in tropical regions where environments are most favorable for plant growth. The least diverse occur in severely cold environments in arctic and alpine areas and the extremely hot and dry desert regions of the world. The temperate regions with intermediate environments, typically fall somewhere in between. Within temperate regions, those areas that offer the greatest diversity of habitats also have the greatest floristic and vegetation diversity.

California's Diverse Flora

California, because of its remarkable variety of habitats, has the most diverse flora and vegetation in all of temperate North America (Fig. 1-2). According to *The Jepson Manual*, the known vascular plant flora of California consists of 5867 species. **Native plants** comprise about 82.5% of the flora (4844 species) and alien species comprise about 17.5% (1023 species). Native plants are those that existed in California before visitation or colonization of the state by Europeans. **Alien or introduced plants** have been brought into the state in the past few centuries by humans. **Naturalized plants** are alien species that have become established and are spontaneously reproducing in California. Some are weeds on disturbed sites; others have become integral parts of

Table 1-2. Comparison of native floras in various areas of the world.

Region	Area (square kilometers)	Area (square miles)	Number of species	Percent difference from California
California	411,000	159,000	4844	100
Northwest Territories (Canada)	3,380,000	1,310,078	1055	21
Great Plains	1,600,000	620,155	2496	52
Quebec	1,540,687	597,166	1803	37
Alaska	1,479,000	571,000	1366	28
British Columbia	948,600	367,674	ca. 1900	39
Texas	692,397	268,270	4498	93
Greenland	326,000	126,357	427	9
Sonoran Desert	310,000	120,155	2441	50
North and South Carolina	263,709	102,212	2890	60
New England	172,676	66,929	1995	41
Cuba	114,525	44,390	5790	120
Guatemala	109,000	42,248	7817	162
Hawaii	16,764	6,498	956	20

various plant communities and are now an obvious part of the California flora. Some of these aliens were accidentally introduced by humans whereas others are garden or landscape plants that escaped and became a naturalized part of the flora. **Crop plants** and **ornamentals** that have not spread into the natural vegetation of California are not included in the Jepson Manual. The total number of these non-naturalized species in the state is unknown. The significance of alien species to the California flora and vegetation are discussed in Chapter 22.

Comparisons with other geographical areas (Table 1-2) illustrate California's rich native flora. For example, floristic diversity in California is significantly higher than that of Alaska and other northern lands (even though Alaska has over four times the land area of California). The biggest contrast is between California and the very harsh, cold environments of Greenland;. Although this island's land area is about 80 percent of that of California, its has less than 10 percent as many species. California also has a higher number of species than that of the hot, arid Sonoran Desert. The flora of California is also more

Table 1-3. Composition of California's flora by major taxonomic groups.

Group	Number of Species	Percent of Flora
Dicots	4646	79.3%
Monocots	1058	18.0%
Gymnosperms	60	1.0%
Ferns and Allies	103	1.7%
Total	5867	100%

diverse than temperate regions in the east such as the Great Plains, New England states and the Carolinas. Even Texas, though significantly larger in area, has fewer species mostly because of the greater diversity of habitats found in California. Only tropical areas of the world (e.g., Cuba, Guatemala) show a greater floristic diversity within a similar or smaller area than California.

The majority of plant species in California are flowering plants, especially dicots (Table 1-3). However, all taxonomic groups are significant components of California's vegetation. For example, gymnosperms (coniferous trees) comprise only about 1% of the plant species found in the state but are the dominant

Table 1-4. Eight largest families of vascular plants in California

Family	# of Species	% of California Flora
Asteraceae (Compositae) The sunflower family)	748	12.8%
Poaceae (Gramineae) (Grass family)	428	7.3%
Fabaceae (Leguminosae) (Pea or legume family)	385	6.6%
Scrophulariaceae (Figwort family))	291	5.0%
Brassicaceae (Cruciferae,) (Mustard family)	260	4.4%
Polygonaceae (Buckwheat family)	219	3.7%
Liliaceae (Lily family)	218	3.7%
Cyperaceae (Sedge family)	207	3.5%
TOTAL	**2756**	**47.0%**

Table 1-5. Seven largest genera of vascular plants in California.

Genus	Family	Common Name	Species
Carex	Cyperaceae	sedges	133
Eriogonum	Polygonaceae	buckwheats	112
Astragalus	Fabaceae	locoweeds	96
Phacelia	Hydrophyllaceae	phacelias	94
Lupinus	Fabaceae	lupines	71
Mimulus	Scrophulariaceae	monkeyflowers	63
Penstemon	Scrophulariaceae	penstemons	53
TOTAL			**622**

plants in the montane forests of California. Monocots (grasses) are the dominant plants in California's grasslands and in the understory vegetation of oak woodlands in the state.

Although California's flora consists of species in many families and genera, it is noteworthy that 47% of the species are from eight families and 10.6% are from just seven genera (Tables 1-4 and 1-5).

Floristic Provinces

California is divided into a western and eastern portion by an extensive chain of tall mountains that extends from the Oregon border to Mexico. This mountain chain is composed of the Cascade, Sierra Nevada, Transverse and Peninsular Mountain Ranges (listed in order from north to south). The portion of California west of the crests of these mountains comprises about 75% of the state and is referred to as cismontane California. The areas east of the high mountain crests comprise transmontane California (Fig. 1-3). Cismontane California has an extremely variable range of environmental conditions from high mountains to interior valleys and mild coastal habitats. Transmontane California is largely desert except for some mountain peaks that are high enough to support montane forests. The crests of these mountain chains divide the state so that drainages originating in cismontane California flow to the Pacific coast while those originating in the transmontane portion flow into desert regions.

The separation of California into distinct western and eastern geographical regions by a high mountain barrier contributes significantly to the diversity in the plant life of the state. This variation is sometimes divided and discussed in terms of floristic provinces: geographic areas that have plant communities com-

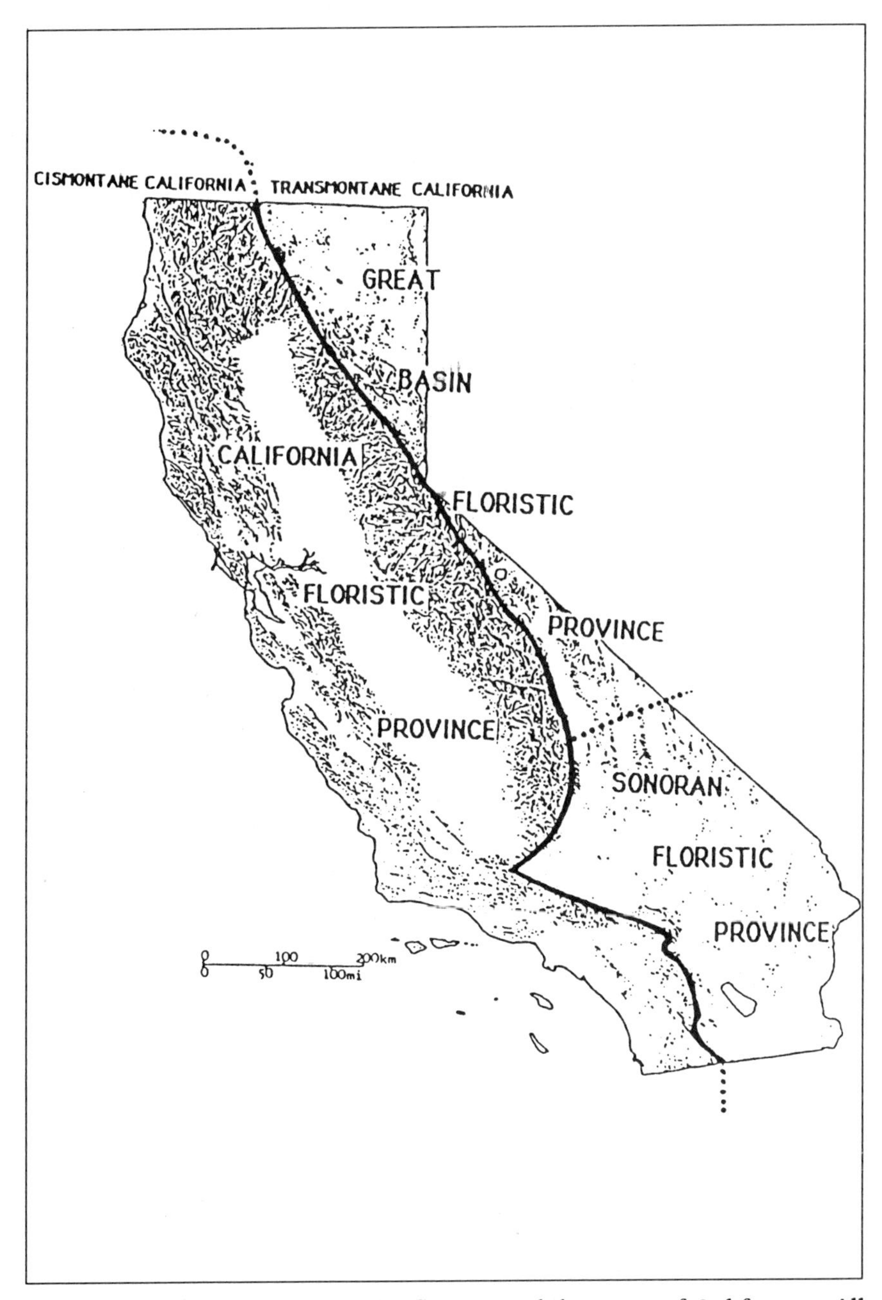

Fig. 1-3. Map illustrating the floristic subdivisions of California. All three floristic provinces extend outside the political boundaries of the state. The California portion of the California Floristic Province is located within the Cismontane portion of the state. The California portion of the Great Basin and Sonoran Floristic Provinces are in Transmontane California.

posed of plant species that are characteristic of and best developed in the particular areas or regions. Three floristic provinces, which are characterized on the basis of climate, flora and history, are included within California's boundaries: the California Floristic Province, the Great Basin Floristic Province and the Sonoran Floristic Province (Fig. 1-3). **The California Floristic Province** consists of that part of the state west of the Cascade-Sierra Nevada and southern California mountain axes and includes all of cismontane California and the Channel Islands (about 75% of the state). Additionally, on the basis of common flora and climate a small portion of southwestern Oregon and the northwestern corner of Baja California are included in the California Floristic Province. The **Great Basin** and **Sonoran Floristic Provinces** include the desert regions of transmontane California east of the mountain crests and in their rain shadows (about 25% of the state). These desert areas are more extensive outside of California and extend over large areas of the western states and Mexico.

The **Great Basin Floristic Province** covers a large portion of western United States from the eastern flanks of the Sierra Nevada and Cascades eastward into Nevada, Utah and portions of adjacent states. In California, this floristic province is limited mostly to the transmontane deserts north of the Owens Lake and Death Valley (including the Modoc Plateau in the northeastern corner of the state). However, it should be noted that there are some small local areas of Great Basin vegetation in cismontane California such as around the base of Mt. Pinos and Mt. Abel in the Transverse Ranges of Kern, Ventura and Santa Barbara Counties. Altogether, the Great Basin Floristic Province only occurs in about 6% of the state. (see Chapters 18 and 19 for more discussion).

The **Sonoran Floristic Province** is located south of the Great Basin Floristic Province and covers a large portion of the southwestern United States and northern Mexico. This floristic province extends from the eastern bases of the southern Sierra Nevada, the Transverse and the Peninsular Ranges eastward and southward into southern Nevada, Arizona and northwestern Mexico. All the hot, dry deserts of southeastern transmontane California south of Owens Lake and Death Valley are included in the Sonoran Floristic Province (about 20% of the state). The Mojave and Colorado Desert regions of California are both found in this floristic province (see Chapter 18 for more discussion).

California's flora includes major representation from all three floristic provinces. The diversity of the vegetation and flora in the California Floristic Province is significantly greater and more impressive than that of the other provinces. For example, the California Floristic Province, which covers about 285,000 km^2 (about three-fourths of the state) contains about 85% of the total genera and 81% of the total vascular plant species found in California. By contrast the Great Basin and Sonoran Floristic Provinces combined cover an area of 129,000 km^2 in California (about one-fourth of the state) but have a proportionally lower species diversity. Approximately 6 families, 102 genera and 938 species occur in the deserts of California but not in the California Floristic Province.

Endemism in the California Flora

A unique and fascinating aspect of California's flora, is the large number of species that are endemic to the state. An **endemic species** is one that is confined to a specific location, region or habitat. Thus, a plant that is endemic to California is one that grows naturally only in California. Of the 4,839 native species that occur in the state, about 29% (1416) are endemic to California (Table 1-6). If subspecies and varieties are included, over 30% of the native flora is endemic. It is also interesting that about 3% of the genera native to California are endemic to the state. The endemism of the California flora is unparalleled in other temperate floras in North America and appears even more impressive when one considers the number of endemic plants found only in the California Floristic Province. Of the approximate 4452 species (native and alien) of vascular plants in the California Floristic Province, about 50% of the species and 6.3% of the genera are endemic. By contrast, the Great Basin and Sonoran Floristic Provinces combined have a much lower number of endemics as well as fewer total species of vascular plants. Only about 85 desert species (less than 2%) are California endemics.

Many of California's endemics are narrow in their geographic range whereas others are broadly distributed over the state. For example, *Cupressus macrocarpa* (Monterey cypress) is an endemic species restricted in its natural distribution to the Monterey peninsula. *Quercus douglasii* (blue oak) and *Pinus sabiniana* (foothill pine; Fig. 1-4B) are also endemics, but both are widespread in the state. *Quercus douglasii* covers about 7.5% of the state and is the dominant tree in the foothills of the Coast Ranges and the Sierra Nevada. Some species are not only endemic to the state but also to specific types of habitats. For example, *Calochortus obispoensis* (San Luis Obispo star-tulip; Fig. 1-4C) and *Cupressus sargentii* (Sargent cypress) are endemic to serpentine soils in the Coast Ranges of California, and *Arctostaphylos morroensis* (morro manzanita) is endemic to stabilized sand dunes around the southern end of Morro Bay.

Fig. 1-4. California natives. **A.** *Castilleja exserta* (owl's clover) is a native species but not endemic. Its range extends to Arizona and northern Mexico. **B.** *Pinus sabiniana* (foothill pine) is a California endemic. It is widespread in the foothill woodlands encircling the Central Valley. **C.** *Calochortus obispoensis* (San Luis Obispo star tulip) occurs only in the hills around San Luis Obispo on soils derived from serpentine rock. This species is considered by the California Native Plant Society to be rare and endangered. Photos A and B by Robert F. Hoover. C. Photo by David Keil

Table 1-6. Native and alien species in the California flora.

Native Species	4844 (82.5%)
Non- endemic species	3428 (70.8%)
Endemic species	1416 (29.2%)
Alien Species	1023 (17.5%)
TOTAL	5862 (100%)

Other plant species are found only on saline soils, in vernal pools, on off-shore islands, etc.

The reason California has so many endemics is not entirely clear. In some cases it is because of the variety of specialized climatic and edaphic (soil) conditions that are found in California. The growth rhythms of many species are adapted to the Mediterranean climate that characterizes the California Floristic Province. Conditions elsewhere in North America are unsuitable for these species. Perhaps the best known of these is *Sequoia sempervirens* (coast redwood) which is found only along the California coast (and a small portion of southwest Oregon) where heavy summer fog drip is prevalent and winters are mild with very little frost. Other plant species are restricted to soils derived from particular parent materials such as serpentine, volcanic deposits, limestone, sandstone, etc., as exemplified by Sargent cypress. Other species occur only on unstabilized sand dunes, in seasonally flooded pools (vernal pools) or such restricted habitats as deposits of seabird guano on islands.

Long term evolutionary changes in the California flora as a result of geological and climatic changes over thousands to millions of years have also played a significant role in the large number of endemics. Many of California's endemic species are evolutionarily old species that were widespread in the geological past but are restricted to much smaller ranges today. These species are referred to as **relict species** or **paleoendemics** and some are considered to be on the road to extinction. The coast redwood and Sierra big tree are examples of species that at one time had wider ranges but now are found only in restricted habitats in California. Fossil records demonstrate that the prehistoric range of *Sequoiadendron* included Idaho and Nevada, and that *Sequoia* occurred across the North American continent and in Europe as well. Other examples are *Pinus radiata* (Monterey pine), *Abies bracteata* (Santa Lucia fir), and *Lyonothamnus floribundus* (Catalina ironwood) all of which were much wider spread in the past.

Fig. 1-5. Introduced plants. Both *Silybum marianum* (milk thistle) and *Lolium multiflorum* (annual ryegrass) are native to the Mediterranean region. They were introduced to California during or after the Spanish Mission period. Both are well-established members of the state's modern flora. Photo by Elizabeth Bergen.

At the opposite end of the spectrum are **new endemics** (**neoendemics**), species in genera such as *Gilia*, *Madia* and *Hemizonia* that probably evolved a million years ago or less. These species may be considered relative newcomers to the flora and may not have yet successfully spread into their potential geographic ranges. Whether they ever will or not is unknown. However, there are some species of *Clarkia* found in restricted localities in the southern Sierra Nevada that have apparently differentiated rather recently. These species appear to be spread-

Table 1-7. Increase in exotic plant species in California over time.

Year	Number of exotic species known in California
1824	16
1848	79
1860	134
1925	292
1959	725
1968	874
1993	1023

ing into new areas and may become more widespread in the future. Between the extremes of the relicts and the newcomers, there are many endemic species of intermediate age.

Introduced Plants

Introduction of exotic plants into the California flora has been of concern to Californians for many years. Many of these plants are weeds that were accidentally introduced and have become well established and sometimes locally dominant components of California's vegetation (Fig. 1-5). Others are garden or ornamental plants that are well adapted to California's climate and have escaped and become part of the naturalized flora of the state. Additional exotic species are introduced deliberately or accidentally into California each year, and some of these become established as part of the state's flora. Chapter 22 discusses this topic in more detail.

The introduction of alien plants into California apparently began in about 1769 when Father Junipero Serra founded the first permanent European settlement in San Diego (Table 1-7). At least 16 exotic species became established in California from 1769 to 1824 during this Spanish colonization period. From 1825 to 1848, during the Mexican occupation, this number grew to 79 (an addition of 63), and between 1849 and 1860, another 55 alien species became established resulting in a total of 134. The exotics continued to increase so that by 1925 (when Jepson's *Manual of Flowering Plants in California* was published), there were an estimated 292 naturalized exotics in California. When the first edition of *A California Flora* by Munz and Keck was published in 1959, the number grew to 725, and when Munz published the supplement to the flora in 1968, there were 874 alien species that were considered naturalized and part of the flora of California. *The Jepson Manual* (1993) includes 1023 alien naturalized species, 17.4% of the California Flora.

Some of these are aggressive, noxious weeds that are spreading and replacing native plant species in nature. Others are now an established part of the natural flora but are not spreading and replacing native vegetation. As humans continue to disturb and modify the California landscape, the establishment of new exotics will continue to increase and be a problem in California. Chapter 22 includes a more complete discussion of introduced plants.

Rare and Endangered Species in California

Another fascinating aspect of the California flora is the large number of plant species that are presumed to be extinct or that are rare and/or endangered (Table 1-8). A plant species is presumed extinct if it has not been found after careful search of known or likely habitats, or if it has not been found for a number of years, and its habitat had been destroyed or modified. A rare species is one that is limited in terms of number of individual plants still present in the wild, and also one that has a limited distribution. Usually rare plants are found in only a few highly restricted populations in the state. This distribution is usually determined by the rarity of the habitat in which the plant is able to grow. While these plants are not presently threatened with extinction, they occur in such small numbers over such a limited range that they could be threatened if their remaining habitat is modified. An endangered species is one that is not only rare, but the remaining populations are considered to be threatened with extinction because its survival and reproduction are jeopardized. The main reason that most such plants in California are extinct or rare and endangered is that humans are gradually destroying their habitats through urbanization, forest destruction, agricultural practices and pollution. In addition, some species are being collected from the wild and used in gardens or are exploited commercially. Attempts are being made to eliminate these practices and to protect the rare and/or endangered species in California.

In the 1994 edition of its *Inventory of Rare and Endangered Vascular Plants of California*, the California Native Plant Society (CNPS) listed 0.5% (34 taxa) of the California Native flora as presumed extinct. Approximately 13.6% (857 taxa) are listed as rare and endangered in California and elsewhere, 4.3% (272 taxa) are listed as rare and endangered in California but more common elsewhere, 8.4% (532 taxa) are listed as rare in terms of having a limited distribution and 0.8% (47 taxa) are listed as needing more information before they can be assigned to a specific category. This means that about 27.7% (1742 taxa) of California's native flora is either extinct or in some form of rarity or endangerment. The number of taxa and percentages

Table 1-8. Plants listed by the California Native Plant Society as rare and endangered (Skinner and Pavlik, 1994).

CNPS List	Taxa	Percent of Flora
1A. Presumed Extinct in California	34	0.5%
1B. R/E in California and elsewhere	857	13.6%
2. R/E in California, more common elsewhere	272	4.3%
3. Need More Information	47	0.8%
4. Plants of Limited Distribution	532	8.4%
TOTAL	1742	27.7%

given by CNPS are based on California having 6300 species which is somewhat higher than the number listed in *The Jepson Manual*.

As California continues to grow and plant habitats continue to disappear, more and more of the California flora becomes rare or endangered. For example, in the last ten years the number of rare or endangered plants listed by CNPS has increased from 1399 (1984) to 1742 (1994), an increase of 20% in one decade. In 1984 22% of California's flora was considered rare or endangered compared to 27.7% in 1994. Without a solid conservation effort, these numbers will continue to rise.

The California Department of Fish and Game maintains an inventory of locations of California's rare plants in The Natural Diversity Data Base (NDDB). The information in this data base is available to the public and has recently been made available electronically on computer programs that can be purchased. However, it is well-known that to protect a rare plant species, its habitat must be protected. With that in mind, The Natural Diversity Data Base developed a classification of natural communities to help define habitats. This classification system was designed so that an inventory of rare plant communities could be put into a data base comparable to the NDDB and CNPS rare plant inventories. As of 1994, 280 plant communities are included in this inventory and 135 of these are considered rare and need more study and protection. CNPS, in collaboration with the California Department of Fish and Game, is currently preparing a multi-faceted effort to inventory, define and protect California's natural communities and ecosystems. This is discussed in more detail in Chapter 5.

About 25% of all the rare and endangered species listed for the entire United States occur in California. This is not surprising considering the large number of endemic species, the extreme diversity of specialized habitats, rare plant communities and the rapid destruction of California's wildlands and plant habitats. It appears that California is in danger of losing a large portion of its irreplaceable heritage of native plant species and native plant communities unless the citizenry and political leadership in the state accept the responsibility of protecting this important resource. Concerned botanists must also take active roles in informing the public and political leaders about the severity of this problem.

General References on California Vegetation and Flora

Abrams, L., and R. S. Ferris. 1923–1960. Illustrated flora of the Pacific states. Stanford Univ. Press, Stanford. 4 vols.

Bakker, E. 1972. An island called California. Univ. California Press, Berkeley. 357 pp.

Barbour, M. G., and W. D. Billings. 1988. North American terrestrial vegetation. Cambridge Univ. Press, Cambridge, New York, et al. 434 pp.

———, B. Pavlik, F. Drysdale, and S. Lindstrom. 1993. California's changing landscapes. California Native Plant Society, Sacramento.

———, M. Torn, and J. Harte. 1993. In our own hands: a strategy for conserving biological diversity in California. University of California Press, Berkeley.

Sawyer, J., and T. Keeler-Wolf. 1995. Series level descriptions of California vegetation. California Native Plant Society, Sacramento.

Schoenherr, A. 1992. A natural history of California. University of California Press, Berkeley.

Beauchamp, R. M. 1986. A flora of San Diego County, California. Sweetwater River Press, National City, California. 241 pp.

Benson, L., and R. A. Darrow. 1981. Trees and shrubs of the southwestern deserts. 3rd ed. Univ. Arizona Press, Tucson. 416 pp.

Billings, W. D. 1941. The problem of life zones on Mt. Shasta, California. Madroño 6:49–56.

Bradley, W. G., and J. E. Deacon. 1967. The biotic communities of southern Nevada. Nevada State Mus. Anthrop. Paper 13, part 4.

Brown, D. E. (ed.). 1982. Biotic communities of the American Southwest—United States and Mexico. Desert Plants 4:1–341.

Burchman, L. T. 1957. California range lands. Calif. Dept. Nat. Res., Sacramento.

Campbell, D. H., and I. L. Wiggins. 1947. Origins of the flora of California. Stanford Univ. Publ. Biol. Sci. 10:1–20.

Critchfield, W. B. 1971. Profiles of California vegetation. U.S.D.A. Forest Serv. Res. Pap. PSW-76. 54 pp.

Daubenmire, R. 1978. Plant geography with special reference to North America. Academic Press, N.Y. 338 pp.

Detling, L. E. 1968. Historical background of the flora of the Pacific Northwest. Bull. Mus. Nat. Hist. Univ. Oregon 13:1–57.

Dice, L. P. 1943. The biotic provinces of North America. Univ. Michigan Press, Ann Arbor. 78 pp.

Donley, M. W., S. Allan, P. Caro, and C. P. Patton. 1979. Atlas of California, pp. 146–152. Academic Book Center, Portland, Oregon.

Dunkle, M. B. 1950. Plant ecology of the Channel Islands of California. Publ. Allen Hancock Pac. Exped. 13:247–386.

Elias, T. S. 1980. The complete trees of North America. Field guide and natural history. Van Nostrand Reinhold Co., N.Y.

——— (ed.). 1987. Conservation and management of rare and endangered plants. Proceedings from a conference of the California Native Plant Society. California Native Plant Society, Sacramento. 630 pp.

Ellison, J. 1983. Natural communities—aquatic section. Calif. Dept. Fish and Game Natural Diversity Data Base. 6 pp.

Franklin, J. F., and C. T. Dyrness. 1973. Natural vegetation of Oregon and Washington. U.S.D.A. Forest Serv., Pac. N.W. Forest and Range Exp. Stn. Tech. Rep. NW-8. 417 pp.

Griffin, J. R. and W. B. Critchfield. 1972. The distribution of forest trees in California. USDA Forest Service Pacific SW Forest and Range Exp. Stn. Res. Paper PSW-82. 114 pp.

Harper, K. T., D. C. Freeman, W. K. Ostler, and L. G. Klikoff. 1978. The flora of Great Basin mountain ranges: diversity, sources, and dispersal ecology. Great Basin Naturalist Mem. 2:81–103.

Heckard, L. R., and J. C. Hickman. 1984. The phytogeographical significance of Snow Mountain, North Coast Ranges, California. Madroño 31:30–47.

Hickman, J. C., ed. 1993. The Jepson manual. Higher plants of California. University of California Press, Berkeley, Los Angeles, and London. xvii + 1400 pp.

Holland, R. F. 1986. Preliminary descriptions of the terrestrial natural communities of California. Nongame Heritage Program, California Department of Fish and Game. Sacramento. 156 pp.

Holland, V. L. 1977. Major plant communities of California. Pp. 3–41 in D. R. Walters, M. McLeod, A. G. Meyer, D. Rible, R. O. Baker, and L. Farwell (eds.). Native plants: a viable option. Calif. Native Pl. Soc. Spec Publ 3, Berkeley.

Hoover, R. F. 1970. The vascular plants of San Luis Obispo County, California. Univ. Calif. Press, Berkeley. 350 pp.

Howell, J. T. 1957. The California flora and its provinces. Leafl. West. Bot. 8:133–138.

———. 1970. Marin Flora. 2nd ed. Univ. Calif. Press, Berkeley. 366 pp.

———, P. H. Raven, and P. Rubtzoff. 1958. A flora of San Francisco, California. Wasmann J. Biol. 16:1–157.

Howitt, B. F., and J. T. Howell. 1964. The vascular plants of Monterey County, California. Wasmann J. Biol. 22:1–184.

Jaeger, E., and A. C. Smith. 1966. Introduction to the natural history of southern California. Univ. Calif. Press, Berkeley.

Jensen, D., and G. Holstein. 1983. Natural communities—terrestrial section. Calif. Dept. of Fish and Game, Natural Diversity Data Base. 10 pp.

———, M. Torn, and J. Harte. 1993. In our own hands: a strategy for conserving biological diversity in California. University of California Press, Berkeley.

Jensen, H. A. 1947. A system for classifying vegetation in California. Calif. Fish and Game 33:199–266.

Jepson, W. L. 1925. A manual of the flowering plants of California. Univ. California Press, Berkeley. 1238 pp.

Küchler, A. W. 1977. Appendix: the map of the natural vegetation of California. Pp. 909–938 in M. G. Barbour and J. Major (eds.). Terrestrial vegetation of California. John Wiley and Sons, N.Y.

Latting, J. (ed.). 1976. Plant communities of southern California. Symposium Proc., California Native Pl. Soc., Spec. Publ. 2. 164 pp.

Lewis, H. 1972. The origin of endemics in the California flora. Pp. 179–188 in D. H. Valentine (ed.), Taxonomy, phytogeography and evolution. Academic Press, New York.

Lloyd, R. M., and R. S. Mitchell. 1973. A flora of the White Mountains of California and Nevada. Univ. California Press, Berkeley. 202 pp.

Lowe, C. H. 1964. Arizona's natural environment. Landscapes and habitats. Univ. Arizona Press, Tucson. 136 pp.

Major, J. 1977. California climate in relation to vegetation. Pp. 11–74 in M. G. Barbour and J. Major (eds.). Terrestrial vegetation of California. John Wiley and Sons, N.Y.

Mason, H. L. 1957. A flora of the marshes of California. Univ. California Press, Berkeley. 878 pp.

Mathias, M. 1979. California—the land, climate and plants. Pacific Horticulture 40(2):9–14.

Matyas, W. J., and I. Parker. 1980. CALVEG mosaic of existing vegetation of California. Regional Ecology Group, U.S. Forest Service, San Francisco. 27 pp + large map.

Mayer, K. E., and W. F. Laudenslayer Jr. (eds.). 1988. A guide to wildlife habitats of California. California Dept. Forestry and Fire Protection, Sacramento, CA. 166 pp.

McMinn, H. E. 1939. An illustrated manual of California shrubs. Univ. California Press, 663 pp.

———, and E. Maino. 1935. An illustrated manual of Pacific coast trees. Univ. California Press, Berkeley. 409 pp.

Munz, P. A. 1968. Supplement to A California Flora. Univ. California Press, Berkeley. 224 pp.

———. 1974. A flora of southern California. Univ. California Press, Berkeley. 1086 pp. Univ. California Press, Berkeley. 1238 pp.

———, and D. D. Keck. 1959. A California flora. Univ. California Press, Berkeley. 1681 pp.

Navah, Z. 1967. Mediterranean ecosystems and vegetation types in California and Israel. Ecology 48:445–459.

Niehaus, T. F., and C. L. Ripper. 1976. A field guide to Pacific states wildflowers. Houghton Mifflin Co., Boston. 432 pp.

Ornduff, R. 1974. An introduction to California plant life. Univ. California Press, Berkeley. 152 pp.

Paysen, T. E., J. A. Derby, H. Black, V. C. Bleich, and J. W. Mincks. 1980. A vegetation classification system applied to southern California. Pacific SW Forest and Range Exp.. Stn., Gen. Tech. Rep. PSW-45. 33 pp.

Philbrick, P. N. 1972. The plants of Santa Barbara Island, California. Madroño 21:329–393.

Philbrick, R. W. (ed.). 1967. Proceedings of the symposium on the biology of the California Islands. Santa Barbara Bot. Gard., Santa Barbara.

Raven, P. H. 1963. A flora of San Clemente Island, California. Aliso 5:289–347.

———. 1973. The evolution of "mediterranean" floras. Pp. 213–224 in F. di Castri and H. Mooney (eds.). Mediterranean type ecosystems, origin and structure. Springer-Verlag, N.Y.

———. 1977. The California flora. Pp. 109–137 in M. G. Barbour and J. Major (eds.). Terrestrial vegetation of California. John Wiley and Sons, N.Y.

———, and D. I. Axelrod. 1978. Origin and relationships of the California flora. Univ. Calif. Publ. Bot. 72:1–134.

Rejmanek, M. and J. M. Randall. 1994. Invasion of Alien Plants in California: 1993 summary and comparison with other areas in North America. Madroño 41:161–177

Rzedowski, J. 1981. Vegetacion de Mexico. Editorial Limusa, Mexico City. 432 pp.

Sampson, A. W., and B. S. Jesperson. 1963. California range brushlands and browse plants. Univ. Calif. Div. Agric. Sci. Man. 33, Berkeley.

Sawyer, J., and T. Keeler-Wolf. 1995. Series level descriptions of California vegetation. California Native Plant Society, Sacramento.

Schoenherr, A. A. 1990. Endangered plant communities of southern California. Proceedings of the 15th Annual Symposium. Southern California Botanists spec. publ. 3.

———. 1992. A natural history of California. University of California Press, Berkeley.

Sharsmith, H. S. 1945. The flora of the Mount Hamilton Range of California. Amer. Midl. Naturalist 34:289–367. Reprint edition with introduction and index, 1982. Calif. Native Pl. Soc. Spec. Publ. 6, Berkeley.

Skinner, M. W. 1994. California Native Plant Society's inventory of rare and endangered vascular plants of California. Calif. Native Pl. Soc. Spec. Publ. 1. 5th ed. 336 pp.

Smith, A. C. 1960. Introduction to the natural history of the San Francisco Bay region. Univ. California Press, Berkeley.

Smith, C. F. 1976. A flora of the Santa Barbara Region, California. Santa Barbara Mus. Nat. Hist., Santa Barbara. 331 pp.

Smith, G. L., and A. M. Noldecke. 1960. A statistical report on A California Flora. Leafl. West. Bot. 9:117–132.

Smith, J. P., R. J. Cole, J. O. Sawyer, and W. R. Powell. 1980. Inventory of rare and endangered vascular plants of California. Calif. Native Pl. Soc. Spec. Publ. 1 (2nd ed.). Berkeley, Calif. 115 pp.

———, and K. Berg. 1988. Inventory of rare and endangered vascular plants of California. Calif. Native Pl. Soc. Spec. Publ. 1 (4th ed.). Berkeley, Calif. xviii + 168 pp.

———, and J. O. Sawyer. 1988. Endemic vascular plants of northwestern California and southwestern Oregon. Madroño 35:54–69.

———, and R. York. 1984. Inventory of rare and endangered vascular plants of California. Calif. Native Pl. Soc. Spec. Publ. 1 (3rd ed.). Berkeley, Calif. 174 pp.

Stebbins, G. L. 1978. Why are there so many rare plants in California? I. Environmental factors. Fremontia 5(4):6–10.

———. 1978. Why are there so many rare plants in California? II. Youth and age of species. Fremontia 6(1):17–20.

———. 1982. Floristic affinities of the high Sierra Nevada. Madroño 29:189–199.

———, and D. Major. 1965. Endemism and speciation in the California flora. Ecol. Monogr. 35:1–35.

Sudworth, G. B. 1908. Forest trees of the Pacific slope. U.S. Dept. Agric., For. Serv., Washington. Reprint edition, 1968. Dover Publ. Co.

Thomas, J. H. 1961. Flora of the Santa Cruz Mountains of California. Stanford Univ. Press, Stanford. 434 pp.

Thorne, R. F. 1967. A flora of Santa Catalina Island, California. of the California Islands. Santa Barbara Bot. Gard., Santa Barbara. Aliso 6:1–77.

———. 1969. The California islands. Ann. Missouri Bot. Gard. 56:391–408.

———. 1976. The vascular plant communities of California. Pp. 1–31 in J. Latting (ed.), Plant communities of southern California. Symposium Proc., California Native Pl. Soc., Spec. Publ. 2.

———. 1993. Phytogeography. Pp. 132–153 *in* Flora of North America Editorial Committee, Flora of North America, Vol. 1. Introduction. Oxford University Pres, New York.

Twisselman, E. C. 1956. A flora of the Temblor Range and the neighboring parts of the San Joaquin Valley. Wasmann J. Biol. 14:161–300.

———. 1967. A flora of Kern County, California. Wasmann J. Biol. 25:1–395.

Whittaker, R. H. 1960. Vegetation of the Siskiyou Mountains, Oregon and California. Ecol. Monogr. 30:279–338.

Chapter 2

Topography, Geology, and Soils

California's natural landscape provides spectacular topographic contrasts and biotic habitats that are unmatched elsewhere in the United States and that rival those of any other temperate portion of the world. The lowest and highest elevational points in the lower 48 states are found in California: 85 m (276 ft) below sea level at Death Valley and 4421 m (14,495 ft) above sea level on top of Mount Whitney. The diversity and contrasts begin with the 1600+ km (1000+ miles) of coastline along California's western boundary, continue eastward over the series of mountain ranges that front the coast, across the broad flat central valley, up and over the glaciated crests of the Sierra Nevada-Cascade axis and into the diverse desert areas of eastern California.

Physiographic Regions

California's landscape is dominated by a combination of major topographic features. Several major mountain systems dissect the state into a series of physiographic regions. These are listed in Table 2-1 and illustrated on Fig. 2-1. The most prominent of the major topographic features of California is the nearly continuous high mountain axis that divides the state into a comparatively moist western portion and a very arid eastern portion. This high mountain crest is derived from several different mountain systems, described below. The term cismontane, introduced in Chapter 1 (Fig. 1-3), describes that portion of California to the west of the crest of the Cascade Range, the Sierra Nevada, and the high peaks of the Transverse and Peninsular Ranges. Approximately 75% of the state is cismontane. The other 25% of the state, east of the mountain crest, is transmontane California (Fig. 1-2). Within each of these areas are additional major topographic features.

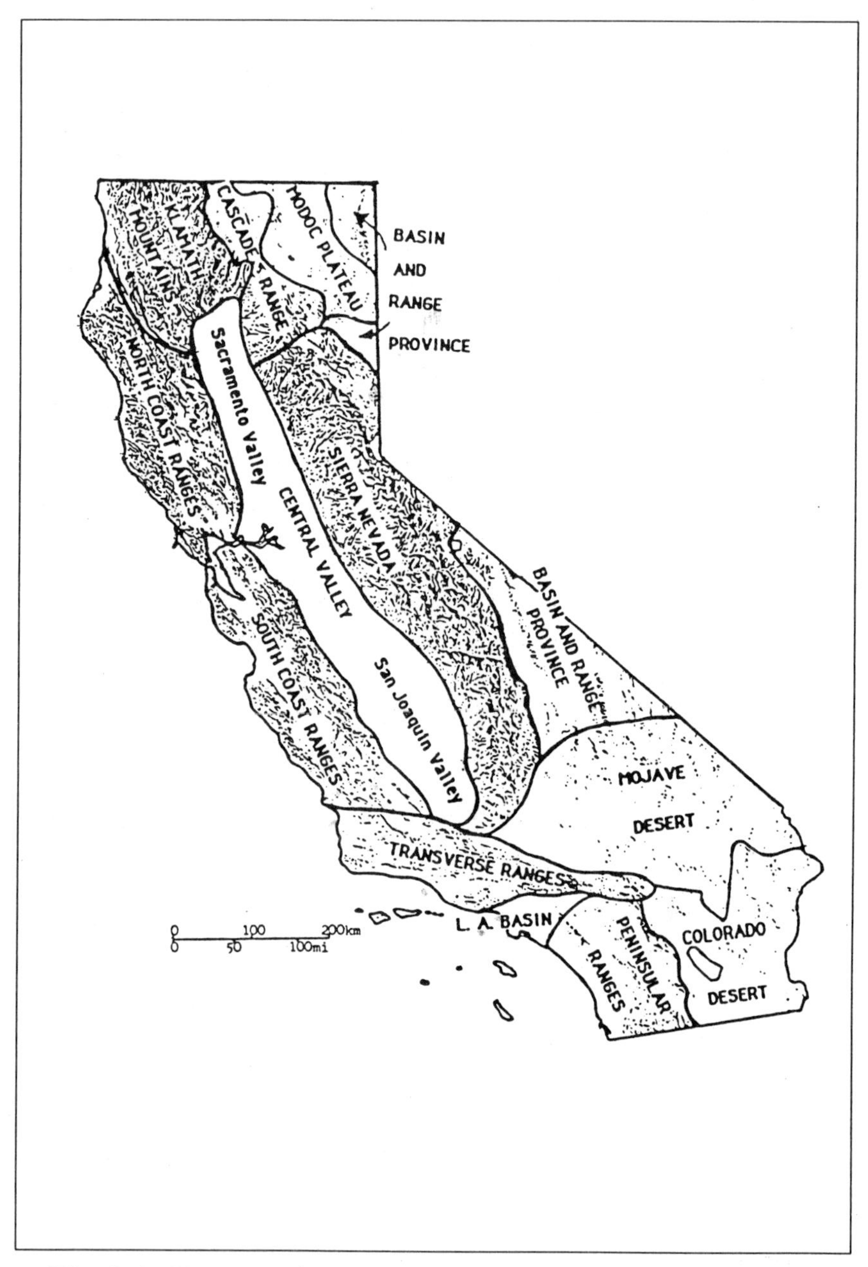

Fig. 2-1. Physiographic regions of California. Refer to the text for discussion of these features of California's landscape.

Table. 2-1. Physiographic regions of California.

CISMONTANE CALIFORNIA	TRANSMONTANE CALIFORNIA
Klamath Mountains	Modoc Plateau
Coast Ranges,	Basin and Range Province
Transverse Range,	Mojave Desert
Los Angeles Basin	Colorado Desert
Peninsular Ranges	
Cascade Range	
Sierra Nevada	
Central Valley	

The Klamath Mountains

The Klamath Mountains occupy the northwestern corner of California and adjacent portions of Oregon. They are bordered by the North Coast Ranges on the south, the Cascade Range and the northern part of the Central Valley on the east. They are among the oldest geological features in the state, composed primarily of ancient metamorphic and granitic parent materials. The Klamath Mountains are folded mountains, like the Coast Ranges, and differ in this respect from the fault-block of the Sierra Nevada and the volcanoes of the Cascades. The Klamath Mountains are dissected by canyons and valleys and are among the most rugged and beautiful areas in California. The highest peaks extend above 2745 m (9000 ft) elevation. The western part of the range receives very heavy precipitation and includes some of the wettest and most heavily forested areas in California. These high western peaks cast a rain shadow that results in the eastern portions of the range being hotter and drier. This rain shadow also extends to the lower elevation slopes of the Cascades and over to the Modoc Plateau. As a result, these areas are mostly vegetated with woodlands (oak and juniper woodlands) and shrublands (chaparral in the Cascade foothills and Great Basin sagebrush on the Modoc Plateau).

The Coast Ranges

The Coast Ranges are a series of northwest-trending parallel ridges extending from Santa Barbara County in the south to Humboldt County in the north. Together they constitute a nearly continuous ridge system, broken only by the lowlands of the San Francisco Bay and Sacramento River delta. From west to east, the Coast Ranges consist of a series of outer (coastal) and inner (interior) ranges separated by north-south trending valleys of various sizes such as the Napa, Sonoma and Salinas Valleys and the Carrizo Plain. The Coast Ranges are bordered on the north by the Klamath Mountains, on the south by the Transverse Ranges and on the east by the Central Valley.

The San Andreas fault and other parallel northwest-trending faults extend through the Coast Ranges and contribute to their geological complexity. The parent materials include sedimentary formations, metamorphics, as well as both intrusive igneous rock and extrusive volcanics. The ranges have been created by folding, uplifting and faulting along the western continental margin. Portions of the mountains were formerly sea bottom sediments or offshore islands that have been accreted (added) to the continent through geological time. Some of the Coast Ranges represent displaced terranes that originated hundreds or thousands of kilometers from California and have been moved by faulting and continental drift to their present locations. Faults in some areas have juxtaposed rock formations of very different origins, creating very diverse habitats. These various parent materials have eroded and weathered into chemically and structurally very different soils.

The Coast Ranges are somewhat naturally divided by the San Francisco Bay and its river delta. The mountains north of San Francisco Bay constitute the North Coast Ranges and those south of San Francisco Bay are the South Coast Ranges. The Coast Ranges are mostly relatively low in elevation but vary from south to north. Portions of the Coast Ranges have been eroded into a very rugged landscape and others are lower and gently rolling. The North Coast Ranges have higher peaks (some over 2440 m or 8000 ft) and a cooler, moister climate than the South Coast Ranges. As a result, the South Coast Ranges differ vegetationally from the cooler, wetter North Coast Ranges that contain significantly larger areas of coastal and montane coniferous forests. Only small areas of South Coast Ranges, near the coast or on top of isolated peaks or high ridges, are moist enough to support coniferous forests. Lower elevation, or drier zones, of both the North and South Coast Ranges are vegetated by foothill woodlands, chaparral, coastal scrub and grasslands.

The climate and vegetation of the Coast Ranges vary on an east-west basis as well south to north. The mountains of the outer Coast Ranges, close to the Pacific, have a moister climate than those of the inner Coast Ranges. Because most rain storms come from the west, the outer Coast Ranges cast a rain shadow on the inner Coast Ranges and the Central Valley. As a result, there is a marked decrease in precipitation from west to east. The innermost South Coast Ranges (e.g., the Temblor Range and the Caliente Range) are so dry that their lower slopes are occupied by desert vegetation similar to that of adjacent interior valleys (e.g., the Carrizo Plain). The west Central Valley Desert extends along the eastern edge of the inner Coast Ranges and Central Valley from Taft to the San Luis Reservoir (Pacheco Pass) area.

The Transverse Range

The Transverse Range is a series of mountain ridges that, unlike most other California ranges, are oriented east-west. The Transverse Range extends from the Santa Ynez Mountains and the northern Channel Islands (Anacapa, San Miguel, Santa Rosa and Santa Cruz Islands) [some of the range is submerged] east to the San Bernardino and Little San Bernardino Ranges (the easternmost extent of the Transverse Range). The Transverse Range is bordered to the north by the South Coast Ranges, the southern San Joaquin Valley, the Tehachapi Mountains (the southernmost portion of the Sierra Nevada), and by the Mojave Desert. To the south the Transverse Range is bordered by the Los Angeles Basin, the San Jacinto Mountains (the northernmost portion of the Peninsular Ranges) and by the northern part of the Colorado Desert. The highest peak, Mount San Gorgonio, is about 3500 m (11,485 ft) high.

The lower-elevation portions of the Transverse Range are located in the western section and are composed largely of sedimentary parent materials. The higher elevation areas to the east are dominated by granitic and metamorphic formations. Many of the mountains are very steep and rugged. Numerous faults dissect the Transverse Range, including the San Andreas Fault that extends across the range between the San Bernardino and the San Gabriel Mountains.

The western and southern flanks of the Transverse Ranges have higher precipitation than do the northern and eastern slopes. The lower elevation cismontane slopes of the ranges are vegetated by chaparral, coastal scrub, grasslands, foothill woodlands, and small areas of mixed evergreen forest. The higher ridges support montane coniferous forests, but these are less extensive than in the Sierra Nevada. Alpine and subalpine conditions occur on top of only the highest peaks. The Transverse Ranges block storms moving into California from the southwest and cast a significant rain shadow on the southern part of the San Joaquin Valley, the Mojave Desert, and the northern part of the Colorado Desert. The transmontane slopes are vegetated by desert woodlands and desert scrub communities. An interesting extension of Great Basin type vegetation occurs in the cismontane portion of this range around the bases of Mt. Pinos and Mt. Abel northwest of Los Angeles.

The Los Angeles Basin

The Los Angeles Basin is a low-lying area adjacent to the ocean, flanked to the north by the Transverse Ranges and to the south and east by the Peninsular ranges. It is underlain by thick layers of sedimentary deposits. Much of the year it is under the strong climatic influence of ocean air masses that moderate its temperature. At times coastal fog extends inland as far as the San Bernardino region. The surrounding mountains tend to hold the ocean air masses in the Los Angeles Basin. This has significant consequences. Atmospheric inversion patterns develop that prevent atmospheric pollutants from escaping from the basin. The air pollution that now characterizes the Los Angeles Basin is seriously damaging forest trees on the slopes of the Transverse and Peninsular Ranges.

At certain times of the year high pressure atmospheric cells settle over the deserts east of California and hot dry Santa Ana winds sweep westward from the desert passes into the Los Angeles Basin. These winds dry the vegetation and greatly increase the potential for brush fires. Prior to urban development much of the basin was vegetated by coastal scrub, chaparral, grassland and southern oak woodland communities. Fires in these communities occurred in prehistoric times, but did not have the devastating economic effects they have today. Much of the present-day vegetation of the Los Angeles Basin is managed in one way or another for human interests.

The Peninsular Ranges

The Peninsular Ranges are a series of northwest trending mountain ridges that are aligned roughly parallel to the San Andreas Fault. They are a northern extension of the mountain backbone of Baja California, and like that peninsula, were rifted from the Mexican mainland by tectonic forces millions of years ago. Their composition is largely granitic and metamorphic parent materials with comparatively small areas of sedimentary rock. Like the Transverse Ranges, they extend offshore with the high peaks emerging as the south Channel Islands (San Clemente, San Nicolas and Santa Catalina Islands). The onshore portion of the Peninsular Ranges extends eastward from the comparatively low-elevation Santa Ana and Laguna Mountains that parallel the Pacific shoreline to the high peaks of the San Jacinto and Santa Rosa Mountains and southward into Baja California. Although the highest peaks extend above 3000 m (9800 ft), the Peninsular Ranges do not form a continuous high-elevation barrier as does the Sierra Nevada. Nevertheless in southern California, because of the lower overall precipitation, the

Peninsular Ranges create an effective rain shadow over the lowlands of the Colorado Desert in southeastern California.

The vegetation of the cismontane slopes of the Peninsular Ranges is similar to that of other mountains of southern California. Chaparral, coastal scrub, grasslands and southern oak woodland communities occupy the lower slopes. Extensive urbanization has greatly modified the vegetation of much of the low-elevation cismontane portion of the Peninsular Ranges. Coniferous forests occur on the higher peaks and ridges. Only Mount San Jacinto, the northernmost high peak of the Peninsular Ranges is high enough to have an alpine zone. Transmontane slopes are vegetated by desert woodland and desert scrub communities.

The Sierra Nevada

The Sierra Nevada forms the interior backbone of California. The range is a northwest trending fault block with a gently sloping western face and a steep eastern escarpment. Much of the Sierra Nevada is a massive granitic batholith. Volcanic deposits cover portions of the northern half of the range. Metamorphic and sedimentary rocks outcrop in scattered localities. The Sierra Nevada is approximately 660 km (400 miles) long, extending from south of Mount Lassen (in the southern Cascades) to Tejon Pass where the Tehachapi Mountains (the southernmost ridge of the Sierra Nevada) meet the Transverse Ranges. The Sierra is bordered on the west by the lowlands of the Central Valley and on the east by the Basin and Range Province. A nearly continuous ridge of the Sierra about 280 km (175 miles) long rises to above 3000 m (9800 ft) with peaks rising over 4000 m (13,000 ft). The rain shadow of this high ridge extends across the western third of North America from the Sierra crest to the Rocky Mountains.

The gently rising cismontane flank of the Sierra Nevada is well watered by Pacific storms and large amounts of snow accumulate at high elevations. Numerous rivers have their headwaters in the Sierra Nevada and most flow westward into the Central Valley. The zonation of vegetation from the low foothills of the western slope to the high peaks reflects the increase in precipitation that accompanies the increase in elevation. In the foothills and on lower mountain slopes the vegetation gradually changes in broad bands from grassland to oak-pine woodlands, mixed hardwood forests and chaparral. Coniferous forests of several kinds grade from one type to another on the upper slopes. At the highest elevations are treeless alpine areas surrounded by dwarfed trees of the subalpine.

The steep transmontane face of the range is much drier and forest vegetation zones are compressed together at high elevations (Fig. 16-11, p. 313). Desert woodlands and desert scrub vegetation extend high up the eastern slopes. The valleys and plateaus that border the Sierra Nevada to the east average 1000–2000 m (3300–6600 ft) elevation, much higher than the Central Valley to the west. The grasslands, chaparral, foothill woodlands, and hardwood forests that cloak the low-elevation western flanks of the Sierra are absent from the eastern slopes.

The Cascade Range

The Cascade Range is a chain of volcanoes that extends from southern British Columbia to northern California. The tallest of these volcanoes in California are Mount Shasta (4317 m; 14,154 ft) and Mount Lassen (3187 m; 10,449 ft). Unlike the Sierra Nevada, the Cascades do not present a continuous high-elevation ridge system. Instead there is a plateau of volcanic deposits above which the high volcanic peaks rise as isolated cones. Mount Shasta, the second-highest peak in California, is much more impressive from a distance than is the slightly taller Mount Whitney (in the Sierra Nevada). Mount Shasta arises alone to tower over the surrounding plateaus and valleys (Fig. 3-1, p. 47), whereas Mount Whitney is surrounded by other peaks that are nearly as tall as it is.

The southern portion of the Cascade Range abuts the northern portion of the Sierra Nevada where volcanic deposits from the Cascades overlie the granitic rocks of the Sierra. To the west the Cascades are bordered by northern part of the Sacramento Valley and by the Klamath Mountains. To the east lies the Modoc Plateau. Vegetational zonation is similar to that of the northern parts of the Sierra Nevada and is considered simply an extension of the vegetation of the Sierra Nevada.

The Central Valley

The Central Valley is an essentially unbroken lowland plain that lies between the Coast Ranges and the Sierra Nevada-Cascade axis. Most of the valley lies below 200 m elevation (655 ft) and sits in the rain shadow of the Coast Ranges. Numerous rivers arising in the surrounding mountains flow into the Central Valley where most drain into one of two major rivers. The northern part of the Central Valley is drained by the Sacramento River and the southern part by the San Joaquin River. Several basins have no natural outlets to the sea and until drained for agriculture, they were occupied by shallow lakes. These are the remnants of a much larger lake that until about 10,000 years ago occupied most of the San Joaquin Valley. Still earlier in geological time the Central Valley was a lobe of the sea that extended to the base of what is now the Sierra Nevada. Underlying the Central Valley are layers of sediments thousands of meters deep. Over millions of years these sediments were eroded by water and ice from the surrounding mountains and deposited in the basin that was to become the Central Valley.

Prior to the agricultural development of the 19th and 20th centuries the low-lying areas of the Central Valley supported extensive marshlands and riparian woodlands. Adjacent areas were wooded by valley oaks. Drier areas were vegetated by grasses, forbs and low shrubs. The present-day vegetation has been altered by the removal of most of the trees, draining of the wetlands, and the conversion of large areas to irrigated cropland. The native grasslands have been modified by introduced grasses forming the new valley grasslands that are now a characteristic feature of the uncultivated portions of the Central Valley.

Because they are not as tall as the higher mountains to the east, the Coast Ranges do not have as great a rain shadow as do the taller mountains of the state. The Central Valley, which is in the rain shadow of the Coast Ranges and receives much less precipitation than the mountain slopes to the west, is, for the most part, moister than the transmontane desert areas. The rain shadow of the South Coast Ranges in combination with the Transverse Range is severe enough to bring about desert formation only in the southern San Joaquin Valley and adjacent smaller valleys. Precipitation in these areas is below 12 cm (4 in) and summers are hot and dry making the southern San Joaquin Valley similar to the Mojave Desert both in terms of climate and vegetation.

The Modoc Plateau

The Modoc Plateau is a comparatively flat area covered by extensive basaltic lava flows and other volcanic deposits. It is a southern extension of the vast lava floods that spread across the Pacific Northwest during the Tertiary Period. The plateau is dotted by small volcanic cones and in the eastern portion has been broken by the faulting that has formed the many small ranges of the Basin and Range Province. The plateau lies mostly between 1000 and 2000 m elevation (3300–6600 ft). The highest mountain ridges of the area rise to 2500–3000 m (8200–9800 ft). The Modoc Plateau is located in the northeastern corner of California in the rain shadow of the Cascades and northern portion of the Sierra Nevada. Much of the plateau is vegetated by desert shrublands and woodlands of the Great Basin Desert. Higher elevation areas support montane coniferous forests.

The Basin and Range Province

The Basin and Range Province is an extensive area that occupies much of Utah, Nevada and Idaho and extends into adjacent states. It comprises numerous more-or-less parallel north-trending fault-block mountain ranges and intervening valleys. In California the basin ranges occur east of the Sierra Nevada (itself a large fault block). The basins vary in elevation from below sea level at Death Valley to over 2000 m (6500 ft) and the ranges include peaks several peaks over 3000 m (9800 ft). The highest peak in the White Mountains is over 4200 m (14,000 ft) high. Elevations tend to decrease from north to south.

Much of the Basin and Range Province is characterized by interior drainage. Streams that originate in the mountains flow into valleys that have no outlet to the sea. During the Pleistocene ice ages these basins were filled by numerous lakes. About 10,000 years ago the climate of the area became warmer and drier and the lakes began to dry out. Most of the basins today are occupied by the dried beds of the ancient lakes (e.g., Death Valley) or by their saline remnants (e.g., Mono Lake).

Because the basins and ranges are located in the rain shadow of the Sierra Nevada-Cascade axis they are dry, even at high elevations. Forest or woodland vegetation is sparse and is restricted to high elevations. Most of the Basin and Range Province is vegetated by Great Basin desert scrub. Creosote bush scrub extends northward into Death Valley and the southern part of the Owens Valley. The saline valleys are occupied by barren salt flats surrounded by alkali sink scrub and saltbush scrub.

The Mojave Desert

The Mojave Desert is located in the rain shadows of the southern Sierra Nevada and Transverse Ranges. It is a comparatively flat region broken here and there by low dry mountains. The mountains are mostly fault-block ranges like those of the Basin and Range Province but lack a consistent north-south orientation. The basins average about 300–1000 m (1000–3300 ft) in elevation and the scattered mountain ranges seldom exceed 2000 m (6600 ft) elevation and are often much lower. Many of the basins of the Mojave Desert, like those further north, contain the dried vestiges of ancient lakes. The easternmost part of the Mojave Desert in California is drained by the Colorado River. A lobe of the Mojave Desert extends westward between the Transverse Ranges and the Tehachapi mountains as the Antelope Valley.

Much of the Mojave Desert is vegetated by desert scrub communities, especially creosote bush scrub and saltbush scrub.

Desert woodlands occur in scattered locations. The peaks of the desert mountains are mostly not high enough to support forests.

The Colorado Desert

The Colorado Desert is a flat, lowland [portions are below sea level] area that includes a lobe of the Gulf of California that was cut off from the sea by the sediments of the Colorado River delta. It is situated in the rain shadow of the Peninsular and Transverse Ranges. In the center of the Colorado Desert is the Salton Sea, a saline lake accidentally filled in 1905 and 1906 by runaway irrigation water diverted from the Colorado River. Studies have shown, however, that this basin has filled naturally and dried out several times without man's assistance. Mountains of the Colorado Desert are low in elevation and do not occupy large areas.

The desert plains are vegetated largely by creosote bush scrub. The desert scrub communities of desert mountains and the transmontane flanks of the Transverse and Peninsular Ranges are often diverse. Freezing weather is infrequent and various frost-intolerant plants reach their northern limits in the Colorado Desert.

The Interaction of Topography and Climate

Topographic variation influences the climate in many ways. The interaction of elevation and temperature are discussed in detail in Chapter 3. Atmospheric temperatures decrease with increasing elevation. This is at times an oversimplification, however. Actually, different slopes of the same canyon or mountain may be located at precisely the same elevation but have very different temperatures. Several factors may be responsible for this variation, including slope aspect, slope angle and location relative to the prevailing winds.

Slope aspect and slope angle interact to determine the angle of incidence of incoming solar radiation. A horizontal surface in California receives solar radiation at angles approaching perpendicular in summer to comparatively low angles in winter. In summer when the sun is high overhead, solar radiation passes through less atmosphere than in winter when the sun is close to the southern horizon. Because the atmosphere absorbs some of the incoming radiation, less reaches the surface. In winter a given amount of solar radiation is also spread over a wider area on a horizontal surface. Much of California is not at all horizontal, however.

Topographic variation greatly affects the amount of solar radiation absorbed at a particular site. A south-facing slope (inclined toward the south) will absorb more incoming sunlight per unit area (insolation) during the winter months when the sun is close to the southern horizon than will a horizontal surface, and still more than will a north-facing slope. A north-facing slope receives sunlight at a very low angle of incidence and a given amount of solar radiation is spread over a wider area. As a consequence of this, the south-facing slope absorbs more heat and warms the air above it more than does a horizontal surface or north-facing slope. As a result of increased insolation, the rate of evaporative loss of soil moisture and the rate of transpiration are greater on a south-facing than on a north-facing slope during the months when the greatest amount of moisture is available. If soil moisture is a limiting factor for the plants of an area, it will become limiting more quickly on south-facing than on north-facing slopes During summer months when the sun is higher overhead, the difference between south-facing and north-facing slopes is less pronounced.

Few slopes are precisely north-facing or precisely south-facing. Most are inclined to some extent toward the east or west. An east-facing slope receives direct sunlight in the early morning hours when temperatures are rather mild., and a west-facing slope is exposed at the time of day when temperatures are usually much higher. On most slopes the microclimate varies continuously from site to site with changes in slope aspect.

Topographic features have the further effect of casting shadows. A site located in a deep canyon receives direct sunlight for a shorter period than does a site in an open area away from canyon walls or other slopes. A site located to the east of a tall mountain may be sheltered entirely from the afternoon sun. The position of the shadow of a major terrain feature shifts with the seasons and may be much more important at one time of the year than at another.

Sites at high elevation receive more solar radiation because there is less atmosphere to intercept incoming sunlight. This is particularly significant to plants with limited moisture supplies. Soil temperatures and leaf temperatures may rise rapidly during the day. At night the temperatures may drop just as rapidly by re-radiation because the thin atmosphere holds little heat. These effects are mitigated at times by clouds that form in mountainous regions and block sunlight or heat loss. The clouds are transparent or nearly so to ultraviolet light, however, and this potentially damaging high energy radiation reaches the ground at high elevations even in cloudy weather.

The location of a slope relative to the prevailing winds can be very important in determining the temperature and other climatic variables. The orographic effect, described in Chapter 3, results in some slopes being exposed to moisture-laden rising air masses and others to dry, descending winds. The western flanks of California's mountain ranges tend to be moister than the eastern flanks. Winds are often channeled through valleys and mountain passes. Just as water rushes rapidly through a narrow canyon and slows down in a broad valley, wind velocity is increased in mountain passes and decreases in open areas. Although temperatures of adjacent sites may be the same, the windier slope will have a higher rate of evapotranspiration.

Topographic features often create their own air currents. Convection causes wind currents that are independent of the prevailing winds. In mountainous terrain the movement of air is influenced by the temperature differential of slopes of varying aspect. A slope exposed to direct sunlight will warm the air above it, causing the air to expand and to rise. The rising air is replaced by cooler air from adjacent sites. As the air rises its temperature again cools and it descends. The storm cells that form in the summer in mountain regions of the state result from this kind of atmospheric convection. The nature of the local topography can be very important in determining local air currents and microclimate. At night, air cooled at high elevations often descends through canyons and valleys, displacing warmer air into adjacent slopes. The effects of this cold-air drainage are discussed in Chapter 3.

Parent Materials and Soil Development

California has a remarkable diversity of different types of soils which form a complex mosaic in the state. Soils develop largely in response to the interaction of climate, topography, parent material (rocks or sediments), and living organisms (especially plants) through time. Topographic variables include elevation, aspect (compass orientation), and slope. Climatic influences on soil development include temperature, precipitation (rainfall, snowfall, fog drip), humidity, and wind. Directly and indirectly, these factors dynamically affect the parent material, weathering it into mineral or organic soil. Interacting with these environmental variables are plants whose roots, litter, and cover contribute significantly to soil development.

California's geology, like its soils, is very complex. As discussed previously, the major terrain features are often composed of parent materials of differing age, structure and chemical composition. Some parent materials are hard, igneous, crystalline rocks. Granites and similar intrusive rocks developed deep in the earth's crust and have been exposed by uplifting and erosion. Extrusive volcanic rocks of various kinds, like basalt, have been deposited on the earth's surface. Sedimentary rocks of various types occur in the state, ranging from coarse sandstones and conglomerates to fine-grained shales, diatomaceous earth, and unconsolidated mudstones. Some types of rock have been formed initially as igneous or sedimentary deposits and subse-quently altered chemically and structurally deep within the earth's crust. These metamorphic rocks include the serpentinite deposits that outcrop in scattered locations in the mountains of California.

Parent materials are exposed to physical and chemical weathering that gradually results in soil formation. The weathering processes include complex interactions involving physical factors such as repeated freezing and thawing, running water, wind, rain, and gravitational slumping. In addition chemical weathering dissolves parts of the parent material and produces subsequent chemical changes. Living plants and animals influence the process of weathering by physically altering the soil or parent material, releasing chemicals into the soil that alter its chemical properties, and removing other chemical substances from the soil. Partially decomposed organic remains of plants and animals that accumulate in the soil modify its water-holding capacity and chemical properties.

The location of a particular rock type in California is important in determining the type of soils that will develop from it. Weathering proccsses vary depending upon the elevation and topography, the amount and type of precipitation and other climatic factors, and the nature of the vegetative cover. The vegetative cover is in turn dependent on the area's climate, its topography, the parent materials and the soils that develop from the parent materials. Just as plant communities undergo successional changes, soils also change through time. Often the changes that take place in the soil are accompanied by and interact with the changes of the plant communities.

The chemical composition of a soil is influenced by the parent material and the nature and degree of weathering. Certain parent materials are lacking or deficient in some mineral nutrients and others have some substances in excess. The chemical weathering of soils may leach away some nutrients, thereby decreasing its fertility. Other chemicals become available only after the soil is thoroughly weathered. Some substances, particularly mineral salts, may accumulate in locations such as desert basins. In dry areas calcium salts may form a cement-like hardpan layer (caliche or petrocalcic horizon) below the surface layers of the soil that restricts water and root penetration.

Soils that develop in place from the weathering of outcrops of a particular parent material retain many of the chemical charac-teristics of that parent material. If a mineral nutrient is deficient or in surplus, the soils will reflect the chemical imbalances.

Plants intolerant of these soil conditions will be unable to grow on these soils. For example, in California serpentinite outcrops occur in various areas. Serpentinite has a much higher ratio of magnesium to calcium than most other parent materials and also usually contains relatively high concentrations of heavy metals such as chromium, nickel and mercury, which are poisonous to many kinds of plants. The vegetation of serpentine soils is often quite different from that of adjacent soils with a more favorable mineral balance. Many plants are unable to grow on serpentine soils and others are able to tolerate the soil's unusual chemical composition but do not grow as well on non-serpentine soils. The juxtaposition of very different parent materials by faulting, uplift and erosion has created some striking contrasts in the plant communities of the state.

Soils that develop from transported parent materials occur in valleys, on alluvial fans and in sand dunes. Rocks and finer particles that have been removed from their place of origin by water, wind or ice often represent a diversity of parent materials and usually have a favorable nutrient balance. The soils of the Central Valley, the Imperial Valley and other agricultural regions have developed from parent materials transported from the surrounding mountains. These soils are often very deep and fertile.

Various soil characteristics are important to plants. Soil depth varies from site to site depending on the local topography, the history of the site and the age of the soils. Soils are deeper and better developed in valleys than on steep slopes. Some parent materials erode very slowly and soil development is consequently slow as well. Other parent materials break down more readily, allowing deeper root penetration. Particle size is very important in determining soil development. The relative proportions of clay, silt, sand, gravel and large rocks influence the soil's aeration, its ability to hold water, its cohesiveness, and its fertility. Clay particles are important because their colloidal surfaces adsorb mineral nutrients and gradually release them into the soil water.

Soils contain a diversity of organisms that interact with the plants in various ways. Some interact with a plant directly. Soil vertebrates, invertebrates and microorganisms may use roots and other underground plant parts as a food source. Many plants establish symbiotic mycorrhizal relationships with soil fungi. The fungal associate in such a relationship is involved in decomposition of soil organic matter and serves as an extended absorptive surface along which water and dissolved nutrients are transported to the roots of the plant. Some plants, particularly members of the legume family, have colonies of nitrogen-fixing bacteria in tumor-like nodules along the roots. The addition of nitrates to the soil by these microorganisms is very important to the mineral nutrition of other plants of a community. Most plants are unable to fix atmospheric nitrogen by themselves.

Chemicals are available to plant roots only when dissolved in water. Coarse unweathered soil particles, living organisms, and undecomposed plant or animal remains may contain chemicals of potential use to a plant, but as long as these substances are not in solution they remain unavailable. The decomposition of organic matter is dependent on a wide array of invertebrates, fungi and bacteria. Dead leaves, twigs and other organic remains contain nutrients that are not directly available to the roots of a plant. These organic remains must be broken down into soluble chemicals before they can be absorbed by the plant. Decomposers in turn release chemicals into the soil that interact with soil particles and the roots of plants. Bacteria are involved in many complex chemical cycles involving sulfur, nitrogen, phosphorus, and other inorganic nutrients. The proportion of organic matter in a soil varies from 0 to 100 percent and can affect the soil's ability to hold water and nutrients.

Substances released by decomposers or by weathering affect the pH of the soil water. This in turn affects the solubility of various nutrients. Fertile soils usually are slightly acidic to neutral. Strongly acid soils sometimes develop as a result of the breakdown of some kinds of plant remains, such as the needles of conifers. The decomposition of these leaves releases organic acids that lower the soil pH. Some mineral nutrients are highly soluble in acidic solutions and are leached from acid soils, thereby reducing soil fertility. The acid conditions also inhibit bacterial decomposition and some of the important nutrient cycles that are mediated by the bacteria. The soils of coniferous forests are often rather infertile because of the long-term leaching that has depleted the soil of soluble nutrients. Acid rain, a result of air pollution, can bring about the same effects. Alkaline soils, on the other hand, occur in some desert and semidesert regions where evaporation has concentrated carbonates and other alkaline salts. Under alkaline conditions certain mineral nutrients, like iron, are essentially insoluble and are unavailable to plants. The salts themselves further inhibit plant growth by the osmotic effects of dissolved salts on plant roots.

Soil Classification

Soils of the United States have been officially classified by the Natural Resources Conservation Service Soil Survey Staff of the U. S. Department of Agriculture. Their taxonomic system is based on natural soil properties found in the field that can be measured quantitatively and uses a nomenclature that is quite similar to plant classification. The nomenclature is designed to fit

Table 2-2. Comparison of plant and soil classification systems.

Plant Classification		**Soil Classification**	
Division	Magnoliophyta	Order	Mollisol
Class	Magnoliopsida	Suborder	Xeroll
Subclass	Asteridae	Great Group	Argixeroll
Order	Asterales	Subgroup	Typic Argixeroll
Family	Asteraceae	Family	fine, montmorillonitic thermic Typic Argixeroll
Genus	*Stephanomeria*	Series	Los Osos
Species	*Stephanomeria exigua*	Type	Los Osos clay loam
Subspecies	*Stephanomeria exigua* ssp. *deanei*	Phase	Los Osos clay loam, 0 to 2% slopes

into any modern language using words coined mainly from Greek and Latin roots with connotations to help users remember the names of mnemonic devices. Eleven soil orders are subdivided into 47 suborders, 225 great groups, 970 subgroups, 4500 families, 10,500 series, and many thousands of soil types and phases.

Table 2-2 shows a comparison between the plant classification of a *Stephanomeria* species with the soil classification of a Los Osos clay loam. *Stephanomeria exigua* identifies a specific kind of plant, and the Los Osos clay loam identifies a specific type of soil. *Stephanomeria exigua* is further divided into a 4 subspecies (e.g., *Stephanomeria exigua* spp. *deanei.*) and Los Osos clay loam is divided into several phases, e.g., Los Osos clay loam, 0 to 2 percent slopes. Several soil phases are grouped into types which are grouped together in a soil series. A soil family is made up of groups of related soil series just as the Aster family includes many different, but taxonomically related genera. California soils include representation of all 11 soil orders, although some are rare and highly localized (Table 2-3).

Soils develop characteristic layers or horizons in response to their environment. These horizons are often distinct and can be differentiated by looking at a soil profile, a section down through a soil to the bedrock. The surface soil just below the litter layer is the zone of maximum organic accumulation and also the zone of eluviation. Materials leached from surface soil move downward and accumulate in subsoil horizons. As a result, subsoil horizons are referred to as zones of illuviation. Illuvial materials include clays, humus and aluminum and iron compounds. Subsoils are also zones of alteration because significant changes occur soil structure. The surface and subsoil characters are determined by the nature of soil-forming factors and are useful in soil classification. In California, high rates of erosion in hills and mountains, coupled with relatively low rainfall in the lowlands tend to produce thin soils that are poorly horizoned.

Due to California's arid and slow soil-forming environments, the dominant soils in the state are **Aridisols, Entisols, Inceptisols, Mollisols,** and **bedrock soils. Aridisols** develop under arid conditions and include the light colored desert soils; **Entisols** lack genetic horizons and form on transported materials in valleys and flood plains; **Inceptisols** are young soils with poorly developed horizons; **Mollisols** are thick dark soils typical of grassland areas; and **bedrock soils** are found on dunes and granite and lava rocklands that have little soil development. **Alfisols** and **Ultisols** typically develop under a forest cover and are also relatively common in California mountain regions. They differ from each other in that Ultisols are more highly weathered and acidic than Alfisols and their bases have been more completely depleted. **Vertisols** are expansive soils that contain more than 30% fine clay and produce deep cracks (>20 inches) when dry, giving rise to partial inversion of the soil. These occur mostly in lowlands and valley areas.

The other soil orders, **Histosols**, **Oxisols**, **Spodosols** and **Andisols**, are rare in California and are found only in localized areas. **Histosols** are soils that develop in wetland areas and typically develop where organic matter retains much of the original plant tissue form. These are bog, peat and muck soils that are saturated at least part of the year. Freshwater and saltwater marshes, which are scattered throughout the state, are characterized by these soils. **Oxisols** have a deep subsurface horizon that is usually very high in clay particles dominated by iron and aluminum. These are very rare in California, known only from a small area near Ione. This unique soil is a paleosol (buried soil) formed in an ancient tropical type climate, and no longer undergoing soil forming processes. It is a highly localized soil heading for extinction because it is being actively mined as a source of kaolin clay to make bricks and ceramics. **Spodosols** are also highly localized soils found in the pygmy forests of Mendocino County. These are old, sandy soils that are highly acidic and very low in nutrients. **Andisols** are young, poorly developed soils that form on volcanic ash, pumice, or cinders. These soils occur only in areas downwind from volcanoes, such as near Mt. Shasta, Mt. Lassen, Long Valley, and on the Modoc Plateau. Table 2-3 shows the soil orders with their descriptive characteristics. Table 2-4 correlates the major vegetation types in California with the types of soils that usually occur there.

Table 2-3. Soil orders and their descriptive characteristics.

Soil Orders (mnemonic device and formative elements)	**Descriptive Characteristics**	**General Vegetation Types**
Aridisols Latin: *aridus*, dry (id)	Older, fertile desert soils; light colored surface horizon due to low organic matter with >50% basic cation saturation (Ca, Mg, K, Na); subsoil variable. Used primarily for grazing, watershed and urban development.	Piñon-juniper and Joshua tree woodlands, some desert shrublands
Entisols English: from the word rec*ent* (ent)	Young mineral soils low in organic matter with poorly developed horizons. Found over a wide range of climates. Formed on transported materials in valleys, flood plains and areas of active erosion or recent deposition.	Desert and semi-desert shrublands, dunes and alpine vegetation. Also includes riparian areas.
Inceptisol Latin: *inceptum*, beginning (ept)	Young soils with poorly developed horizons and light or dark surface. Subsurface weakly developed with very slight clay movement. Shows bright colors.	Forests and foothill wood-lands, montane coniferous forest, chaparral
Mollisols Latin: *mollis*, soft (oll)	Deep, dark colored surface with >1% organic matter, making it soft and friable, and >50% basic cation saturation (Ca, Mg, K, Na). California's most fertile and important agricultural soils.	Grasslands, chaparral, foothill wood-lands, coastal scrub; some Great Basin sagebrush,.
Alfisols nonsense syllable (alf)	Surface horizon light colored due to low organic matter; subsurface with strong clay accumulation and >35% basic cation saturation 2:1 clays dominate; relatively high fertility.	Montane coniferous forest, , foothill woodlands, chaparral
Ultisols Latin: *ultimus*, last (ult)	Old soils of red and yellow colors and 1:1 clays; subsurface horizon consists of clay accumulation with <35% basic cation saturation; not naturally as fertile as Alfisols or Mollisols.	Coniferous, mixed evergreen forests, and woodlands

Table 2-3 (continued).

Vertisols Latin: *verto*, turn (ert)	Fine textured mineral soils characterized by high content clays that shrink and swell , producing deep cracks (>20 In) when dry and usually no subsurface horizon because the soil churns. Sticky and plastic when wet.	Some grasslands and associated lowland vegetation
Spodosols Greek: *spodos*, wood ash (od)	Old, sandy parent material; highly leached; very acidic; very low fertility. Subsurface may be nearly white, strongly leached and bleached.	Rare in CA, localized in Mendocino pygmy forest Humid, stable, sandy areas
Oxisols French: *oxide*, oxide (ox)	Most highly weathered soil group. Subsurface horizon high in kaolinite clay and iron and aluminum oxides (often red) with silica leached out. Very low fertility, and often toxic levels of aluminum and phosphorus.	Rare in CA, highly localized paleosol (buried soil) near Ione
Histosols Greek: *histos*, tissue (ist)	Organic soils that are saturated at least part of the year unless artificially drained; >30% organic matter to 40+ cm. Deeper layers may be become peat material with very high water-holding capacity, and high CEC (cation exchange coefficient).	Freshwater marshes, salt marshes, estuaries, bogs, and fens
Andisols Japanese: *ando*, black soil or volcanic ash (and)	Young, slightly developed soils formed from volcanic ash, pumice, or cinders (localized and associated with areas down-wind from volcanoes). High amount of acid soluble aluminum and iron, high CEC and high P retention capacity.	Grasslands and shrublands around Mt. Shasta, Mt. Lassen and Modoc Plateau

California soils are generally divided into upland and lowland soils. Upland soils are mostly residual soils that develop in place as the bedrock weathers and are considered agriculturally non-productive (e. g. alfisols and ultisols). They are best used in their natural condition as woodlands, forest lands, watershed protection, and are also used for rangeland. Lowland soils are transported in origin and are geologically young sediments including valleys, floodplains, alluvial fans, and consolidated sand dunes (e. g. mollisols and entisols). Most valley soils are under cultivation today. However, large areas of these prime soils have been lost to rapid urban development. A few low, poorly drained valley areas where salts accumulate are not agriculturally useful.

Table 2-4. Soil orders common to major vegetation types in California.

Vegetation	Soil Orders (in order of abundance)
Coastal salt marsh communities Coastal estuarine communities	Histosols
Coastal sand dune and beach communities	Entisols, Mollisols, Alfisols in stable dunes, small pockets of Histosols in some dune slacks
Coastal scrub communities	Entisols, Mollisols, Inceptisols
Chaparral communities	Mollisols, Alfisols, Entisols, Inceptisols
Grassland communities	Mollisols
Closed-cone coniferous forest communities	Alfisols, Ultisols, Spodosols (in pygmy forest)
Coastal coniferous forest communities	Ultisols, Alfisols
Mixed evergreen forest communities	Ultisols, Inceptisols (Sierra)
Oak woodland communities	Alfisols, Mollisols, Inceptisols
Montane coniferous forest communities	Alfisols, Inceptisols, Entisols (steep, eroded mountains)
Alpine meadows	Mollisols, Histosols (in wet depressions)
Rocky alpine communities	Entisols, Inceptisols
Desert alpine communities	Entisols, Aridisols
Desert woodland communities	Aridisols, Entisols
Desert scrub communities	Aridisols, Entisols
Riparian communities	Entisols, Mollisols (on stream terraces)
Freshwater wetland communities	Histosols in freshwater marshes, bogs and fens; Mollisols (Aquolls) in meadows and vernal pools
Anthropogenic communities	Arents (artificial Entisols) are soils physically mixed by humans destroying soil horizons. Only remnant fragments of pre-existing horizons are present.

Because the major soil forming factors are physically and spatially diverse throughout the California, soil types and soil patterns in the state are also varied and complex. Like plant communities, soils have many transitional areas where one soil type grades into another, making mapping and classification difficult (see Chapter 5). However, there is often a direct correlation between the type of vegetation and type of soil that develops in a given area, and soils can often be indicative of vegetation type and vice versa. Relationships of soils and vegetation are fascinating but not completely understood.

References on Topography, Geology and Soils

Bailey, E. H. (ed.). 1966. Geology of Northern California. Calif. Div. Mines and Geol. Bull 190.

Barbour, M. G., J. H. Burk, and W. D. Pitts. 1987. Soil. Pp. 407–433 *in* Terrestrial plant ecology, 2nd ed., Benjamin/Cummings Publ. Co., Menlo Park, CA.

Bassett, A. M., and D. H. Kupfer. 1964. A geological reconnaissance of the southeastern Mojave Desert, California. Calif. Div. Mines and Geol. Spec. Rep. 83.

Brady, N. C. 1974. The nature and properties of soils. 8th ed. MacMillan Publ. Co., N.Y. 639 pp.

Brouillet, L., and R. D. Whetstone. 1993. Climate and physiography. Pp. 15–46 *in* Flora of North America Editorial Committee, Flora of North America, Vol. 1. Introduction. Oxford University Pres, New York.

Daubenmire, R. F. 1974. The soil factor. Pp. 3–73 in Plants and environment: A textbook of plant autecology. John Wiley & Sons, N.Y.

Ernst, W. G. (ed.) . 1981. The geotectonic development of California. Prentice-Hall, Englewood Cliffs, N.J. 706 pp.

Hill, M. 1975. Geology of the Sierra Nevada. Nat. Hist. Guide 37, Univ. California Press.

———. 1984. California landscape. Nat. Hist. Guide 48, Univ. California Press.

Hornbeck, D., P. Kane and D. L. Fuller. 1983. California patterns: A geographical and historical atlas. Mayfield Publ. Co., Palo Alto, California. 111 pp.

Irwin, W. P. 1960. Geologic reconnaissance of the Northern Coast Ranges and Klamath Mountains, California. Calif. Div. Mines and Geol. Bull. 179.

Jahns, R. H. (ed.). 1954. Geology of Southern California. Calif. Div. Mines and Geol. Bull. 170.

Lantis, D. W., R. Steiner, and A. E. Karinen. 1963. California: land of contrast. Wadsworth Publ. Co., Belmont, California. 509 pp.

Miller, C. S., and R. S. Hyslop. 1983. California: the geography of diversity. Mayfield Publ. Co., Palo Alto, California. 255 pp.

Norris, R. M., and W. W. Webb. 1976. Geology of California. John Wiley and Sons, N.Y. 365 pp.

Steila, D. 1993. Soils. Pp. 47–54 *in* Flora of North America Editorial Committee, Flora of North America, Vol. 1. Introduction. Oxford University Pres, New York.

Trott, K. E., and G. Frantz. 1986. An index to soil surveys in California, 2nd ed. Calif. Dept. Conservation, Sacramento. 84 pp.

U.S.D.A. Soil Conservation Service. 1976. Soil association maps.

———. 1977. Classification of soil series of the United States.

Chapter 3

California Climate

From a botanical point of view, climate can be defined as the sum total of atmospheric conditions that influence plant growth and development. Although there is not always a direct correlation between climate and vegetation, climate is usually considered to be one of the most significant features in the complex of factors that determine the vegetation of an area.

California has greater climatic diversity than any other state, and its vegetational diversity is reflective of this varied climate. For example, California has climatic regions in the northwest that receive over 250 cm (100 in) of precipitation per year and support moist temperate rain forests. In contrast, some arid regions in the southeastern portion of the state get less than 5 cm (2 in) per year and are among the hottest, driest deserts in the world (e.g., Death Valley). California has areas along the coast that seldom, if ever, are subjected to freezing temperatures and areas on the peaks of high mountains that can experience frost any day of the year. Locations only a few kilometers apart may have enormously different environments. For example, areas along the immediate coast like Morro Bay and San Luis Obispo in south-central California and Berkeley in the central part of the state have mild summer temperatures. High temperatures average between 16 and 25 °C (in the 60's and 70's in °F). Just a few km inland (e.g., Atascadero and Paso Robles in south-central California, and Walnut Creek and Concord in central California) summer high temperatures average 31–35 °C (in 90's in °F) and are often over 38 °C (100 °F).

The factors that combine to form what we refer to as climate, such as precipitation, temperature, fog, and wind, are complex and interdependent. California's regional climates are mostly a result of its extremely diverse topography, its wide range of elevations, its wide range in latitude (32.5° to 42° north), and its proximity to the ocean,

Table 3-1. Comparison of mean August and January temperatures, growing seasons, and mean annual precipitation for selected coastal sites from southern to northern California.

Location	Mean January Temperature	Mean August Temperature	Number of Days in Growing Season	Mean Annual Precipitation
San Diego	13° C (55° F)	22° C (71° F)	365	25 cm (10 in)
San Francisco	11° C (51° F)	15° C (60° F)	356	51 cm (20 in)
Eureka	8° C (47° F)	14° C (57° F)	331	107 cm (42 in)
Crescent City	7° C (44° F)	13° C (56° F)	320	193 cm (76 in)

The Influence of Latitude on California's Climate

As a general rule, latitude is a major factor influencing climate in the United States. However, in California physiography is also a major determinant of climate. This is illustrated by noting that isotherms (lines of equal temperatures) tend to run north-south, following the topographic contours of the mountain ranges, rather than east-west as in most other areas of the United States. This is not to say that latitude is unimportant in terms of its influence on California's climate. California extends over a greater range of latitudes than any other state except Alaska, and various climatic features do change from south to north. As a general rule, temperatures decrease and precipitation increases from southern California to northern California. This can be illustrated by comparing a series of areas along the immediate coast of California from the southern to the northern border of the state. The sites used, San Diego, San Francisco, Eureka and Crescent City, all are situated just above sea level adjacent to the Pacific Ocean. From a comparison of the climates of these sites, it is clear that there is a gradual change in climate along a gradient from San Diego (32.5° north latitude) near the southern California border to Crescent City (42° north latitude) near the northern border (Table 3-1). Mean winter and summer temperatures decrease, the length of the growing season (number of frost-free days per year) decreases and the mean annual precipitation increases along this south-north gradient. Similar trends also occur along south to north gradients elsewhere in the state such as in the Central Valley from Bakersfield to Redding, and from the Sonoran to the Great Basin Deserts.

Fig. 3-1. Mount Shasta in the Cascade Range. High mountains illustrate the elevational zonation characteristic of California's climate and vegetation. Areas in the foreground are vegetated by chaparral and lower elevation montane mixed coniferous forest The upper slopes extend above treeline into an alpine tundra. Photo by V. L. Holland.

Vegetation responds to these climatic trends. In the areas around San Diego grassland and coastal scrub predominate along the coast. As one travels northward along the coast, the vegetation gradually changes from coastal scrub through coastal live oak woodland and various other communities to coastal redwoods and northern coniferous forests around Eureka and Crescent City. A similar gradient occurs in the Central Valley. The dominant vegetation in the southern end of the valley around Taft and Bakersfield is the arid West Central Valley Desert. As one travel north, the vegetation changes to valley grassland and eventually to foothill woodland in the extreme northern end of the valley around Redding.

Influence Of Topography and Elevation on California's Climate

Because of the state's complex topography, the precipitation and temperature patterns also vary greatly with respect to elevation and major topographic features (Fig. 3-1). Precipitation increases with elevation due to the "orographic effect". Moist marine air masses are carried by prevailing westerly winds over the western slopes of the California mountains. As this air rises over the mountains, it cools, reducing the air's ability to hold water vapor which condenses and falls as precipitation (rain or snow) mostly on the western slopes. As a general rule, precipitation increases at a rate of 5–10 cm per 100 m increase in elevation (2–4 in per 300 ft). Maximum precipitation is usually reached between 1200 and 2100 m (4000–7000 ft) elevation depending on location and latitude, higher in the south, lower in the north. Highest precipitation usually corresponds with upper montane coniferous forests. Above this elevation, precipitation decreases because most of the water has already fallen at lower elevations. Temperatures also decrease along elevational gradients at an average rate of 0.6° C per 100 m increase in elevation (3.5° F per 1000 ft). This generalized trend, called the "lapse rate", can be modified by other environmental factors. For example, air passing over a California mountain range in summer is usually dry, and the lapse rate is much higher than in winter when air masses are moist. Canyons, where cold air settles and flows downslope, can also affect the lapse rate.

Mountains tend to create "rain shadows". An air mass that has passed over a mountain heats up as it descends in elevation down into valleys at about the same rate as the generalized lapse rate. As air warms, its moisture holding capacity increases, and it can absorb more moisture from the land and from plants living there. Consequently, there is much less precipitation, and the climate is drier on lee slopes of a mountain chain than on windward slopes. Because storms in California move eastward from the ocean and perpendicular to the north-south trending mountains, rain shadows occur to the east of the mountains. The deserts of the state are all on the lee side of one or more mountain ranges. The Central Valley of California is in the rain shadow of the relatively low elevation Coast Ranges. Taller mountains tend to cast a more effective rain shadow. For example, the high Cascade–Sierra Nevada axis blocks moisture from about a third of the United States. Death Valley, which receives 5 cm or less (2 in) rain per year, is in the double rain shadow of two tall mountain ranges, the Sierra Nevada and the Panamint Range.

Effects of the north-south trending mountains ranges on precipitation and temperature can be illustrated by examining a transect across central portion of the state from the ocean, over the coast ranges, across the Central Valley, up and over the Sierra Nevada and ending in the desert regions in eastern California (Fig. 3-2). Beginning in the Morro Bay–San Luis Obispo area along the immediate coast, moist air from the Pacific Ocean hits the land and rises over the Coast Ranges. Precipitation in this area, which is just above sea level (0 to 100 m [330 ft]), is about 40–60 cm (15–23 in). As the air mass continues to rise over the crest of the Coast Ranges, precipitation increases

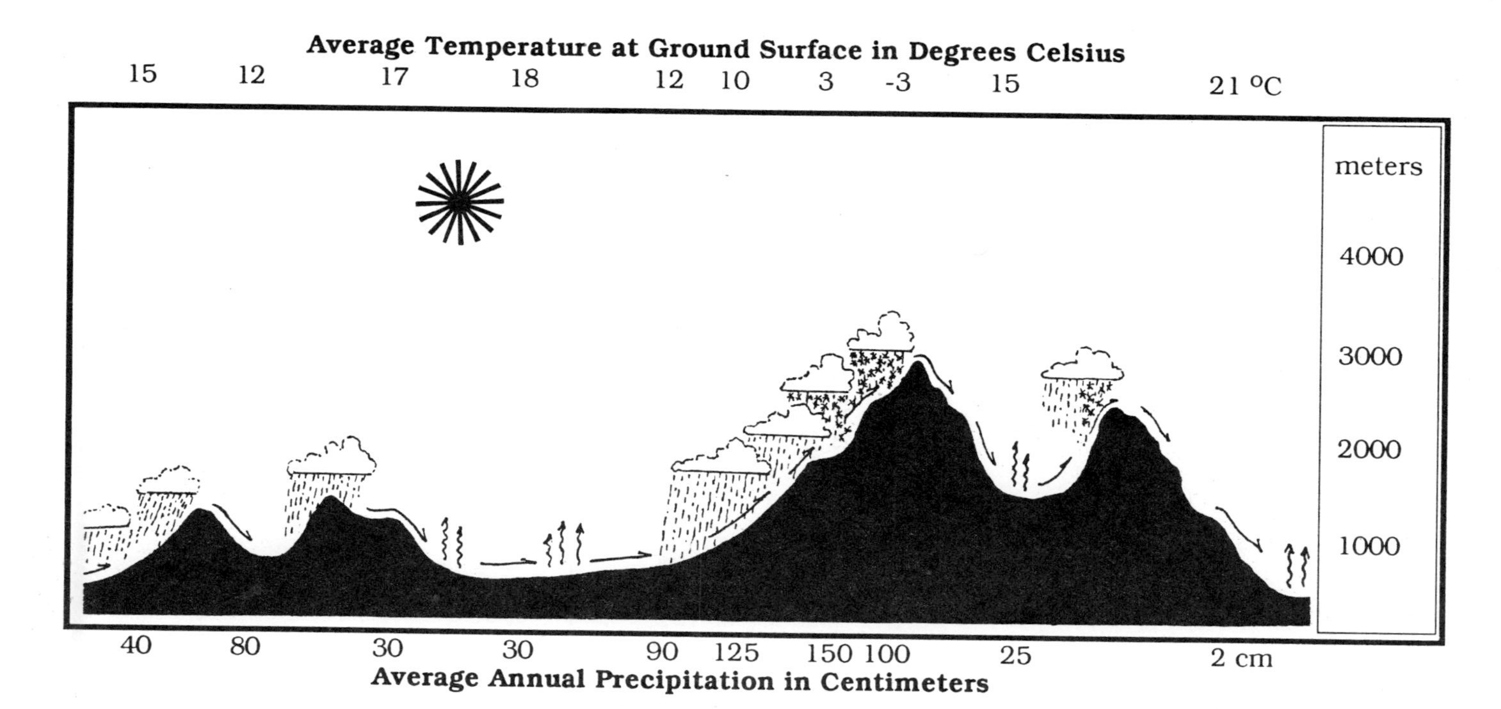

Fig. 3-2. Generalized transect across California illustrating variation in average temperature and precipitation with respect to topography and elevation. The precipitation increases as the temperature decreases with elevation. Rain shadows develop on the eastern side of the Coast Ranges, the Sierra Nevada, and the Desert Ranges. These climatic changes play an important role in the vegetational changes that occur along the same transect. See Figures 16-11 (p. 313) and 16-22 (p. 335).

to about 80 cm (over 30 in) at elevations of 610 to 910 m (2000–3000 ft) on the moist, western slopes. On the eastern slopes, precipitation decreases because much of the moisture has already fallen on the seaward slopes. As the air descends into the Central Valley, it gets hotter and drier and can take up moisture. This area in the western portion of the Central Valley (e.g., Coalinga, Taft) is often called the West Central Valley Desert and receives about 13–17 cm (5–7 in) of precipitation annually. As an air mass moves across the Central Valley, it often picks up some moisture from the valley. Areas on the western side of the valley (at the base of the Sierra Nevada) get slightly more precipitation than those on the east side (e.g., Fresno and Visalia average about 28 cm [11 in]) and support valley grassland.

As the air mass rises up the western slopes of the Sierra Nevada, precipitation increases at the general rate of 5–10 cm per 100 m increase in elevation (2–4 in per 300 ft) as discussed previously. In the foothills from about 305 to 610 m (1000 to 2000 ft), the precipitation is about 50–75 cm (20–30 in); from about 610 to 1524 m (2000–5000 ft) it is about 77–104 cm (30–40 in). For example, Yosemite Valley at an elevation of about 1220 m (4000 ft) averages about 94 cm (37 in) of rain and snow annually. From 1524 m to 2440 m (5000 to 8000 ft), precipitation reaches its maximum level of 102 to 153 cm (40 to 60 in). For example, Giant Forest in Sequoia National Park at about 2012 m (6600 ft) averages 116 cm (46 in) of precipitation. Above about 2440 m (8000 ft), precipitation decreases because most of the water has already fallen at the lower elevations. As a result, alpine and subalpine zones (3280 to 4000 m [10,000–13,000 ft]) receive about 77 cm (30 in) of precipitation, mostly as winter snow. Flora and vegetation gradually change along this elevational and climatic gradient. Above the Central Valley grassland, vegetation changes from foothill (oak) woodland to montane coniferous forests to subalpine forests and eventually to alpine fell fields on the highest peaks and crests.

After an air mass passes over the crest of the Sierra Nevada, it descends abruptly into the Owens Valley, becoming hotter and drier; this results in an extensive rain shadow. Precipitation in the Owens Valley averages between about 13 and 26 cm (5–10 in). For example, Independence receives about 13 cm (5 in) of precipitation annually. East of the Owens Valley, the air mass rises again, over the western slopes of the White and Inyo Mountains where maximum precipitation is about 38 cm (15 in) at about 2748 to 3050 cm (9000–10,000 ft). In the rain shadow of these mountains, precipitation drops to the lowest levels in the state, about 5 cm (2 in) at Greenland Ranch and Death Valley. Vegetation in these dry desert regions east of the Sierra Nevada is characterized by various types of desert shrublands at lower elevations and desert woodlands at higher elevations.

Finally, it should be noted that other characteristics of precipitation patterns are also influenced by elevation and topography. For example, seasonal precipitation patterns change along elevational gradients, e.g., the length of the summer dry season decreases with increasing elevation. Summer thunderstorms, which are not common at low elevations, are frequent at mid- and high elevations during July and August. Because of gradual release of water from the winter snowpack, summer thunderstorms, and lower seasonal temperatures the summer drought for montane plant communities is shorter than that at lower elevations. In addition, the percentage of orographic precipitation that falls as snow or ice increases with elevation as air temperature decreases. Although almost all precipitation in lowlands and foothills occurs as rain, over 80% in the subalpine and alpine zones may fall as snow. In mid-elevation zones, about 50% of precipitation falls as rain and 50% as snow.

The Influence of the Ocean on California's Climate

Along the immediate coast of California, the climate is cool and mild and does not display much daily or seasonal temperature fluctuation. This maritime climate is a result of the ocean serving as a giant heat reservoir. The ocean moderates the temperature of air masses passing over it by absorbing heat in the summer and releasing heat in the winter. Consequently the air temperature of areas along the coast is generally cooler than inland areas in the summer and warmer in the winter. Additionally, areas with a maritime climate have smaller diurnal temperature fluctuations than do inland areas. Maritime climates occur on the lower seaward side of the coastal mountains.

Inland from the ocean's influence, maritime climates give way to more continental climates from the inner coastal mountains across the Central Valley and into the interior mountain ranges. Hot dry summers and cold winters characterize the continental climates of the Central Valley and nearby foothills which is one reason many people from the Central Valley enjoy visiting coastal areas both in summer and winter. Montane climates are found at high elevation in the Sierra Nevada and other interior mountain ranges. These climates are more similar to the continental climates in the central portions of the United States where the winters are harsh and most of the winter precipitation falls as snow. Both continental and montane climates have distinct seasons and show much greater daily and seasonal temperature fluctuations than do maritime climates.

Climatic Regions of California

California is often said to have a Mediterranean climate with mild, wet winters and warm, dry summers. This statement is a generalization that is applicable to only a portion of the state. It does not accurately describe all of the different climates that actually occur in California. Climates in the state include hot, dry deserts and severely cold arctic-alpine environments. The climatic map of California (Fig. 3-3) shows the general distribution of California's regional climates and provides an excellent indication of the variety of climates found within its boundaries. The system used for the map, which is based on temperature and precipitation factors, was developed by Wladimir Köppen, a German climatologist. It is the most popular classification system for climates and conveniently illustrates California's great diversity of climates. By examining the map (Fig. 3-2), one can see that California has representation of four of Koppen's five climatic types including hot dry (B) climates, temperate rainy (C) climates, cold snowy (D) climates and cold arctic/alpine (E) climates. Only tropical rainy (A) climates are not represented somewhere within the state.

The dry (B) climates are found in areas where evaporation, for the most part, exceeds precipitation throughout the year because of the combination of low precipitation and warm to hot temperatures. There are two major subdivisions of the dry climate (1) Desert (BW) which includes both hot desert (BWh) and very hot desert (BWhh) and (2) Steppe (BS) which includes the semi-arid regions. The hot desert climates are characterized by having mean temperatures above 0 C (32 F) in the coldest months and mean annual temperatures over 18 C (64.4 F). The very hot desert is similar except at least 3 months have maximum average temperatures over 38 C (100.4 F). The hot desert climate is characteristic of high deserts such as the Mojave Desert and the West Central Valley Desert in southwestern San Joaquin Valley. The very hot desert climate is found in the low deserts like Death Valley and the Colorado Desert located around the Salton Sea in the southeastern section of California. The steppe climates are similar to the desert climates but are typically cooler and moister. Steppe climates are characteristic of higher elevations of the Great Basin area east of the Sierra-Cascade Mountains. Vegetation in this area is mostly Great Basin Sagebrush with scattered grasslands. Steppe climates are also found in the northern and western portions of the San Joaquin Valley and some of the surrounding foothills that are too dry to support trees. Vegetation in these areas is typically California grasslands. A coastal form of the steppe climate (with fog) occurs in the extreme southern end of the state around San Diego where semi-

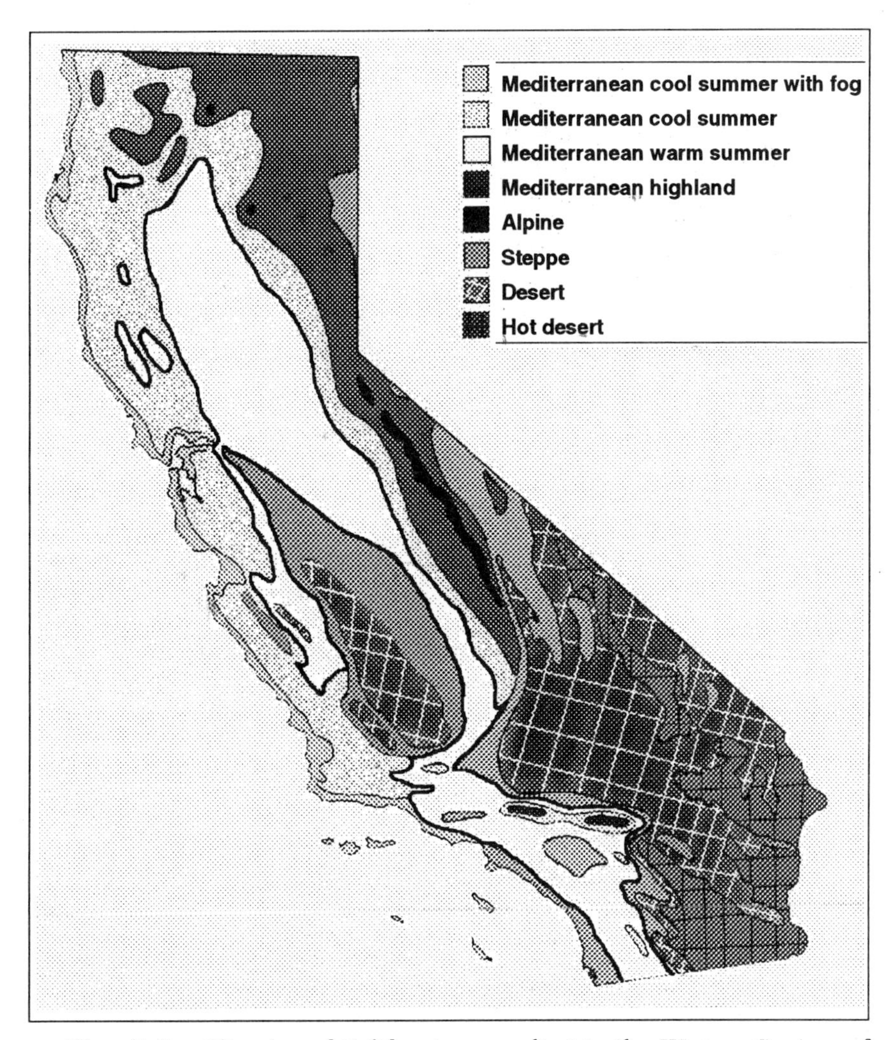

Fig. 3-3. Climates of California according to the Köppen System of climate classification. See the discussion in the text for details (redrawn from James, 1966).

desert coastal scrub and grasslands are the predominant vegetation types.

The western two-thirds of the state (cismontane California) is dominated by various types of Mediterranean climates. Mediterranean climates are included in Koppen's C- and D-type climates. The C-type climates are referred to as temperate rainy (= summer subtropical or mesothermal climates). The D-type climates are referred to as Mediterranean highland (= snowy forest or microthermal climates). Mediterranean climates are found throughout the earth's continents wherever there is a west coast between about 30° and 40° north or south latitude. In the northern hemisphere, they occur in areas around the Mediterranean Sea (from which they take their name) and in California and northern Baja California; in the southern hemisphere the occur in Chile, South Africa, and southern Australia. In all, Mediterranean climates, including California's, occupy only two percent of the earth's land surface.

The C-type climates are characterized by warm, dry summers and mild, wet winters. The coldest month has an average temperature under 18° C (64.4° F) but above -3° C (26.6° F) and at least one month has an average above 10° C (50° F). At least 70 percent of the precipitation occurs during the 6 winter months, November to April, and almost no precipitation occurs in the summer months, June to September. California has three C-type Mediterranean climates: (1) Mediterranean warm (hot) summer (Csa), (2) Mediterranean cool (warm) summer (Csb), and (3) Mediterranean cool (warm) summer with fog (Csbn). The main differences among these three climates is the summer temperatures and presence of summer fog.

In Mediterranean warm summer climates, the average temperature of the warmest month is over 22° C (71.6° F), whereas in the Mediterranean cool summer climates, the warmest month averages below 22° C (71.6° F). Mediterranean warm summer climates are found in the dry Central Valley and interior foothills, especially in southern California. Mediterranean cool summer climates are associated with higher elevations in the mountains and along the immediate coast. The Mediterranean warm summer climates average between approximately 31 to 64 cm (12 and 25 in) of precipitation annually, almost all of which occurs as rainfall. The Mediterranean cool summer climates typically average between 64 and 140 cm (25 and 55 in) annually. In coastal areas this occurs mostly as rainfall; however, at mid-elevations in the mountains, half or more of the precipitation usually falls as winter snow.

The Mediterranean cool summer with fog climate is differentiated from Mediterranean cool summer by the presence of

Table 3-2. Types of vegetation found in the major climatic regions of California. Some vegetation types occur in more than one climatic region.

Climate	Vegetation Types
Mediterranean cool summer with fog (Csbn)	Coastal redwood forests, Coastal closed cone conifer forests, North coast coniferous forests, Coastal scrub
Mediterranean cool summer (Csb)	Chaparral, Foothill woodland, Lower montane coniferous forests, Mixed evergreen forests,
Mediterranean warm summer (Csa)	Grasslands, Lower foothill woodland, Valley oak woodland
Mediterranean highland (Dsb)	Montane coniferous forests, Red fir forests, Lodgepole pine forests, Subalpine forests, Piñon-Juniper woodlands.
Alpine (EH)	Alpine
Steppe (south coast) (BSh, BShn)	Grasslands, Semi-desert coastal scrub
Steppe (interior) (BSh, BSk)	Grasslands, Great Basin sagebrush scrub
Desert (BW)	Joshua tree woodlands, Desert shrublands of Mojave Desert
Hot Desert (BWhh)	Desert shrublands of the (Sonoran Desert)

more that 30 days per year of dense fog. This climate is characteristic of the cool coastal fog belt of California, one of the foggiest coasts anywhere in the world. Fog in this climate is most important in the summer months because it provides moisture to the vegetation during an otherwise dry season of the year. The wettest areas in the state fall within this climate. These areas, located in the northwestern part of the state, average over 250 cm (100 in) of precipitation per year and support some of the lushest temperate rain forests in the world. The remainder of the Mediterranean C-type climates support a diversity of vegetation types typical of the California Floristic Province, such as dense forests, woodlands, chaparral, coastal scrublands and grasslands.

The Mediterranean highland (Dsb) climates also have the dry summer and wet winter typical of C-type Mediterranean climates. However, these climates have the colder temperatures that characterize interior highlands. Within this climatic type, the mean temperature of the coldest month is 0° C (32° F) or less, and the warmest summer months range between 10 and 21° C (50 and 72° F). These climates occur at elevations higher than the regions with Mediterranean cool summer climates and generally receive less precipitation. The annual average precipitation, which falls mostly as winter snow, may be less than 50 cm (20 in) in some areas (e.g., the Modoc Plateau), but is generally about 75 cm (30 in) on western slopes of the Cascade and Sierra Nevada Mountains. Plant communities in this climate range from high elevation montane forests to the northern juniper woodland of the Modoc Plateau region.

The Alpine (EH) Climate (= polar, tundra, arctic climates) is California's coldest climate and does not have a true summer. The warmest summer months have mean temperatures below 10° C (50° F) and the minimum mean of the coldest months is about -14° C (7° F). Frost may occur on any day of the year. This climate in not only severely cold, but also dry. As a result, many ecologists refer to alpine areas as extremely cold, high elevation deserts. Alpine areas are characterized by poorly formed, rocky soils that hold very little moisture. As a result the vegetation consists of low growing herbs and shrubs with many of the adaptations found in desert plants (e.g., reduced leaf surfaces).

As discussed above, climate plays a major role in determining regional vegetation types. Often specific groups of plant communities are associated with the climates discussed above on a regional basis. The general association of vegetation with climates is summarized in Table 3-2.

Regional Climates, Local Climates and Microclimates

The various types of climates shown on the climatic map (Fig. 3-3) are the prevailing, regional climates for California. These are very useful in predicting general vegetational patterns for the state as shown on Table 3-2. However, a local area may have a diversity of local or microclimates that fall within one regional climate. The vegetation in the area may show considerable variation because of local site to site differences.

Local climates are a result of the local variation in temperature, moisture, wind, humidity, etc., within a regional climate (= macroclimate). These local differences are often attributable to such variables as slope aspect and angle, minor topographic features and influence of the vegetation (Chapter 2). Local topography may strongly influence local temperature and moisture conditions. For example, a canyon located within a given regional climate can be expected to have very different local climates on the north and south facing slopes. The north facing slope will be cooler and moister than the south facing one and will usually support a more mesic type plant community. For example, within a regional Mediterranean warm summer climate a north facing slope may have foothill woodland whereas the opposing south facing slope may have grassland or chaparral because it is too hot and dry to support trees.

As discussed previously, canyon bottoms may also have a different local climate than the slopes as a result of cold air drainage. At night, air on mountain slopes cools. As it does, its density increases, and the heavier, cooled air flows down slope where it displaces warmer air near the ground surface. Nocturnal cold air drainage commonly occurs in canyons and valleys and the displaced warmer air moves upward onto slopes and ridges. This results in a temperature inversion with lower elevation areas cooler than adjacent higher elevation sites. Canyon and valley sites are consequently often cooler than are adjacent slopes and may support vegetation that includes species mostly found at higher elevations. The cool temperatures and higher moisture availability in a canyon bottom often allow montane coniferous trees such ponderosa pines (*Pinus ponderosa*) to extend down canyons into the hotter, drier regions where adjacent slopes are covered by foothill woodland or chaparral. In the southern part of their range, coast redwoods are mostly restricted to cool, moist coastal canyons Adjacent north facing slopes are usually vegetated by mixed evergreen forests and hot, dry south facing slopes by chaparral.

Within local climates there are also microenvironmental variations that result in microclimates. For example, the microclimate under a tree may be quite different from that of an adjacent open area within the same regional and local climate. The temperature will be more moderate (not as cold at night, not as hot during the day), the humidity and soil moisture will be higher, the light intensity will be less and the wind may be reduced. As a result, the plants growing under trees may be different than those in adjacent open areas. Even large rocks create microclimates that support plants not found away from them. For example, on a dry slope, it is not uncommon to find ferns and other mesophytic species in the shade of rocks where moisture and temperature conditions are favorable. By contrast, only very hardy, drought tolerant plants can grow on the exposed surfaces of these rocks (e.g., lichens).

Plants respond to those climatic conditions to which their surfaces are exposed. The climate just a few cm from a plant, for example, may differ from that impinging directly on the plant's leaf surfaces in terms of wind velocity, temperature and humidity.

For this reason, an understanding of the interrelationships of regional climates, local climates and microclimates is essential in explaining distribution of plants and plant communities in nature.

Other important factors to consider are unpredictable variations in seasonal and long-term climatic trends. For example, San Luis Obispo averages about 50 cm (22 in) of precipitation annually. However, there have been years in which precipitation was less than 25 cm (10 in) and other years that it has been over 140 cm (55 in). Although in most years rainfall comes during winter months, there have been years when the rains came during the fall, and the winters were comparatively dry. Although winter temperatures are typically cooler than summer temperatures, some months or years have exceptionally high or low temperature conditions that greatly affect plants. January temperatures may even reach the 90's in parts of southern California in some years. Obviously, these seasonal and annual variations in climate have a significant effect on plants. For example, there can be great variation in the number and size of plants and in species composition of annual wildflower displays as a result of seasonal and yearly climatic variation within the same area. If one were to visit the same site year after year, it would be evident that the wildflower displays are never the same from year to year. One of the main reasons appears to be the interaction of temperature and precipitation with seasons. For example, a cool, wet fall may result in a very different display of wildflowers than a hot, dry fall. This seems especially noticeable in desert regions of California where wildflower displays are so completely different from year to year.

Day to day and diurnal variations further complicate attempts to fully understand the influence of climate on vegetation. A single climatic event such as a severe storm, a killing frost, or a fire whipped by howling Santa Ana winds may be of only a few hours' duration, yet it may have long-term effects on vegetation.

Precipitation

As discussed above, precipitation in California varies in a complex fashion in response to latitude, topography and seasonal factors. Along California's latitudinal gradient, annual precipitation ranges from over 250 cm (100 in) in coastal regions of the north to less than 25 cm (10 in) in the south. The Death Valley area is the driest in the state and averages less than 5 cm (2 in). The rainy season in California, which occurs predominantly during the winter months, begins earlier and lasts longer in the north than in the south. More Pacific storms come ashore in northern California than in the south, and those storms that reach the southern part of the state are often of lesser intensity and duration than those of the north.

Large cyclonic storms that move off the Pacific Ocean are the primary source of precipitation in California. These storms are most frequent and most dependable during the winter months. Those winter storms that move out of the north Pacific are accompanied by cold winds and often by high snowfalls in the mountains. Storms that move onshore from tropical or subtropical latitudes are warmer though, and the snow level may be several thousand feet higher.

The summer in California is mostly rather dry. Occasional summer thunderstorms occur in the southwestern deserts and in the higher mountains, but for the most part lowland areas are dry for several months. During the summer, and at times during the remainder of the year, a high pressure atmospheric cell forms off the coast of the state and deflects moisture-bearing air masses away from the California coastline. The dry period begins earlier and lasts longer in southern California, sometimes lasting for six months or longer.

The condition of winter storms and summer drought is paralleled in the other Mediterranean climate areas of the world (southern Africa, southern Chile, southern and southwestern Australia, and the regions bordering the Mediterranean Sea). Plants of California are generally either adapted in various ways to surviving or avoiding the dry season, or are restricted to localized habitats where summer moisture is available. Plants and the communities that they comprise in other Mediterranean climatic areas of the world often have similarities in physiognomy to those that occur in California. For example, oak woodlands and chaparral areas in Spain are very similar ecologically to those in California and look very similar as well, although most of the plants in the two areas are unrelated taxonomically. Plants of other Mediterranean climatic regions are often grown in California as ornamentals, and many of the weedy species found in California originated in areas with Mediterranean climates. There are also differences among these widely separated areas with Mediterranean climates. They vary in such important features as intensity of summer drought, fire frequency, and other related factors that influence the vegetation.

Season to season variation adds to the complexity of precipitation patterns. Some years have much higher than average precipitation. This can result in floods, landslides, and other destructive effects and also in deep penetration of water into the soil, enhanced plant growth, and in some cases, beautiful wildflower displays. When this lush vegetative growth dries in the summer it can provide fuel for extensive fires. At high elevations, years with higher than average precipitation can result in a heavy snowpack that may not melt until mid-summer in some areas. This reduces the growing season for some plants to a few weeks. Stream-flow in such years is much greater than in dry years and this affects riparian vegetation at various elevations.

Drought years put stress on perennial plants, allow for less vegetative growth, and decrease the effective life spans of annual plants. Less potential fuel accumulates but vegetation may dry out to a greater extent than in moist years. The snowpack at high elevations may melt off early in a dry year, exposing plants to an early spring and a long, dry summer. Montane forests become very susceptible to fire in such years. These overall patterns are complicated still further by variations in temperature and distribution of precipitation within a single season.

The quality of precipitation is very important. Water must be available to a plant before it can be of use to the plant. Most plants can absorb water only when the water has penetrated into the soil and is in contact with the roots. Gentle rainfall spread over a long period of time allows greater penetration of water into the soil than does a brief, intense storm. If a storms drops a large quantity of rainfall in a short period, much of the water may run off without being absorbed into the soil. This rapid runoff may cause flooding, landslides, and physical damage to the plants.

Ice and snow are not directly usable by a plant; melting must occur first. As long as snow covers the ground, little plant growth can occur. However, snow is a rather effective insulator and often shelters plants from the extreme temperatures of high elevations. Once melting has begun, a snowbank may serve as a water reservoir that supplies nearby plants for several weeks. However, a sudden thaw will allow rapid runoff and comparatively little penetration. These high-elevation snowfields are also the sources of water for species of canyons, riverbanks, and other similar areas during the summer months. They are crucial in supplying water for human use as well.

Snow and ice can have various negative effects on plants. Branches can be snapped off or bent by an accumulation of snow or ice. On steep slopes, saplings are often bent at the base by slowly creeping masses of snow. Heavy accumulations of snow at high elevations can result in avalanches that can destroy forests. Windblown snow and ice can abrade plant surfaces, stripping away bark and foliage on the windward side of a tree resulting in "wind flagging". In high elevations near timberline, trees are often severely pruned and deformed by the combined effects of severe winds and ice. Plants completely buried by snow are protected from this type of damage.

Potential Evapotranspiration

Water is essential to the development of plants and communities. However, the amount of water that falls onto an area as rain or snow is not the most significant factor. More important than precipitation are the availability of water and the length of time in which it is available. Water that falls to the ground can be unavailable for a number of reasons. Runoff can transport water from snow or rain to lower elevation sites or to the sea. Water that percolates through the soil below the level that is accessible to plant roots is also lost (though it may emerge at some other lower-elevation site and add to the available moisture at that site). A major loss of water to all plant communities is through evaporation, either directly from the surface of soil, rocks, or open water or as transpiration from plant surfaces. These are often combined as evapotranspiration.

The amount of water actually lost through evapotranspiration may be greater than, less than, or equal to the amount received as precipitation. The **potential evapotranspiration** is the amount of water that could be lost from a site if moisture were continually available. Potential evapotranspiration (PE) is more important than precipitation (P) in determining whether a climate is wet or dry. Areas receiving equal amounts of precipitation may have very different climates because the amount of water lost as vapor differs greatly. Where P markedly exceeds PE the climate tends to be humid or very humid. Where P and PE are equal or nearly so the climate is subhumid. In semiarid or arid climates PE greatly exceeds P. These can be calculated as an **Index of Moisture** (I_m):

$$I_m = 100\left(\frac{P}{PE} - 1\right)$$

The Index of Moisture can have any value from -100 to more than 100. Some sites in the northwestern portion of California have very humid climates with an I_m of more than 200. Very dry desert sites in the southeastern part of the state may have an I_m below -95. Table 3-3 lists climatic types as determined by I_m, and Table 3-4 lists some California localities, their annual precipitation, I_m values, and vegetation types. This is a much more accurate way of looking at the relationship of climate to vegetation than that discussed and shown in Table 3-2.

The rate of water loss through evapotranspiration varies greatly on an annual basis. At times of high precipitation the water falling on a site may be much greater than the rate of evapotranspiration. Under these conditions there is a water surplus and excess water flows away as runoff or percolates

Table 3-3. Climatic Types as Determined by Index of Moisture .

Climatic Type	Index of Moisture (I_m)
Very humid	100 and above
Humid	20 to 100
Subhumid	-33 to 20
Semiarid	-66.7 to -33
Arid	-100 to -66.7

toward the water table. During the summer drought, however, the rate of evapotranspiration much exceeds precipitation, if their is any at all, and there is a water deficit. Under conditions of water deficit the soil and plants growing in it lose moisture. Some areas of California have a large water surplus and only a small summer deficit. High temperatures promote a greater water deficit. Desert regions with little or no surplus at any time in the year together with high summer temperatures have a large water deficit.

Water Retention in Soil

Another important feature to consider when examining available water for plants is the texture and composition of the soils. Soils have significant differences in water holding capacities and permeability. Heavy, clayey soils have the highest water holding capacity and will retain water for extended periods of time whereas sandy and rocky soils have low water holding capacities and only retain water in upper soil horizons for a short time. If these different soil types occur side by side in an area with the same precipitation, the vegetation they support will still be significantly different. For example, along a small section of coastline, one can find sand dunes, marine terraces and rocky headlands next to each other. Obviously, these are all in the same climatic area with the same amount of precipitation yet support very different types of plants. This is partly because of different water holding abilities of the soils. Water will be available to plants on the fine textured soils of the marine terraces for a longer period of time, than on the rocky headlands and dunes. There is a great diversity of soil types in California, and all have different water holding capacities that affect plant growth. Of course, we must remember that soil water holding is just one feature of a whole complex of soil factors that affect the growth and distribution of plants and plant communities in California.

Table 3-4. Annual Precipitation, Potential Evapotranspiration, Index of Moisture, and Vegetation Types for selected California Locations. Data from Major (1965).

Location	P (mm)	PE (mm)	I_m	Vegetation Type
Branscomb	2091	673	211	redwood forest
Crescent City	1910	658	190	coastal coniferous forest
Giant Forest	1092	575	115	montane mixed coniferous forest
Ellery Lake	760	397	91	subalpine forest
Huntington Lake	775	509	52	red fir forest
Fort Bragg	965	649	49	closed cone conifer forest
Placerville	1056	715	48	ponderosa pine forest
Eureka	918	645	42	northern coastal scrub
Yosemite Valley	839	659	27	montane mixed coniferous forest
Point Reyes	759	638	19	coastal dunes
Los Gatos	777	743	5	mixed evergreen forest
Sonora	847	825	3	foothill woodland
Mt. Tamalpais	701	695	1	coastal live oak woodland
San Francisco	551	702	-22	coastal live oak woodland
Pismo Beach	423	708	-40	southern coastal scrub
San Joaquin Experimental Station	513	874	-41	foothill woodland
Bridgeport Dam	267	530	-50	Great Basin sagebrush
Sacramento	416	846	-51	valley oak woodland
San Diego	262	785	-67	chaparral
Palmdale	259	875	-70	Joshua tree woodland
Fresno	236	905	-74	valley grassland
Barstow	104	933	-89	creosote bush scrub (Mojave Desert)
Indio	78	1275	-94	creosote bush scrub (Colorado Desert)
Cow Creek	53	1381	-96	alkali sink

Fog and Cloud Cover

Fog and clouds cannot be fully separated in their effects. We perceive a cloud as fog when it is low enough to affect surface features. Fog and clouds are common in California. Fog and low stratus clouds are a frequent feature of the California coastline, particularly in summer. Clouds that form in mountainous areas as a result of orographic effects often cling to the slopes or ridgelines as fog layers. In winter months, thick fog layers form in the Central Valley; these valley fogs are referred to locally as tule fog.

Fog and clouds affect plant communities in several ways. The most noticeable of these effects is the reflection of radiation. Clouds reduce the incoming solar radiation by reflecting or absorbing much of the incoming light energy. This results in a decrease in photosynthesis and transpiration rates. The latter reduction is particularly important in areas where fog and clouds are frequent and of considerable duration. This is the case in some mountainous areas and is particularly true along much of the California coastline. During the summer months coastal fog and stratus cloud layers overlie the immediate coast, sometimes for days at a time, and often extend into the outer Coast Ranges. The fog/cloud layers together with the proximity of the ocean greatly ameliorate the climate of coastal areas. The length of time during the day when plants are exposed to direct solar radiation is reduced when fog or clouds persist through parts of the daytime. Clouds and fog also tend to prevent rapid night-time heat radiation from land surfaces and plant bodies. Frost is much less likely when there is a cloud cover.

Another major influence of fog is a result of condensation on plant surfaces and soil. Tall trees that grow in areas with frequent fog, such as the coast redwoods (Chapter 12), serve as condensation surfaces. Moisture condenses on leaves and branches and fall to the ground, adding to the effective precipitation of the area. Some botanists have hypothesized that this extra precipitation is responsible for maintaining some redwood populations by providing the trees with summer moisture. A second effect of the condensation is the reduction of transpiration due to the wetting of the foliage. In much of California, high atmospheric humidity and fog contribute to growth of both epiphytic and terrestrial lichens, algae and mosses.

Temperature

As discussed previously, temperature variations are correlated with factors that also influence precipitation. Temperatures decrease along both latitudinal and elevational gradients. Areas in the northern part of California generally have lower average temperatures than those in the south, and higher elevations are usually cooler than lower elevations. Also, coastal areas tend to have milder climates than those away from the ocean (Fig 3-2 to 3-4).

At times high pressure atmospheric cells develop over the high deserts of California and adjacent regions. If there is a strong pressure differential between the coastal region and the dry uplands, air begins flowing from the deserts toward the coast. As the air descends in elevation it increases in temperature. The hot, dry Santa Ana winds of southern California are the result. These warm winds can rapidly dry out vegetation, and it is in these situations that the most severe brush and forest fires tend to occur. Whipped through the parched vegetation by fierce hot winds, the fires can rapidly spread across large areas, charring vegetation and expensive homes as the flames race out of control.

Air Density

Air becomes thinner as elevation increases. At an elevation of 3050 m (10,000 ft), a given volume of air contains about 70% of the gas molecules contained in the same volume at sea level. Some people who travel to high elevations suffer from altitude sickness because of the reduced oxygen content of the air. Plants and animals that live in high elevation environments must be adapted to thin mountain air. However, little scientific work has been carried out to date to see how this factor affects plants along an elevational gradient.

Heat capacity of air decreases with increasing elevation. Thin air at high elevations is unable to hold as much heat as denser air at lower elevations. The atmosphere warms rapidly in the daytime and cools rapidly at night at high elevations. Because fewer photons of incoming light or heat energy are absorbed by the atmosphere, insolation is greater at high elevations. At night thin air holds little of the heat re-radiated from the ground surface or from plants, and the temperature of the soil and plant surfaces decreases rapidly after dark. If temperatures drop below freezing, frost results. Frosts are more common at high than at low elevations as a consequence of lower overall air temperature and rapid temperature fluctuations.

Ultraviolet light, which is potentially harmful to plants and animals, is mostly absorbed by the atmosphere and only a small amount reaches the ground surface at low elevations. At high elevations, however, the intensity of ultraviolet radiation is greater because there is less atmosphere between the sun and the ground. The degree to which the increased level of ultraviolet radiation affects plants at high elevations has not been determined. Much of the incoming ultraviolet radiation is absorbed by the epidermis and does not penetrate very far into leaf tissues. Some damage to photosynthetic pigments may occur. Many plants at high elevations also produce large amounts of anthocyanin pigments that harmlessly screen out the ultraviolet radiation.

Snow skiers are very aware of the increased level of radiation at high elevations. Even though the temperature may be cold, high levels of incoming radiations plus that reflected from the white snow can cause serious sun burns. The role this increased radiation plays on plant growth and distribution is not completely understood.

The Growing Season

The growing season for plants in California is determined by a combination of factors. These include moisture availability, freedom from killing frosts, and amount and duration of snow cover. In many regions of the world the growing season is determined as the period between killing frosts. In California's Mediterranean climate the end of the growing season for many plants is marked not by frost but by summer drought. This is most applicable to shallowly rooted plants at low and mid elevations. Spring ephemerals, for instance, die when they can no longer extract water from the soil. At high elevations plants may be exposed to frost on almost any day of the year and must be able to grow in spite of frost. The time of snow-melt is for many plants the factor that determines when a growing season begins. Although air temperatures may be suitable for growth, snow prevents the plants from growing. For some sites the snow may not melt at all in some years. Consequently plants in these sites may not have a growing season at all during some years.

Wind

Winds are important to plants in several ways. Wind that occurs predominantly from a particular direction may determine the shape that a plant takes by bending the plant while it is growing. Strong, gusty winds may snap branches or tear off leaves or flowers. Strong winds also tend to increase the rate of transpiration. The combined effects of wind and fire are described above.

Many of the effects, however, are a result of the capacity of moving air to carry small objects. These small objects range from a salt-spray aerosol to airborne sand and dust, raindrops, snow and ice crystals. The effects of these wind-borne objects may be either direct, through physical damage to a plant, indirect through latent damage that occurs after the particles have settled on the plants, or through modification of the environment such as the movement of dune sands or snow drifts.

Strong winds occur under several different circumstances. The causes of hot, dry Santa Ana winds are described above. Topography is also very significant. Topographic features that parallel the direction of prevailing winds tend to channel and locally strengthen the winds. This is particularly notable in certain valleys and mountain passes. Coastal sand dunes tend to form where the shoreline is at a right angle to the direction of the prevailing winds. Blowouts in the dunes (Chapter 7) develop parallel to the wind direction.

Winds generally increase with elevation and reach their maximum in alpine and subalpine zones. In winter these winds move snow from place to place, often piling so much onto some slopes that avalanches are the result. Winds can break limbs, strip away foliage, and subject plants to desiccation and abrasion. Plants at high elevations, especially near treeline, may be severely wind-flagged or dwarfed by wind borne ice particles. Winds play a less significant role in overall habitat conditions at lower elevations except where particular topographic features occur.

Winter storms are often accompanied by severe winds, particularly in coastal areas. These winds can create high ocean waves that are responsible for coastal erosion and flooding. The combination of wind and waves can in a few hours bring about long-term changes in coastal plant communities.

Winds affect some plants as a dispersal agent for pollen, seeds, or fruits. Wind pollination is particularly common in some kinds of trees (e.g., oaks, pines), in grasses, sedges, rushes, and various other groups. Some communities are dominated by wind-pollinated species (grasslands, coastal salt marshes, coniferous forests). The wind-borne pollen of some species is the major cause of hay fever. The seeds or fruits of various plants bear wings or tufts of bristles or hairs that can be borne for short or long distances by winds. Such structures are particularly common in the Asteraceae. Entire plants of a few aggressive weedy species (e.g., *Salsola iberica* [tumbleweed or Russian thistle], *Sisymbrium altissimum* [tumble mustard]) are dispersed as tumbling rounded spheres that drop seeds as they roll across the landscape.

Atmospheric Pollution

One of the major byproducts of our civilization is the atmospheric pollution that clouds the skies and obscures distant features in much of California. The enormous population growth of California plus the proliferation of automobiles and other pollution sources have added a new variable to the state's climate. These airborne substances have various effects on plants, not all of which are fully understood. Components of the pollution clouds such as ozone, sulfuric acid, oxides of nitrogen, and various other compounds are highly reactive and can cause cell death or leaching of nutrients from plant cells. A significant effect is the death or diseased condition of large numbers of coniferous trees in the mountains of southern California. Because the intensity of the pollution has risen so rapidly and has been present at an elevated level for a comparatively short period, the long-term effects are unknown. If short-term effects are an indication, the effects will be severe and wide-ranging.

A component of air pollution that has recently received publicity is the suspended acid aerosols that are washed from the skies as acid rain. Because the major mountains of California are mostly granitic or volcanic, there is not a high level of basic ions available to buffer local waters against pH decrease. Lakes, bogs, and other moist upland areas are likely to become progressively more acidic. This will undoubtedly affect the community composition of these areas. Polluted air in coastal regions as well as in mountains becomes a major problem in the form of acid fog. The acidity of the droplets of acid fog may exceed that of vinegar.

Summary

As you can see from reading this chapter, California's diversity of plant communities is reflective of an equally diverse array of climates. All general climatic types except tropical ones occur in the state. Climates are complex and are created by the interaction of many factors and features. To really understand the role climate plays in plant distribution, one must look carefully at the diversity of local and microclimates that exist within one regional climate. Each of these localized climates is an integral part of the environmental complex that creates the mosaic of plant habitats and communities one encounters within a region. It is essential that local and microclimates are examined carefully when investigating distributions of plants and plant communities. However, also remember that climate is just one of the features of the environmental complex and by itself climate does not explain the distribution of plants in California.

References on California Climate

Bailey, H. P. 1966. The climate of southern California. Univ. Calif. Pr., Berkeley.

Barbour, M. G., and R. A. Minnich. 1990. The myth of chaparral convergence. Israel J. Bot. 39:453–463.

Brouillet, L., and R. D. Whetstone. 1993. Climate and physiography. Pp. 15–46 *in* Flora of North America Editorial Committee, Flora of North America, Vol. 1. Introduction. Oxford University Pres, New York.

Carter, D. B., and J. R. Mather. 1966. Climatic classification for environmental biology. Publ. Climatol. 19(4):305–395.

Donley, M. W., S. Allan, P. Caro, and C. P. Patton. 1979. Atlas of California, pp. 130–141. Academic Book Center, Portland, Oregon.

Halbwachs, G. 1983. Effects of air pollution on vegetation. Pp. 55–67 in W. Holzner, J. A. Werger, and I. Ikusima (eds.). Man's impact on vegetation, Dr. W. Junk Publ., The Hague, Boston London.

Hawkins, L. 1980. Gardener's guide to the Mediterranean climate. Pacific Horticulture 41(4):21–26.

James, J. W. 1966. A modified Koeppen classification of California climates. Calif. Geographer 7:1–12.

Major, J. 1965. Potential evapotranspiration and plant distribution in the western states with emphasis in California. Pp. 93–129 *in* R. H. Shaw (ed.), Ground level climatology. American Association for the Advancement of Science, Washington, D.C.

———. 1977. California climate in relation to vegetation. Pp. 11–74 *in* M. G. Barbour and J. Major (eds.). Terrestrial vegetation of California. John Wiley and Sons, N.Y.

Marotz, G. A., and J. F. Lahey. 1975. Some stratus/fog statistics in contrasting coastal plant communities of California. J. Biogeogr. 2:289–295.

Mather, J. R. 1985. The water budget and the distribution of climates, vegetation and soils. Publ. Climatol. 38(2):i–iii + 1–36.

———, and G. A. Yoshioka. 1968. The role of climate in the distribution of vegetation. Ann. Assoc. Amer. Geogr. 58:29-41.

Miller, D. H. 1955. Snow cover and climate in the Sierra Nevada of California. Univ. Calif. Publ. Geogr. 11. 218 pp.

Miller, P. C., D. K. Poole, and P. M. Miller. 1983. The influence of annual precipitation, topography, and vegetative cover on soil moisture and summer drought in southern California. Oecologia 56:385–391.

Minnich, R. A. 1983. Fire mosaics in southern California and northern Baja California. Science 219:1287–1294.

Patterson, M. T., and P. W. Rundel. 1995. Stand characteristics of ozone-stressed populations of *Pinus jeffreyi* (Pinaceae): extent, development, and physiological consequences of visible injury. Amer. J. Bot. 82:150–158.

Russell, R. J. 1926. Climates of California. Univ. Calif. Publ. Geogr. 2(4):73–84 + map.

Thornthwaite, C. W. 1948. An approach toward a rational classification of climate. Geogr. Rev. 38:55-94.

———, and J. R. Mather. 1955. The water balance, Publ. Climatol. 8(1):1–104.

Wolfe, J. A. 1979. Temperature parameters of humid to mesic forests of eastern Asia and relation to forests of other regions of the northern hemisphere and Australasia. U.S. Geol. Survey Prof. Pap. 1106. 37 pp. + 3 maps.

Chapter 4

Vegetational History of California

California's vegetation appears as it does because of a combination of modern-day and historical factors. Each plant in California is able to live within a particular range of environmental variables. These variables include climate, soils, topography and other living organisms. The precise interaction among these variables is not known in full for any species, but much is known or can be inferred about the modern-day factors that influence the range of a species. Each plant's ancestors also responded to a wide array of ecological factors. Evidence from geology and paleoclimatology indicates that the California of various times in the past differed in many ways from the California of the present. Fossils indicate that some plants occurred in the past in locations far from where their relatives live today, that some kinds of plants formerly lived in California that no longer occur in the state, and that many plants occur today in locations that at times in the past would have been completely unsuitable habitats. The fossil record and present-day distributions also indicate that many species in California have evolved as the climate and topography of the state have changed.

A detailed survey of California's vegetational history is beyond the scope of this text, but a brief overview is necessary to provide a background for the descriptions of communities in the chapters that follow. In-depth discussions of the history of California's vegetation can be found in some of the references listed at the end of this chapter. Not all of the sources are in full agreement, and the discussion that follows will differ from some of them in various ways.

The history of the state's vegetation is the sum of the individual histories of all of the component species. Each species in California's flora has its own unique history. Each has its own genetic background and its own ways of interacting with its environment. Unfortunately the fossil record provides clues to the history of only a very small sample of the many species that occur in the state. Those plants that have left fossils had to have died in the right place at the right time for fossils to have formed. To be useful to us those fossils that did form must have survived subsequent geological processes (uplift, erosion, metamorphosis, etc.) and must be found and studied by someone who can correctly interpret them. The fossil record is biased towards those kinds of plants that leave behind recognizable hard parts. Delicate herbaceous plants are very poorly represented by fossils. Often we can only speculate regarding the ecological interactions of the plants of the past. Our knowledge of these plants is a very incomplete picture, and much must be inferred from modern distributions, present-day ecological interactions and known geological events.

Some paleobotanists recognize communities of plants of the past as geofloras. A geoflora is defined as an assemblage of plants that has maintained itself with only minor changes in composition for millions of years. The geoflora concept is based on the idea that communities of the past remained relatively stable, unified, and recognizable, although composed of several to many individually varying component species. The geoflora concept is not accepted by all paleobotanists. Some have argued that the geoflora concept is an oversimplification that requires a degree of organization and unity in community structure not observed in present-day communities. Others have qualified the use of the concept, acknowledging that there was undoubtedly much variation, but nevertheless admitting the concept as a useful generalization. The geoflora concept is useful to the extent that it summarizes observable similarities among communities of the past and helps to explain the vegetational changes that have brought about the formation of modern-day plant communities. The fossil record is too discontinuous and fragmentary for us to trace the history of all the individual lineages through space and time. Nonetheless, it should be remembered that the geoflora concept is a simplification of the actual history of the vegetation.

The history of the plant life of California is closely entwined with the history of the state's climate (Chapter 3) and its topographical features (Chapter 2). The climate of the state has changed from wet and tropical, with high precipitation throughout the year, to strongly seasonal with a cool damp winter and warm dry summers. The terrain of California has been modified over millions of years by enormous tectonic forces. Uplift, volcanism and faulting have raised up mountain ranges, plateaus and other topographical features. Erosion by water, wind, ice and gravity have sculpted canyons and valleys, worn down mountains and ridges, and deposited the weathered fragments as layer upon layer of sediments. Soils of various structure and chemical composition have developed on the diverse terrain features. All the while plants, animals and other organisms have lived and died over the surface of the land. Over geological time the plant life has changed as the face of California changed.

The discussion that follows is based upon the Cenozoic history of California (Table 1). This era of approximately 65 million years has been a time of great climatic, geological and vegetational changes throughout the world. Although plants and animals occupied California in still earlier eras of the earth's past, their history is so remote from the present that their influence on the present is in many cases not readily measurable.

Table 4-1. Subdivisions of the Cenozoic Era.

	Years Before Present
Quaternary Period	
Holocene Epoch	0–12,000
Pleistocene Epoch	12,000–1,800,000
Tertiary Period	
Pliocene Epoch	1,800,000–5,000,000
Miocene Epoch	5,000,000–23,000,000
Oligocene Epoch	23,000,000–35,000,000
Eocene Epoch	35,000,000–53,000,000
Paleocene Epoch	53,000,000–65,000,000

Vegetational Changes of the Tertiary

The California of the early Tertiary Period was a land very different from the modern state. Many of the major terrain features were not yet in existence. The western half of the state had not yet formed. The Pacific Ocean extended to the foot of low mountain ranges that occupied the sites of the modern-day Sierra Nevada and Klamath Ranges. The slice of modern-day California that is at present located west of the San Andreas Fault was far to the south, part of the western coast of mainland Mexico. Volcanic activity had not yet raised the Cascades or the lava plateaus of the Pacific Northwest. The climate of California was moist and warm.

Evidence from fossils indicates that the vegetation of the state was tropical, with rain-forests and savannas extending well to the north along the Pacific Coast. Rainfall was plentiful and there was apparently no prolonged summer drought. There were no deserts in the state, and areas that are now in the rain-shadow of the Sierra Nevada and other high mountains were well-watered and forested. The plant communities of California included many plants whose modern relatives occur only in frost-free areas of the tropics. The fossil remains of these ancient tropical plants are grouped by some paleobotanists with other similar fossils as the Neotropical-Tertiary Geoflora. These tropical communities included palms, tree ferns, cycads and many other plants of tropical affinities. The southern third of North America was tropical at this time and tropical plants of California resembled those of such areas as New Mexico and the southeastern United States. The low mountain ranges of California were probably vegetated by coniferous forests.

Climatic fluctuations occurred during the early Tertiary but the state remained largely tropical until the end of the Eocene Epoch. Near the end of the Eocene the climate became more strongly seasonal and a gradual cooling and drying took place over much of the state. With this change of climate came major vegetational shifts. Tropical elements in California's flora were progressively shifted from the interior of the state and concentrated toward the coast and to the south. Gradually these tropical plants were eliminated from the state's flora. The elements of the Neotropical-Tertiary Geoflora did not disappear as a group, but declined one by one.

The forests that replaced the tropical vegetation in California were part of a northern forest that, according to some paleobotanists, extended from eastern Asia to Alaska and through the northern half of North America and across Greenland and the narrow Atlantic Ocean into Europe. The fossils of this great forest have been recognized as the Arcto-Tertiary Geoflora. It included numerous deciduous hardwoods as well as conifers. Some modern-day California plants are apparently descended directly from Arcto-Tertiary ancestors. Many of the forest trees of the state have such an ancestry. However, many genera of the Arcto-Tertiary Geoflora that once were represented in California no longer occur in the state. Some are completely extinct. Others are represented at present by modern-day species in the eastern part of the United States, in eastern Asia, or in both regions. These include the majority of the deciduous hardwoods that once grew in California and some of the conifers.

The western portion of North America underwent many changes during the mid-Tertiary. Extensive volcanic activity in the Pacific Northwest raised up great plateaus of lava along with the precursors of the Cascade Range. Interactions between the great North American and Pacific crustal plates and the smaller Farallon Plate crushed together and uplifted oceanic sediments, coastal islands and continental shelf to form the California Coast Ranges. The Central Valley was still a lobe of the ocean into which sediments were being deposited by erosion from the surrounding mountains. Toward the end of the Miocene Epoch the Baja California peninsula began to separate from the Mexican mainland and to move northwest along the fault system that includes the modern San Andreas fault.

Accompanying these vast geological changes were climatic shifts. The climate of California was becoming more seasonal, with a progressively less tropical character. The mountains being uplifted began to cast rain shadows on the interior of western North America. At the same time the climate was also becoming cooler with a more pronounced seasonality. The changing climate affected the vegetation in many ways. The cooling of the climate that began early in the Tertiary and that culminated in the great ice ages of the Pleistocene Epoch brought about latitudinal and altitudinal shifts in the vegetation. Many species changed evolutionarily as the climate and topography were modified. In California the drought-intolerant species were progressively restricted to mountain slopes and coastal environments. Many of the deciduous hardwoods and some of the conifers were entirely eliminated from the state's flora as the climate changed. A major cause of this decline was probably the shift from a summer-moist to a summer dry climate. Species unable to adapt to the dry summers or to shift to areas that remained moist became extinct.

As the climate of the state changed to the detriment of some kinds of plants, other species adapted to the new conditions. An assemblage of species of warm-temperate or dry tropical affinities began to expand, particularly in the comparatively warm, dry, mountainous areas of the American southwest and northern Mexico. The members of this assemblage, the Madro-Tertiary Geoflora, were characterized by comparatively small, often thick leaves. Some were drought-deciduous and others had evergreen, sclerophyllous leaves. Some of these formed subtropical thorn-forests in the southern parts of California and adjacent regions. Others contributed to the changing vegetation of the remainder of California. Assemblages of plants similar to those of the Madro-Tertiary Geoflora occur today in the Sierra Madre mountains of Mexico, hence the name, *Madro*-Tertiary.

Plants derived from Madro-Tertiary ancestors are very prominent in California's modern flora. Many species from the chaparral, woodland and grassland communities apparently had their origin in this assemblage of plants. The modern representatives of this geoflora in California do not represent the full spectrum of variation that formerly occurred in the state. With the ever more pronounced summer drought that characterized the end of the Tertiary and the Quaternary Periods, species requiring summer moisture were restricted to localized moist sites or eliminated from the state's flora.

Vegetational Changes of the Quaternary

By the end of the Pliocene Epoch many components of California's modern flora and vegetation were well-established, and many of the archaic elements had been eliminated, but more great changes were yet to come. The Quaternary has been a period of enormous changes both in California and throughout the rest of the world. Commencing about 1.5 million years ago, the climate of the earth underwent a series of major cold episodes during which giant ice sheets formed on the northern continents and mountains glaciers expanded in highlands throughout the world. The shifting climate changed both latitudinal and elevational zonation of vegetation. The climate of California became much cooler. The ice sheets came and went repeatedly. Each of the cold glacial episodes lasted for many thousands of years. Mountain glaciers scoured the slopes of the high mountains of the state, carving deep, U-shaped valleys and depositing vast piles of rubble at their tips and on their flanks. Meltwater from glaciers flooded low-lying areas creating large lakes. The San Joaquin Valley was flooded, as were basins of the Basin and Range Province (Chapter 3). So much water was tied up in the form of ice on the land that the level of the sea was depressed by as much as 100 meters, exposing coastal shelves and increasing the size of islands, sometimes forming dry-land bridges connecting island to mainland or island to island.

In addition to the great climatic shifts there were also major geological changes in California. The Pleistocene was a period of great mountain building. The Cascade volcanoes rose to their present height on the platform built by earlier eruptions. The Sierra Nevada, the Transverse Ranges, the Coast Ranges and the Peninsular Ranges underwent much uplift as well. The rising high mountain crest enhanced the rain shadow effect over the areas to the east. High mountain habitats became available during interglacial periods for colonization and speciation.

Cold-intolerant species were forced to shift latitudinally or elevationally, to adapt evolutionarily to the changed conditions, or to go extinct. Some species expanded their ranges as suitable habitats were opened up in new areas. Forest communities spread from mountains into adjacent valleys. Species spread across previously unforested valleys to formerly inaccessible mountains. The coastal climate was milder than that of the interior and provided a refuge from the ice and cold. The exposed coastal shelf was invaded by plants of many kinds. Forest communities were more widespread in coastal areas than they are today, extending far to the south of their present distribution.

The glacial episodes were followed by warming periods of shorter duration during which the ice sheets and mountain glaciers melted, releasing vast floods of water. Sea levels once again rose, flooding coastal areas and land-bridges. Plants reinvaded previously glaciated terrain as the climate warmed and elevational and latitudinal shifts once again took place. Species that had spread across forested valleys retreated up the slopes into mountain regions where they were isolated from other conspecific populations. Again and again the cycle was repeated. The Holocene (or Recent) Epoch apparently represents merely the most recent interglacial period. The retreat of the last ice sheets began about 12,000 years ago.

The end of the last glaciation was accompanied by an unprecedented dying off of many species of large mammals. Little is known about the ecological relationships of these large animals with the plant communities of the state. Such animals as mammoths and giant ground sloths were a part of the Pleistocene fauna of California. Their demise coincided with the arrival of humans in the area, a correlation that perhaps is more than coincidental. Whatever the cause of their extinction, their disappearance must have had an effect on the plants that they used for food and on the communities in which they lived.

At the end of the last glacial period the ice cap of the Sierra Nevada and the mountain glaciers of other high mountains began to melt off, leaving behind deep scarred valleys and piles of glacial debris. Only a few tiny remnants of these glaciers persist in the northern and central Sierra Nevada today. The retreat of the glaciers was followed by a period warmer and drier than the present climate of California. This Xerothermic Period lasted from approximately 8000 to 4000 years ago. During this prolonged drought major changes took place. Most of the lakes that had occupied valleys in Transmontane California and the Great Basin region dried out, leaving behind barren, salt-encrusted plains and lakebeds. Desert communities expanded into rain-shadow areas. Forest communities retreated up into the mountains and in coastal areas were progressively fragmented and isolated in local moist sites or restricted to northern areas. Chaparral, coastal scrub and drought-tolerant woodland communities became established in previously forested areas. Grasslands expanded.

Following the Xerothermic Period the climate moderated somewhat, but plants that had been eliminated from parts of their previous range often were unable to return to their former habitats. Man was well-established on the scene by this time and human activity was influencing the vegetation in various ways. Perhaps the greatest influences were a result of man's association with fire. Various of California's Indian tribes burned the vegetation to open it up and increase its value as a habitat for game. Human-caused fires probably eliminated fire-susceptible species from many areas. Fire-tolerant communities such as chaparral expanded at the expense of other communities.

Vegetational Changes of the Past 300 Years

The coming of Europeans to California and the subsequent colonization, agricultural and urban development of the state have probably changed the vegetation as much as did the climatic shifts of previous times. Large areas have been fundamentally altered. The original vegetation has been stripped away and replaced by farms, mines, roads, towns and cities. Forests have been cut to the ground. Species of plants have been introduced into California for use as crops or ornamentals or inadvertently as weeds. Agriculture has changed the state in many ways. Plowing has changed the character of the vegetation, disturbing the soil and opening it up for invading weedy species. Grazing of domesticated animals has affected the vegetation in many ways. Highly palatable native plants have been reduced or eliminated. Introduced species tolerant of grazing pressure, particularly annual grasses of Eurasian ancestry, have displaced the native grasses, creating a new kind of grassland community. The changes brought about by civilization continue today at an accelerated pace. Human-induced changes in California's vegetation are discussed in subsequent chapters and detailed in Chapter 22.

References on the History of California's Vegetation and Flora

Anderson, R. S., and S. L. Carpenter. 1991. Vegetation change in Yosemite Valley, Yosemite National Park, California, during the protohistoric period. Madroño 38:1–13.

Axelrod, D. I. 1950. Evolution of desert vegetation in western North America. Carnegie Inst. Wash. Publ. 590:215–260.

———. 1958. Evolution of the Madro-Tertiary geoflora. Bot. Rev. 24:433–509.

———. 1959. Geological history. Pp. 5–9 *in* P. A. Munz and D. D. Keck. A California flora. Univ. California Press, Berkeley.

———. 1967. Evolution of the California closed-cone pine forest. Pp. 93–149 *in* R. N. Philbrick (ed.). Proceedings of the Symposium on the Biology of the California Islands. Santa Barbara Bot. Gard., Santa Barbara.

———. 1967. Geologic history of the California insular flora. Pp. 267–315 *in* R. N. Philbrick (ed.). Proceedings of the Symposium on the Biology of the California Islands. Santa Barbara Bot. Gard., Santa Barbara.

———. 1973. History of the Mediterranean ecosystems in California. Pp. 225–277 *in* F. di Castri and H. Mooney (eds.). Mediterranean type ecosystems, origin and structure. Springer-Verlag, N.Y.

———. 1975. Evolution and biogeography of Madrean-Tethyan sclerophyll vegetation. Ann. Missouri Bot. Gard. 62:280–334.

———. 1976. Evolution of the Santa Lucia fir (*Abies bracteata*) ecosystem. Ann. Missouri Bot. Gard. 63:24–41.

———. 1976. History of the conifer forests, California and Nevada. Univ. Calif. Publ. Bot. 70:1–62.

———. 1977. Outline history of California vegetation. Pp. 139–193 *in* M. G. Barbour and J. Major (eds.). Terrestrial vegetation of California. John Wiley and Sons, N.Y..

———. 1978. Origin of coastal sage vegetation, Alta and Baja California. Amer. J. Bot. 65:1117–1131.

———. 1980. History of the maritime closed-cone pines, Alta and Baja California. Univ. Calif. Publ. Geol. Sci 120:1–143.

———. 1980. Contributions to the Neogene paleobotany of central California. Univ. Calif. Publ. Geol. Sci. 121:1–212.

———. 1982. Age and origin of the Monterey endemic area. Madroño 29:127–147.

———. 1989. Age and origin of chaparral. Pp. 7-19 *in* S. C. Keeley (ed.). The California chaparral. Paradigms reexamined. Natural History Museum of Los Angeles County, Science Series no. 34.

———, and P. H. Raven. 1975. Evolution and biogeography of Madrean-Tethyan sclerophyll vegetation. Ann. Missouri Bot. Gard. 62:289–334.

Campbell, D. H., and I. L. Wiggins. 1947. Origins of the flora of California. Stanford Univ. Publ. Biol. Sci. 10:1–20.

Chabot, B. F., and W. D. Billings. 1972. Origins and ecology of the Sierran alpine flora and vegetation. Ecol. Monogr. 42:163–199.

Daubenmire, R. 1978. Evolution of North American Vegetation. Pp. 48–65 *in* Plant Geography with special reference to North America. Academic Press, N. Y.

Delcourt, P. A., and H. R. Delcourt. 1993. Paleoclimates, paleovegetation, and paleofloras during the late Quaternary. Pp. 71–94 *in* Flora of North America Editorial Committee, Flora of North America, Vol. 1. Introduction. Oxford University Pres, New York.

Detling, W. R. 1961. The chaparral formation of southwestern Oregon with considerations of its postglacial history. Ecology 42:348–357.

Graham, A. 1993. History of the vegetation: Cretaceous (Maastrichtian) —Recent. Pp. 57–70 *in* Flora of North America Editorial Committee, Flora of North America, Vol. 1. Introduction. Oxford University Pres, New York.

Major, J., and S. A. Bamberg. 1967. Some cordilleran plants disjunct in the Sierra Nevada of California and their bearing on Pleistocene ecological conditions. Pp. 171–178 *in* H. E. Wright and W. H. Osborn (eds.). Arctic and alpine environments. Indiana Univ. Press, Bloomington.

Mason, H. L. 1947. Evolution of certain floristic associations in western North America. Ecol. Monogr. 17: 201–210.

Nowak, C. L., R. S. Nowak, R. J. Tausch, and P. E. Wigand. 1994. Tree and shrub dynamics in northwestern Great Basin woodland and shrub steppe during the Late-Pleistocene and Holocene. Amer. J. Bot. 81:265–277.

Raven, P. H. 1973. The evolution of "mediterranean" floras. Pp. 213–224 *in* F. di Castri and H. Mooney (eds.). Mediterranean type ecosystems, origin and structure. Springer-Verlag, N.Y..

———, and D. I. Axelrod. 1978. Origin and relationships of the California flora. Univ. Calif. Publ. Bot. 72:1–134.

Robichaux, R. 1980. Geologic history of the riparian forests of California. Pp. 21–34 *in* A. Sands (ed.). Riparian forests of California. Their ecology and conservation. Univ. Calif Div. Agric. Sci., Davis.

Schaffer, J. P. 1993. California's geological history and changing landscapes. Pp. 49–54 *in* J. C. Hickman, ed., The Jepson manual. Higher plants of California. University of California Press, Berkeley, Los Angeles, and London.

Sharp, R. P. 1960. Pleistocene glaciation in the Trinity Alps of northern California. Amer. J. Sci. 258:305–340.

Stebbins, G. L. 1978. Why are there so many rare plants in California? II. Youth and age of species. Fremontia 6(1):17–20.

———. 1982. Floristic affinities of the high Sierra Nevada. Madroño 29:189–199.

———, and J. Major. 1965. Endemism and speciation in the California flora. Ecol. Monogr. 35:1–35.

Tidwell, W. D., S. R. Rushforth, and S. Simper. 1972. Evolution of floras in the intermountain region. Pp. 19–39 *in* A. Cronquist, A. H. Holmgren, N. H. Holmgren and J. L. Reveal. Intermountain Flora. Vol. 1. New York Bot. Gard. and Hafner Publ. Co., N.Y.

Wells, P. V. 1983. Paleobiogeography of montane islands in the Great Basin since the last glaciopluvial. Ecol. Monogr. 53:341–382.

———, and R. Berger. 1967. Late Pleistocene history of coniferous woodland in the Mohave desert. Science 155:1640–1647.

———, and J. H. Hunziker. 1976. Origin of the creosote bush (*Larrea*) deserts of southwestern North America. Ann. Missouri Bot. Gard. 63:843–861.

Wilken, D. 1993. California's changing climates and flora. Pp. 55–58 *in* J. C. Hickman, ed., The Jepson manual. Higher plants of California. University of California Press, Berkeley, Los Angeles, and London.

Wolfe, J. A. 1969. Neogene floristic and vegetational history of the Pacific Northwest. Madroño 20:83–111.

———. 1978. A paleobotanical interpretation of Tertiary climates in the Northern Hemisphere. Amer. Scientist 66:694–703.

———. 1980. Neogene history of California oaks. Pp. 3–6 *in* Plumb, T. R. (tech. coord.). Proceedings of the symposium on the Ecology, management, and utilization of California oaks. Pacific S.W. Forest and Range Exp.. Stn., Berkeley.

Chapter 5

Community Classification: the Need and the Problems

Even a cursory examination of California's natural landscape shows that plant species are distributed and aggregated into more-or-less definite plant communities. A **plant community** is an assemblage of plants that interact among themselves and with their environment within a space-time boundary. Plant communities usually do not form vegetational belts, but instead form a mosaic pattern on the landscape. Questions often posed are: "Why is grassland here and chaparral there?" Or, "Why is chaparral restricted to this area and coniferous forest to another area?" Questions such as these are not always easy to answer. Partial answers will be given in the chapters that follow, but we must emphasize that many 'why' questions cannot be fully answered. A plant community with its component species exists where it does as a result of evolution and past dispersal events coupled with complex interactions among the following factors: climate, soils, topography, biota, fire, and man. Each of these factors is internally complex and interrelated with the others and changes through time.

Community Structure: Dominance

Communities comprise individuals of varying sizes, physical structure, abundance, distribution and taxonomic affinities. These plants have varying roles in the structure of the community. Not all plants occurring in a community are of equal ecological importance. Usually some species, the **dominants**, greatly influence the environment of all the other species of the community. Some communities have only a single dominant species; other communities have several co-dominant species. Occasionally communities are so diverse that no dominance is apparent.

A community is almost always composed of many more **subordinate species** than dominant species. Subordinates have a smaller influence than the dominants. There are two general types of subordinates, dependent and tolerant. **Dependent subordinates** are those that are dependent on the conditions that are created and maintained by the dominants. Mistletoe is a good example. It is dependent on the host plant being in the community; without it the mistletoe would not be able to occur. *Oxalis oregona* (redwood sorrel) is perhaps another good example. It is restricted to heavily shaded, moist conditions created by tall coastal trees, especially coastal redwoods. **Tolerant subordinates** are those that tolerate conditions imposed on the habitat by the dominants but are not restricted to these areas. For example, *Vaccinium ovatum* (huckleberry) occurs in the shade of the coastal redwood forest and also in the adjacent highly exposed chaparral areas.

Dominance has been ambiguously defined by plant ecologists. Essentially it refers to those plant species whose removal would result in the greatest impact on edaphic, climatic and biotic features of the plant community (ecosystem). These plants are generally the tallest and cover the largest area in the community. However, a plant may dominate its environment in other ways. Dominance is usually a result of size of the plants (coverage), number of the plants (density), and the distribution of the plants in an area (frequency). Large plants usually influence the environment of plants around them in several ways. The largest plants in an area cast shade on the plants growing under or around them. The fallen leaves, branches and other litter derived from large plants can both physically and chemically modify the environment of associated smaller plants. The root systems of large plants may directly compete for water and nutrients with the roots of smaller plants. Generally trees have a larger sphere of influence than shrubs, and shrubs a greater influence than herbs.

In communities where size is not the determining factor, a species may be dominant because of the collective cover of numerous individuals. A species may be a dominant even though it has small stature if it is abundant. For instance, grasses are the dominants in some communities as a result of high density. The overall biomass of grasses may vastly exceed that of all the other members of the community even though individual grass plants are not any larger than the other plants in the community. The other species are simply not as numerous. Larger plants may be found in the community but occur in sufficiently small numbers to have only limited influence on the overall community.

Ecologists recognize a form of dominance called **sociologic dominance.** This is based on the importance of a species in controlling the regeneration of the community as it exists at that time. These plants may be understory species. For example, if an understory species inhibits the establishment of certain tree species by root competition for moisture and nutrients, it would be considered a sociologic dominant because of its influence on the community.

Another important term used by ecologists is **aspect dominance**. This term is applied to species that at first glance are very noticeable and appear to be dominant in a community but on closer inspection turn out to be not as numerous as some of the other, less showy species found in the community. This is common in grasslands and meadows during the peak wildflower season when certain species of bright, showy wildflowers give the appearance of being dominant because they provide such large beautiful, extensive color displays.

Often plants may dominate a small area within a community (where they are very important), but are not common elsewhere in the community. These are often described as **locally common** or **locally abundant**, indicating that they are not widespread, but where they do occur they may be important or dominant members of the community. For example, dominance may vary locally over a small area depending upon slope aspect, soil variation, drainage channels or other factors. A species may be dominant on one site and subordinate on another. Often dominance in a community shifts subtly from species to species in response to minor habitat variations. In communities with much site to site variation, it may not be possible to determine overall dominance.

Perennial plants usually have a more long-lasting influence than do annuals. Perennials tend to occupy the same site year after year whereas the distribution of annuals may be highly variable from year to year. All shrubs and trees are perennials, and once these plants become established they may maintain their status in the community for several to many years. In a community such as an alpine meadow, in which the growing season is short, perennials have the advantage of being able to flower as soon as conditions become suitable. Annuals may not have enough time to grow from seed to reproductive maturity in such an environment. Annuals may, however, dominate if perennials are few and small, and if the growing season is sufficiently long. For instance, most California grasslands are at present dominated by annual grasses.

Community Structure: Physiognomy

The physical structure of the plants dominant in a community together with their distribution in the community are both impor-

Fig. 5-1. A forest dominated by Jeffrey pine (*Pinus jeffreyi*) on the eastern side of the Sierra Nevada in Mono County. The understory is composed of a sparse shrub cover of Great Basin sagebrush (*Artemisia tridentata*) and an herb layer with scattered grasses and forbs.

tant in assessing community structure. On the basis of **physiognomy** (the overall appearance or life-form of a mass of vegetation), we recognize three major community groups: tree-dominated communities, shrub-dominated communities and herb-dominated communities. Some communities may be dominated by members of more than one size class.

Tree-dominated communities can be grouped as forests, woodlands and savannas. **Forests** (Fig. 5-1) comprise fairly dense assemblages of relatively tall trees. Often the canopies of adjacent large trees overlap forming a more-or-less continuous canopy over the ground. The trees of a forest tend to have overlapping areas of influence at the ground level. In some forest communities there is a considerable amount of vertical stratification. The dominant or overstory trees form a leafy canopy that is exposed to direct sunlight. In the shade of the overstory trees one or more layers of understory plants may be developed. Saplings of the dominant trees or successional trees together with shade-tolerant subordinate trees may form a subcanopy below the canopy formed by the dominants. Often the subcanopy is more broken than that of the overstory. Shade-tolerant shrubs and tree seedlings may combine to form a shrub layer. At the ground level an herb layer may or may not be well developed. In some forests one or more of the lower strata may be absent or poorly developed. As an example, the montane mixed coniferous forest around Yosemite has an overstory of various confers such as ponderosa pine, incense cedar, sugar pine and Douglas-fir. Under these dominants is a layer of smaller trees such as black oaks, dogwoods and alders. Below these trees is a scattered bushy layer composed of various shrubs such as ceanothus, gooseberries and manzanitas, and along the ground surface are a variety of scattered herbs.

Woodlands (Fig. 5-2) generally have a less continuous tree cover than do forests and are often composed of smaller individual trees. In a woodland, the area of direct influence of one tree may or may not overlap areas influenced by adjacent trees (Fig. 5-2A). Understory vegetation in woodlands varies markedly depending on habitat conditions and density of the trees. In some cases, the understory is dominated by shrubs with only a sparse herbaceous cover; in other cases it is dominated by a sparse to lush herbaceous cover with few shrubs. Foothill woodland communities are good examples of this type of physiognomy. Here various oak trees form the overstory and shrubs typical of chaparral or coastal scrub communities may form patches or may be scattered in a more or less continuous cover of typical grassland in the understory. **Savanna** communities are similar to woodlands except that the trees are more widely spaced and the understory is almost entirely dominated by various species of grasses and forbs (Fig. 5-2B). The arcas of influence of adjacent trees are usually well separated and seldom overlap. Most plants of the herb layer in a savanna are intolerant of deep shade. A good example of a savanna in California are some of the valley oak woodlands.

All gradations exist between forests, woodlands and savannas, and a stand of trees may vary from open woodland to closed-canopy forest over a short distance. No sharp boundary exists between savannas and woodlands; an open stand of trees may vary from savanna to woodland depending on tree density. In foothill woodland communities, for instance, the trees on south-facing slopes may be widely scattered, whereas those on adjacent north-facing slopes may be clustered together.

California has many communities dominated by shrubby species (Fig. 5-3). Some of these shrub-dominated communities have been given names with a particular connotation. We are treating the terms **shrubland** and **scrub** as essentially interchangeable for any shrub-dominated community [though

A
B
C

Fig. 5-3. Desert scrub community on the eastern side of the Sierra Nevada in Kern County. Dominants include various drought-tolerant shrubs. Photo by David Keil.

some ecologists treat them separately]. The nature of a particular shrubland community is often indicated by a descriptive term indicating the dominants or the ecological position of the community (e.g., salt-bush scrub or coastal dune scrub). The term, chaparral, has a particular connotation in that it describes dense shrubland communities dominated by stiffly branching, mostly evergreen shrubs with thick, leathery leaves (sclerophylls) and hard wood.

Fig. 5-2. Examples of woodlands. **A.** Foothill woodland in southern San Luis Obispo County. Dominants are coast-live oak (*Quercus agrifolia*) and blue oak (*Q. douglasii*). On the openly wooded south-facing slope the understory includes many shrubs. On the more densely wooded north-facing slope herbaceous plants are the principal components of the understory. Photo by David Keil. **B.** Savanna-like foothill woodland near Shell Creek in the La Panza Mountains of San Luis Obispo County. The scattered trees include valley oaks (*Quercus lobata*), blue oaks, and foothill pines (*Pinus sabiniana*). The understory is dominated by grasses. Photo by Paula Kleintjes. **C.** Northern juniper woodland in Siskiyou County dominated by western juniper (*Juniperus occidentalis* var. *occidentalis*). The understory is dominated by Great Basin sagebrush (*Artemisia tridentata*). Photo by V. L. Holland

Fig. 5-4. Grassland community near Mount Shasta.

Shrubland communities vary from dense stands with a continuous shrub canopy to very open stands with much open space separating individual shrubs. There may be a well-developed herb layer, or herbs may be absent or nearly so. Very open shrub communities with a grassy understory are sometimes described as shrub-savannas.

Shrubland or woodland communities that occur in permanently wet soils with standing water are **swamp** communities. California has few areas that could be classified as true swamp communities. Those that occur in the state are found in riparian areas, often in association with freshwater marshes.

California has several kinds of herb-dominated communities. **Grasslands** (Fig. 5-4) are the most widespread of these. These communities (also called prairies and steppes) are dominated by various species of grasses. Such a community often includes various non grassy herbs (called forbs) and sometimes scattered, low-growing shrubs. Grasslands often intergrade with woodland or savanna communities and grassland species may form the herbaceous understory in these communities.

Meadows are open grassy areas in otherwise wooded areas (or above the tree-dominated communities in high mountains). The dominant plants in meadows are usually grasses and sedges, frequently with an assemblage of associated herbaceous species. Meadow soils are usually seasonally moist and frequently are composed of fine-grained sediments.

Marshes are permanently or seasonally inundated communities dominated mostly by sedges, rushes, cattails and other semi-aquatic herbs. Marshes differ from swamps in lacking a significant woody component in the community structure. Marsh soils are generally saturated through most or all of the year. Seasonal marshes also occur and are characterized by being inundated only during the wet season. Species composition in seasonal marshes is somewhat different from that in permanently inundated marshes. The salt content of the water has a very important effect on the species composition and community structure. Marshes are usually grouped as freshwater marshes and saltwater marshes.

Community Characteristics

We perceive plant communities around us. How real are they though? As defined above, a community is "an assemblage of plant species that interact among themselves and with their environment within a space-time boundary." This definition seems quite satisfactory until we examine it closely. The first part of the definition gives us no trouble, until we realize that every plant community is unique at whatever level we choose to measure or describe it. How big does an assemblage have to be to be considered a community? Do a few individuals growing together constitute a community? No two assemblages are exactly alike in ecology, physiognomy, or composition.

The second part of the definition, the spatial boundaries of communities, presents even more difficulty. Where two or more contiguous communities come into contact, they overlap and form transition zones called **ecotones**. Ecotones may be rather abrupt, such as where an aquatic community comes in contact with a terrestrial community, or they may be quite broad such as the change from woodland to forest along an elevational gradient. In many locations vegetation forms a mosaic of shifting plant assemblages, sometimes with sharp boundaries and sometimes with indistinct ecotones (Fig. 5-5). In some of these cases, the ecotonal zones could be considered to be communities in themselves. For example, the transition between desert woodland communities on the eastern slope of the Tehachapi Mountains and foothill woodland communities on the western slope covers a zone several miles wide. Is this an ecotone or are the intermediate assemblages of plants worthy of recognition as communities?

Boundaries in time may be just as difficult to determine. Communities change through **ecological succession** and as a result of various types of disturbances. Ecological succession is a gradual process. While the beginning and end stages of succes-

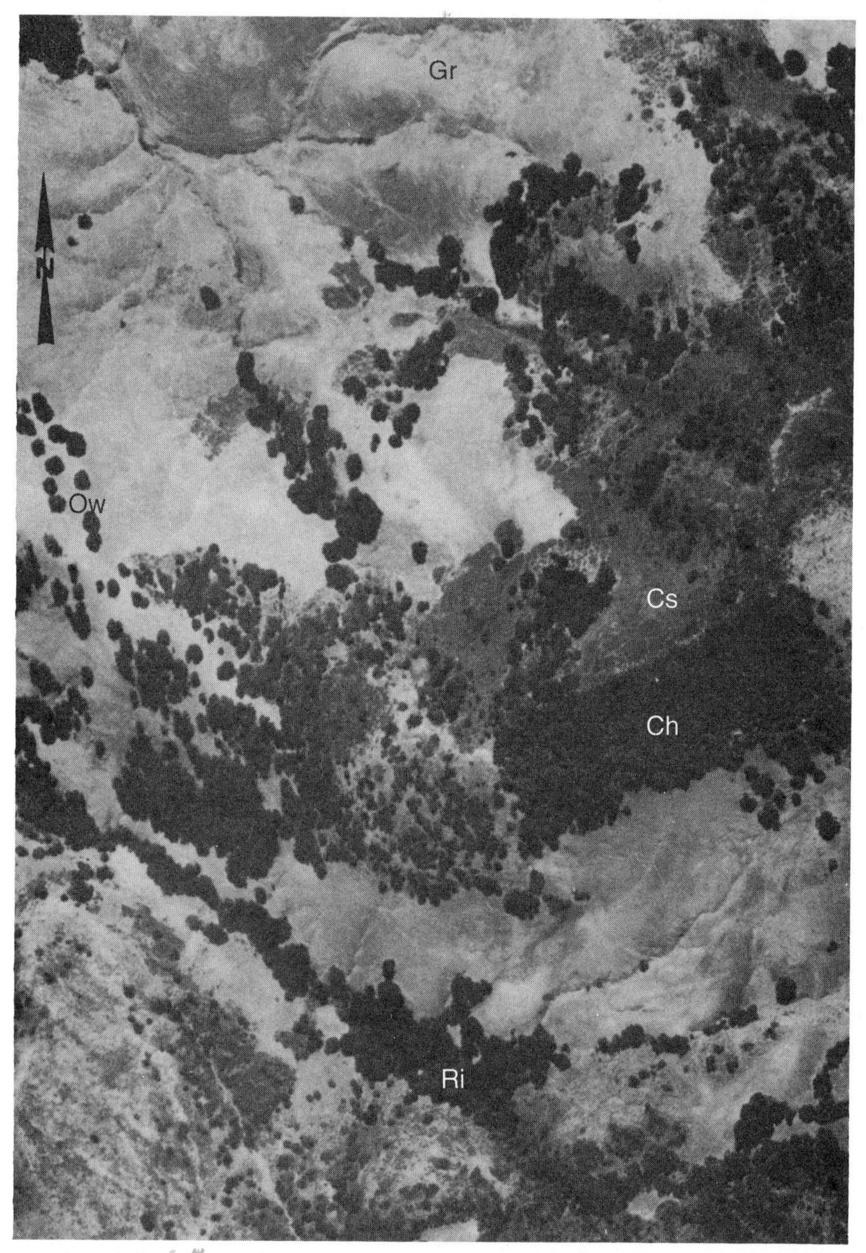

Fig. 5-5. Aerial photograph of coastal hills southeast of San Luis Obispo. The vegetation is a mosaic of grassland (Gr), coastal live oak woodland (Ow), chaparral (Ch), southern coastal scrub (Cs), and riparian (Ri). The distribution of vegetation is dependent on a combination of soils, slope angle and exposure, and moisture availability. Ecotones vary from diffuse to rather sharp.

sion may be very different, there is a continuous gradient of intermediates. For example, following a forest fire, it may take a hundred or more years for the forest to mature and reach the climax forest community. (A **climax community** is one that has reached equilibrium with its environment and is able to reproduce and maintain itself for an indefinite period of time). During the succession back to the climax forest, the area will be dominated by a whole series of **seral** (immature, non-climax) communities that gradually change through time. These in turn may be subjected to disturbances. It is important for an ecologist to recognize whether the vegetation in the area is in a climax stage, or if it will change with time. However, this is sometimes difficult to determine. It is often important to know the precise history of a patch of land in order to recognize whether the community is in a state of succession. Because in many cases ecological succession is so slow, changes may not be readily perceptible during individual human lifetimes.

Communities are not permanent assemblages. They vary in time. They are composed of species, each of which has its own unique evolutionary history. Community history can be observed both in evolutionary time and in successional time. We perceive the latter much more readily than the former because it corresponds more closely with our own lifetimes and with recorded history. However, the species composition of a particular stand is a response to both evolutionary and successional history. In theory at least it is possible to trace the unique evolutionary history of a group of related organisms. In classifying species, genera, and families we group organisms that share a common ancestry and history. Communities do not have this kind of close genealogical relationship, however. When we classify communities, we must realize that the modern-day assemblages we see may represent organisms with very different histories that happen to grow together at present. Their occurrence together at present does not mean that their ancestors grew together, though. Classification of communities is not based on the same kind or strength of evidence as is classification of taxa. An overview of the vegetational history of California and its importance to modern day vegetative units are discussed in Chapter 4.

Classification of California's Plant Communities

The primary purpose of classification systems, whether applied to plant communities, taxonomic groups, or to inanimate objects of our everyday surroundings, is communication. We organize the world around us into groups of items with some perceived similarities, we organize these groups into hierarchies, and we apply names to the groups. In the process of classifying we often must simplify, overlooking or deliberately choosing to ignore minor or even major inconsistencies to make our classification systems useful to us. Any community classification will involve some of these problems. Anyone who wants to look closely enough can find exceptions to any community classification—local aggregations of species that do not "fit" into recognized categories.

California has a diverse and complex vegetation that corresponds to the diversity and complexity of environments present in the state. This makes classification very difficult, and sometimes subjective and arbitrary, but nevertheless important. An analogy may be made with the colors of the rainbow. Everyone readily accepts classification of colors into red, orange, yellow, etc. However, in reality, they are not that clear-cut. In examining a rainbow, one notes that red gradually changes into orange which grades into yellow and so on. If one selects a given portion of this color spectrum or gradient, it is easy to say this is red, or this is orange, but what about the intermediate areas? When does red-orange become orange-red? How many of these color variations should we identify for a classification system of colors? In comparison with communities, colors are simple to classify. Colors are expressions of a single variable, the wavelength of light. Communities are the product of the independent spatial and temporal variations of the all the component species and numerous environmental variables. Classification in biology is not simple because of the natural complexity of living systems. The complexity and variation found in plant communities make it very difficult to categorize them into units. However, it is important to have a classification system so that we can refer to plant communities by name and communicate about community ecology in an understandable way.

Communities may be classified on the basis of habitat characteristics, physiognomy of the component plants, or species composition. An example of classification by habitat characteristics is coastal sand dune communities. These communities show variation in both physiognomy and species composition from site to site, but all share the feature of growing in a unique habitat, coastal sand dunes.

Some communities are commonly distinguished by their physiognomy. Chaparral communities, for instance, may vary from site to site in habitat and/or species composition but are characterized by the morphological features shared by the dominant plants. These features include stiff woody stems, evergreen sclerophyllous leaves, and a shrubby or dwarf tree growth habit. Sometimes phases of chaparral, like some communities, are classified by species composition, e.g., chamisal chaparral where chamise is dominant.

Where one or a few species are overwhelmingly dominant or very conspicuous in a community, we often characterize and name the community by these species. For example, we recognize Joshua Tree Woodlands as identifiable plant communities based on the dominance of *Yucca brevifolia* (Joshua tree). Although the species composition of the associated plants may be extremely varied depending on environmental conditions and site history, the dominance of the Joshua trees makes these communities readily recognizable. Other excellent and well-known examples are coastal redwood forests, named for the dominant species *Sequoia sempervirens* (coast redwood) and red fir forests, dominated by *Abies magnifica* (red fir).

The classification system that we outline in the chapters that follow is a pragmatic one. We have used habitat characteristics where it seemed most useful for conveying information. Similarly we have employed physiognomy and species composition in our system where these aspects appeared to be most appropriate. Wherever possible we have attempted to tie together all three. We admit that this is an artificial classification that suffers, like all other community classifications, from generalizations, over-simplifications and lack of information in some areas. Much remains to be known about California plant communities. Our purpose is communication about an imperfectly known subject.

One aspect of our classification that we wish to emphasize is our use of the plural when dealing with communities. In many references one can find statements such as "The Montane Coniferous Forest is characterized by.." We prefer to word such statements differently: "Montane Coniferous Forests are characterized by .." While such wording differences may seem to be mere semantics, we believe that there is an important underlying philosophical difference between them. The first version defines communities as if they were real entities. The second version shows that we are recognizing groups, not entities. In our treatment we do not often "divide" a community type into "subtypes", but instead group communities into higher level categories. Our approach comes from the bottom rather than from the top.

Some of the plant communities we have recognized are endemic to California, are well known and are of great interest to the public as well as the scientific community (e.g., coastal redwood forests, oak woodlands, chaparral, coastal salt marshes). Some, such as the desert shrublands, chaparral, oak woodlands and montane coniferous forests, are very extensive and cover large portions of the state. Others are restricted to rather limited

Table 5-1. Outline of the community classification used in this text.

Formation Community Type Phase	Approximate percent of California land area
Marine Aquatic Communities	<1.0 %
Subtidal and Intertidal Communities	
Coastal Estuarine Communities	
Coastal Salt Marsh Communities	
Coastal Sand Dune and Beach Communities	<1.0 %
Pioneer Dune Communities	
Dune Scrub Communities	
Dune Wetland Communities	
Coastal Scrub Communities	2.8 %
Northern Coastal Scrub Communities	
Southern Coastal Scrub Communities	
Southern Semidesert Coastal Scrub Communities	
Sea-Bluff Coastal Scrub Communities	
Chaparral Communities	9.0 %
Mixed Chaparral Communities	
Chamisal Chaparral Communities	
Red-Shanks or Ribbon-Bush Chaparral Communities	
Manzanita Chaparral Communities	
Ceanothus Chaparral Communities	
Scrub Oak Chaparral Communities	
Maritime Chaparral Communities	
Island Chaparral Communities	
Serpentine Chaparral Communities	
Montane Chaparral Communities	
Semidesert Chaparral Communities	
Grassland Communities	13.0 %
Native Bunchgrass Grasslands	
Valley Grasslands	
Northern Coastal Grasslands	
Desert Grasslands	
Closed-Cone Coniferous Forest Communities	<1.0 %
Coastal Closed Cone Conifer Forests	
Coastal Cypress Forests	
Coastal Closed Cone Pine Forests	
Pygmy Forests	
Interior Closed Cone Conifer Forest Communities	
Knobcone Pine Forests	
Interior Cypress Forests	
Coastal Coniferous Forest Communities	3.0 %
North Coast Coniferous Forests	
Coastal Redwood Forests	

Table 5-1 (continued).

Community	Percent
Mixed Evergreen Forest Communities	4.7 %
Northern Mixed Evergreen Forests	
Central and Southern Mixed Evergreen Forests	
Sierran Mixed Hardwood Forests	
Oak Woodland Communities	10.0 %
Coastal Live Oak Woodlands	
Valley Oak Woodlands	
Foothill Woodlands	
Northern Oak Woodlands	
Southern Oak Woodlands	
Island Oak Woodlands	
Montane Coniferous Forest Communities	19.4 %
Montane Mixed Coniferous Forest Communities	
Coulter Pine Forests	
Ponderosa Pine Forests	
Jeffrey Pine Forests	
Mixed Conifer Forests	
Giant Sequoia Forests	
White Fir Forests	
Red Fir Forest Communities	
Lodgepole Pine Forest Communities	
Subalpine Forest Communities	
Alpine Communities	<1.0 %
Alpine Meadows	
Rocky Alpine Communities	
Desert Alpine Communities	
Desert Woodland Communities	3.0 %
Piñon Pine and Juniper Woodland Communities	
Piñon-Juniper Woodlands	
Sierran Piñon Pine Woodlands	
Northern Juniper Woodlands	
Mountain Juniper Woodlands	
Southern Piñon and Juniper Woodlands	
Joshua Tree Woodland Communities	
Desert Scrub Communities	29.0 %
Great Basin Sagebrush Scrub Communities	
Saltbush Scrub Communities	
Blackbush Scrub Communities	
Creosote Bush Scrub Communities	
Desert Sand Dune Communities	
Desert Dry Wash Communities	
Alkali Sink Communities	
West Central Valley Desert Scrub Communities	
Riparian Communities	1.0 %
Valley and Foothill Riparian Communities	
Montane Riparian Communities	
Desert Riparian Communities	

Table 5-1 (continued).

Freshwater Wetland Communities	2.5 %
Limnetic Plant Communities	
Freshwater Marsh Communities	
Bog and Fen Communities	
Montane Meadow Communities	
Vernal Pool Communities	
Anthropogenic Communities	not estimated*
Agrestal Communities	
Pastoral Communities	
Ruderal Communities	
Plantations	
The Urban Mix	

*Human activities have modified the original vegetative cover in some parts of the areas occupied by all of the other communities.

unique habitats and, while ecologically very important, do not cover large land areas. Examples of these are the vernal pools, freshwater marshlands, alpine communities, coastal salt marshes and coastal sand dune communities. A number of ecologists and botanists have developed classification schemes for California's plant communities. These schemes have identified from as few as 15 and to many as 280 different plant communities. In our classification system we recognize 68 groups of plant communities which are categorized into 16 formation-type vegetative units based on physiognomy, species composition and/or common habitat requirements. An outline of the classification system used in this book is found in Table 5-1 along with an estimate of the percentage of California that is or could be potentially covered by these major vegetative units.

The percent cover values are approximations, at best, which were modified from the potential areal coverage figures given in *Terrestrial Vegetation of California* by Barbour and Major. No cover estimates are given for areas where the native vegetation has been replaced by urban development, agriculture, etc., or for anthropogenic (human-influenced) communities. Native plant communities continue to disappear at a rapid rate every day as a result of land modifications such as urbanization, land clearing, agricultural practices, etc. This makes the actual area covered by plant communities very difficult to measure or estimate. As an example, grassland communities are estimated to potentially cover 13 percent of the state. However, a major portion of these potential grassland communities is located in the Central Valley where agricultural land use practices and rapid urban growth have already eliminated much of the grassland vegetation and are continuing to eliminate still more of these communities. The cover approximations given in Table 5-1 are provided only to give the reader a general idea of the relative areal coverage of the major vegetation types found or potentially found in California.

A Note about Distribution Maps

In the chapters that follow, maps are often presented that depict the distributions of communities or species. The maps purport to show the areas where a community or species occurs, but the overall area of the map inevitably exceeds that given in Table 5-1 or actually occupied by the community or species. When mapping the distribution of some communities of small overall area in California we have chosen to darken in enough of the map that the community can actually be located. Otherwise, on the scale of the map used in this book, some of the areas would be too small to be noticeable. In almost all cases the actual distribution of vegetation types is patchy. As we have separately mapped the communities, their distributions have broad areas of overlap. The scale of the map is not sufficient to show more than overall patterns of distribution. Even large vegetation maps such as that of Küchler (1977) make many compromises in the presentation of community distributions. A brief examination of the aerial photograph in Fig. 5-5 should be sufficient to indicate the futility of attempting to map community distributions precisely.

The Natural Diversity Data Base Natural Communities Program

We also must acknowledge the work of the California Department of Fish and Game in developing and maintaining The Natural Diversity Data Base (NDDB) Natural Communities program. This program keeps up-to-date records of California's rare communities as part of The Natural Diversity Data Base. This data base describes, classifies and maintains an inventory of California's plant communities which is separate from the inventory of rare plants.

The Natural Diversity Data Base (NDDB) community inventory uses a classification of natural communities originally devised by Cheatham and Haller (1975). Their system recognized about 250 communities which has been modified over the years and now includes about 280 communities. Of these, 135 are considered rare and need more study and protection. This classification system was designed so that an inventory of rare communities could be put into a data base comparable to the NDDB and CNPS rare plant inventories. Currently, CNPS, in conjunction with NDDB, Department Fish and Game, is preparing a multi-faceted effort to inventory, define and protect California's natural communities and ecosystems. Once this inventory is completed and published, it will be used as the classification system of natural communities used by the State Department of Fish and Games Natural Diversity Data Base. It will also be used by The Nature Conservancy in the California section of their national and western regional vegetation classification that they are currently developing.

The new CNPS classification system is hierarchical arrangement using **series** as the organizational level. They define a series as a community that "represents areas of the landscape dominated by similar plants". A series is composed of stands with similar overstory composition. Stands are defined as local examples of plant communities defined by floristic composition and structure. The CNPS classification system will also have dichotomous keys to assist the reader in identifying the various series.

The CNPS system differs from ours in describing a much larger number of series than we do in our communities classification system. We find that California's plant communities are so diverse that attempting to describe all of the variations may be confusing and still may not always fit the vegetation in a given area. Thus, we prefer using a somewhat broader classification system with a detailed description of the local variation found within the community being studied. For example, within a given area of Pioneer Coastal Dune communities, one can find significant variation in species composition from site to site. Some areas may be dominated by beach-bur and sea-rocket, others by beach-bur, beach morning glory, and dundelion, others by sand verbena, others by beach-bur, ice plant and some by European beach grass, all within a short distance of one another. Describing all the variations that occur within such a single dune system would be cumbersome and may not be of value in describing the vegetation of the area. Thus, we prefer describing an area like this as Pioneer Coastal Dunes dominated by various mixtures of sand verbena, beach-bur, beach morning glory, ice plant, dundelion and sea rocket.

We applaud CNPS for taking on such a huge, ambitious undertaking, and we will refer to this classification in our book so our readers can easily go from our discussion of communities to the series, stands and habitats described in the new CNPS system. Appendix 1 compares our system with the most recent draft CNPS system. The CNPS hopes to publish the first edition of their system by mid-1995.

References on Community Classification

Barbour, M. G. 1988. Californian upland forests and woodlands. Pp. 131–164 *in* M. G. Barbour and W. D. Billings (eds.)., North American Terrestrial Vegetation. Cambridge Univ. Press, Cambridge.

———, and W. D. Billings (eds.). 1988. North American Terrestrial Vegetation. Cambridge Univ. Press, Cambridge.

———, and N. L. Christensen. 1993. Vegetation. Pp. 97–131 *in* Flora of North America Editorial Committee, Flora of North America, Vol. 1. Introduction. Oxford University Pres, New York.

Benson, L., and R. A. Darrow. 1981. Overview of the southwestern deserts. Pp. 1–26 *in* Trees and shrubs of the southwestern deserts. 3rd ed. Univ. Arizona Press, Tucson.

Billings, W. D. 1951. Vegetational zonation in the Great Basin in western North America. Pp. 101–122. *in* Comp. rend. du colloque sur les bases ecologiques de regeneration de la vegetation des zones arides, Union Internat. Soc. Biol., Paris.

———. 1941. The problem of life zones on Mt. Shasta, California. Madroño 6:49–56.

Bradley, W. G., and J. E. Deacon. 1967. The biotic communities of southern Nevada. Nevada State Mus. Anthrop. Paper 13, part 4.

Braun-Blanquet, J. 1932. Plant sociology. Translated by G. D. Fuller & H. S. Conard. McGraw-Hill, N.Y.

Brown, D. E., ed. 1982. Biotic communities of the American Southwest—United States and Mexico. Desert Plants 4:1–341.

Cain, S. A., and G. M. Castro. Manual of vegetation analysis. Harper & Row, N.Y.

Cheatham, N. H., and J. R. Haller. 1975. An annotated list of California habitat types. University of California Natural Land and Water Reserve System, unpubl.

Critchfield, W. B. 1971. Profiles of California vegetation. U.S.D.A. Forest Serv. Res. Pap. PSW-76. 54 pp.

Daubenmire, R. 1968. Plant communities: a textbook of plant synecology. Harper & Row, N.Y. 300 pp.

———. 1978. Plant geography with special reference to North America. Academic Press, N.Y. 338 pp.

Dice, L. P. 1943. The biotic provinces of North America. Univ. Michigan Press, Ann Arbor. 78 pp.

Franklin, J. F. 1988. Pacific Northwest forests. Pp. 103–130 *in* M. G. Barbour and W. D. Billings (eds.)., North American Terrestrial Vegetation. Cambridge Univ. Press, Cambridge.

———, and C. T. Dyrness. 1973. Natural vegetation of Oregon and Washington. U.S.D.A. Forest Serv., Pac. N.W. Forest and Range Exp. Stn. Tech. Rept. PNW-8. 417 pp.

Gleason, H. A., and A. Cronquist. 1964. The natural geography of plants. Columbia Univ. Press, N.Y. 420 pp.

Goldsmith, F. B., and C. M. Harrison. 1976. Description and analysis of vegetation. Pp. 85–155 *in* S. B. Chapman (ed.). Methods in plant ecology. Blackwell Scientific Publ., Oxford.

Holland, R. F. 1986. Preliminary descriptions of the terrestrial natural communities of California. Nongame Heritage Program, Calif. Dept. Fish and Game, Sacramento. 156 pp.

Holland, V. L. 1977. Major plant communities of California. Pp. 3–41 *in* D. R. Walters, M. McLeod, A. G. Meyer, D. Rible, R. O. Baker, and L. Farwell (eds.). Native plants: a viable option. Calif. Native Pl. Soc. Spec Publ 3, Berkeley.

Howell, J. T. 1957. The California flora and its provinces. Leafl. West. Bot. 8:133–138.

Jensen, D., and G. Holstein. 1983. Natural communities—terrestrial section. Calif. Dept. of Fish and Game, Natural Diversity Data Base. 10 pp.

Jensen, H. A. 1947. A system for classifying vegetation in California. Calif. Fish and Game 33:199–266.

Keeley, J. E., and S. C. Keeley. 1988. Pp. 165–208 *in* M. G. Barbour and W. D. Billings (eds.)., North American Terrestrial Vegetation. Cambridge Univ. Press, Cambridge.

Kellman, M. C. 1980. Plant geography. 2nd ed. Methuen, London & N.Y. 181 pp.

Küchler, A. W. 1949. A physiognomic classification of vegetation. Ann. Amer. Assoc. Geogr. 39:201–210.

———. 1964. Manual to accompany the map. Amer. Geogr. Soc. Spec. Publ. 36, N.Y. 116 pp.

———. 1964. Potential natural vegetation of the conterminous United States. Amer. Geogr. Soc. Spec. Publ. 36, N. Y. map.

———. 1977. Appendix: the map of the natural vegetation of California. Pp. 909–938 *in* M. G. Barbour and J. Major (eds.). Terrestrial vegetation of California. John Wiley and Sons, N.Y.

———. 1977. Natural vegetation of California. Map in folder *in* M. G. Barbour and J. Major (eds.). Terrestrial vegetation of California. John Wiley and Sons, N.Y.

Latting, J. (ed.). 1976. Plant communities of southern California. Symposium Proc., California Native Pl. Soc., Spec. Publ. 2. 164 pp.

Lowe, C. H. 1964. Arizona's natural environment. Landscapes and habitats. Univ. Ariz. Press, Tucson. 136 pp.

MacMahon, J. A. 1988. Warm deserts. Pp. 231–264 *in* M. G. Barbour and W. D. Billings (eds.)., North American Terrestrial Vegetation. Cambridge Univ. Press, Cambridge.

Matyas, W. J., and I. Parker. 1980. CALVEG Mosaic of existing vegetation of California. Regional Ecology Group, U.S. Forest Service, San Francisco. 27 pp + large map.

Mayer, K. E., and W. F. Laudenslayer Jr. (eds.). 1988. A guide to wildlife habitats of California. Calif. Dept. Forestry and Fire Protection, Sacramento. 166 pp.

McIntosh, R. P. 1967. The continuum concept of vegetation. Bot. Rev. 33:130–187.

Mueller-Dombois, D., and H. Ellenberg. 1974. Aims and methods of vegetation ecology. John Wiley & Sons, N.Y. 547 pp.

Odum, E. P. 1971. Fundamentals of ecology. 3rd ed. Saunders, Philadelphia. 574 pp.

Oosting, H. J. 1956. The study of plant communities: an introduction to plant ecology. 2nd. ed. Freeman, San Francisco. 440 pp.

Paysen, T. E., J. A. Derby, H. Black, V. C. Bleich, and J. W. Mincks. 1980. A vegetation classification system applied to southern California. Pacific SW Forest and Range Exp. Stn., Gen. Tech. Rept. PSW-45. 33 pp.

Phillips, E. A. 1959. Methods of vegetation study. Holt, Rinehart & Winston, N.Y. 107 pp.

Raven, P. H., and D. I. Axelrod. 1978. Origin and relationships of the California flora. Univ. Calif. Publ. Bot. 72:1–134.

Rzedowski, J. 1981. Vegetación de México. Editorial Limusa, Mexico City. 432 pp.

Sawyer, J., and T. Keeler-Wolf. 1995. Series level descriptions of California vegetation. California Native Plant Society, Sacramento.

Shimwell, D. W. 1972. The description and classification of vegetation. University of Washington Press, Seattle. 322 pp.

Sims, P. L. 1988. Grasslands. Pp. 265–286 *in* M. G. Barbour and W. D. Billings (eds.)., North American Terrestrial Vegetation. Cambridge Univ. Press, Cambridge.

Thorne, R. F. 1976. The vascular plant communities of California. Pp. 1–31 *in* J. Latting (ed.). Plant communities of southern California. Symp. Proc., Calif. Nat. Pl. Soc., Spec. Publ. 2.

Vankat, J. L. 1979. The natural vegetation of North America. John Wiley & Sons, N. Y. 261 pp.

West, N. E. 1988. Intermountain deserts, shrub steppes, and woodlands. Pp. 209–230 *in* M. G. Barbour and W. D. Billings (eds.)., North American Terrestrial Vegetation. Cambridge Univ. Press, Cambridge.

Whittaker, R. H. 1975. Communities and ecosystems. 2nd ed. MacMillan, N.Y. 385 pp.

Zippen, D. B., and J. M. Vanderwier. 1994. Scrub community descriptions od the Baja California Peninsula, Mexico. Madroño 41:85–119.

Chapter 6

Plant Names

One of the most remarkable aspects of California's plant life is the state's diversity of plant life. The flora of California comprises over 5000 different named plant species. One of the major challenges is keeping track of the names for all these plants. No botanist alive today is familiar with all the different kinds of plants that occur in California. Fortunately, as a result of the efforts of generations of California botanists, books are available that provide an inventory of the state's flora and means for identification of unfamiliar plants. With a basic knowledge of plant structure and some experience in using keys and botanical descriptions, a person can determine the name of just about any plant found growing wild in California. The most up-to-date manual for identification of California plants is the *Jepson Manual*. Published in 1993 after a ten-year joint effort by almost 26 botanists, the *Jepson Manual* is set to serve as the standard reference on the California flora for years to come.

The names that we use for plants serve as means for communication. They allow us to convey the concept of a plant without necessarily having a sample of the plant in hand. Through its name we can find information that has been published about a plant.

There are two more or less independent kinds of names for plants. These are **scientific names** and **common names**. Scientific names evolved out of the study and reporting of plants in books during the period from the 13th to 18th centuries. This was a period when Europeans were finding out about the world beyond their continent and many unfamiliar plants and animals were being seen for the first time. There was a need to catalog these new finds, some of which had potential economic or medicinal value. Because the language of learned men of this period was Latin, names of plants were written in Latin. During the 1700s the Swedish botanist, Carolus Linnaeus standardized the way in which scientific names of species are constructed, a convention we still use today.

Historically, common names have been created by people doing the everyday living of life. They are words in the language of the layman and thus are easy to understand and use. Common names are widely used and are perceived by most of the general public as "friendlier" and easier to understand than scientific names. There are many different systems of common names that have developed in different cultures and languages. Undoubtedly many of California's plants had names in the languages of the various native American tribes that lived in the state before the arrival of the Spanish colonists. Many of the plants subsequently were given Spanish and then English common names as well.

Scientific names

The application of scientific names is regulated by the *International Code of Botanical Nomenclature*. This book sets forth a series of principles that underlie botanical nomenclature and details the rules for naming plants in a series of Articles and Recommendations. The purpose of the code is to standardize the procedures for naming plants, to rectify nomenclatural problems, and ultimately to stabilize the names of plants. To the extent that our understanding of plant relationships and our interpretation of taxonomic boundaries remain stable, the *Code* has been successful. As discussed below, however, there are situations where name changes become necessary and the scientific nomenclature seems anything but stable.

Early botanists did not have a code of nomenclature. The rules we follow gradually developed through the practices of early taxonomists and were later codified. Not all botanists, of course, did things the same way and certain early practices are now considered to be against the rules. In order to consistently apply the *Code* it has been necessary to retroactively outlaw certain early procedures.

The Classification Hierarchy

The system of classification used for living organisms can be viewed as a set of nested boxes—large boxes with smaller boxes inside (Table 6-1). Each level in the hierarchy is a taxonomic rank. An individual box corresponds to a taxon (plural, taxa) —a recognized group of plants that has been given a name. The names of taxa above the genus level have standardized suffixes that indicate the taxon's rank . The taxonomic rank for these can automatically be determined from the name. These names are *not* underlined or italicized. Some higher level taxa (e.g., subfamily and tribe) are only used for certain families. Additional taxa (e.g., suborder) can be added if needed in a classification scheme.

Table 6-1. Classification hierarchy for the California chicory (in the system of Cronquist, 1981).

Kingdom	Meta**phyta**
Division	Magnoli**ophyta**
Class	Magnoli**opsida**
Subclass	Aster**idae**
Order	Aster**ales**
Family	Aster**aceae**
Subfamily	Lactuc**oideae**
Tribe	Lactuc**eae**
Genus	*Rafinesquia* (no standardized suffix)
Species	*Rafinesquia californica* (no standardized suffix)

Construction of Genus and Species Names

The scientific names we use most frequently in *California Vegetation* are genus[1] (generic) names and species names. A **genus name** is a single word. Regardless of its derivation, it is considered to be a Latin noun and is subject to the rules of grammar of that language. Some generic names were used by the ancients. Many more have been coined as needed for newly discovered genera. The origin of the word that is used to make a new generic name most commonly is Greek or Latin, but a name can originate from any other language if it can be written in the Roman alphabet. The words chosen to be Latinized for use as names usually have some relationship to botany. A generic name may be derived from various sources such as a combination of botanical terms, a person's name, the name of a mythological being, or it may even be a meaningless [but pronounceable] combination of letters (Table 6-2)[2]. Regardless of its origin, a generic name is treated as a singular Latin noun written in the nominative case and may be either masculine, feminine, or neuter. [Latin nouns have a grammatical gender that usually has nothing to do with the actual gender, if any, of the object being named.] A generic name should be capitalized and either underlined or written in italics.

---

[1]The plural of genus is genera. The word species is spelled the same way whether it is singular or plural.

[2]In *The Jepson Manual* (Hickman, 1993) and *A California Flora* (Munz and Keck, 1959) the derivation of each generic name is given at the end of the description of the genus.

Table 6-2. Derivation of some generic names in the California flora.

Allium	Latin for garlic
Muilla	An anagram of *Allium*
Convolvulus	Latin meaning twisted together
Crassula	Latin—diminutive of thick
Cassiope	Greek mythology—Mother of Andromeda
Helianthus	Compound of Greek words meaning sun flower
Xylorhiza	Compound of Greek words meaning woody root
Clarkia	Named for Capt. William Clark of the Lewis and Clark Expedition
Krascheninnikovia	Named for Stephan Kraschenninikov, 18th century Russian naturalist
Pinus	Ancient Latin word for pine tree
Tsuga	Japanese word for hemlock tree
Pseudotsuga	Compound of Greek and Japanese meaning false hemlock

Species names are binomials, two-word names written in Latin. A **species name** consists of two parts. The first word is the name of the genus to which the plant is assigned. The second word, the **specific epithet**, is usually either an adjective or a possessive noun. The specific epithet by itself is *not* a species name. A specific epithet is usually written in lower case and underlined or italicized. [Certain specific epithets may be capitalized (e.g., those derived from a person's name), but the *International Code of Botanical Nomenclature* recommends that specific epithets always be written in lower case]. The specific epithet is usually either a Latin adjective, or a noun in the genitive (possessive) case (Table 6-3). If the specific epithet is an adjective, it must agree with the noun it modifies in grammatical gender and number (singular) and thus often has the same ending as the noun. The specific epithet may be descriptive of the plant, it may refer to its geographical range, or it may honor a person [generally someone who has some connection with botany]. There are exceptional cases in which the specific epithet is a second noun that is not possessive. Consider the examples below.

Cirsium brevistylum Cronq.

Cirsium andersonii (A. Gray) Petrak

In both cases the generic name is *Cirsium*, a genus in the sunflower family. The *-um* ending of *Cirsium* indicates that this word is grammatically neuter. [Generic names ending in *-a* are usually feminine. Those ending in *-us* are most commonly mas-

Table 6-3. Derivation of some specific epithets.

albiflora	Latin compound meaning "white-flowered"
californica	adjective meaning "pertaining to California"
chrysanthus	Greek compound meaning "yellow-flowered"
eastwoodiae	possessive noun honoring Alice Eastwood (1859–1953)
eastwoodiana	adjectival form of the name of Alice Eastwood
flavus	Latin meaning "yellow"
lanatus	Latin adjective meaning "woolly"
maritimus	Latin meaning "by the sea"
nemaclada	Greek compound meaning "thread-stemmed"
nevadensis	adjective derived from the Spanish word for snowy, generally pertaining to Nevada or to the Sierra Nevada
obispoensis	adjective meaning "pertaining to [San Luis] Obispo"
occidentalis	Latin adjective meaning "western"
orientalis	Latin adjective meaning "eastern"
sphaerocephalus	Greek compound meaning "spherical head"
watsonii	possessive noun honoring Serano Watson (1826–1892)

culine though some of the most common tree genera ending in -*us* are feminine (e.g., *Pinus*, *Quercus*). Words ending in -*is* may be masculine or feminine. Those ending in -*e* may be feminine or neuter. Names ending in -*um* are neuter.] The specific epithet of the first species is an adjective that in Latin means "short-styled". The second species was named in honor of Charles Lewis Anderson, a nineteenth century plant collector. The suffix -*ii*, which is the possessive case for the Latinized version of a man's name [or -*iae* for a woman's name], is the equivalent of -'s in English. The name means Anderson's thistle.

Both names have been printed here with their **authors**, the individual or individuals who are responsible for having given the plants their names. The names of these individuals are often abbreviated. The author for the first species is Arthur Cronquist, a leading 20th century botanist and expert on the sunflower family. The history of the second name is a bit more complicated. The species was originally named by Dr. Asa Gray, a prominent 19th century botanist from Harvard as *Cnicus andersonii*. About 50 years later, Franz Petrak, an Austrian botanist, concluded that this species would better be classified as a species of *Cirsium*, and transferred the specific epithet, *andersonii* from *Cnicus* to *Cirsium*. Gray (the **parenthetical author**) gets credit for having published the epithet, *andersonii*. Petrak (the **combining author**) gets credit for transferring the epithet to *Cirsium* and publishing the combination, *Cirsium andersonii*.

In almost all floras and identification manuals species names are written with their authors. Some manuals include a list of the abbreviations of authors' names with the name spelled out and a bit of biographical information provided for each. In *California Vegetation* we have chosen to not include author citations. Readers who wish to look up author citations should refer to the *Jepson Manual.*

Names of Infraspecific Taxa

Taxonomists commonly encounter species that are variable. Individuals may differ in size, flower color, leaf shape, or various other features. If these variations represent geographical races, a botanist may choose to formally recognize them as **subspecies** or **varieties**.[3] Subspecies are more inclusive than varieties. In a highly variable species both subspecies and varieties may be recognized. [This does not happen very often]. In practice, subspecies and varieties are often essentially equivalent taxa. One botanist may recognize infraspecific taxa as subspecies, whereas another botanist might treat the same plants as varieties. If a variant is sporadic in its occurrence and does not have a geographical range of its own, most taxonomists will not formally name it. Some botanists, however, recognize such plants as **forms**. The name of an infraspecific taxon is an epithet similar in construction to a specific epithet and preceded by a word or abbreviation that indicates its rank (e.g., *Arctostaphylos glandulosa* ssp. *crassifolia*; *Cirsium occidentale* var. *venustum*). It should be emphasized here that the horticultural "cultivar" is *not* a part of the formal system of botanical nomenclature.

When a species is divided into infraspecific taxa or when a taxonomist recognizes that two or more taxa that previously had been recognized as separate species are actually geographical races or forms of a single species, names must be adjusted accordingly. One of the infraspecific taxa that results includes the type specimen of the species. Its subspecific (varietal, formal) epithet will be exactly the same as the specific epithet, and the name will be written as in the following example:

Pinus contorta Louden ssp. *contorta*

---

[3]In the example of *Rafinesquia californica* in Table 6-1, no subspecies or varieties have ever been recognized. Therefore, these ranks were not indicated in its classification hierarchy.

Notice that the epithet of the subspecies that contains the type specimen is written without an author. This is often referred to as the "typical" subspecies because it contains the type specimen of the species, but other subspecies may be more common or widespread. The epithets of all other subspecies are written together with their authors:

Pinus contorta Louden ssp. *bolanderi* (Parl.) Critchf.

Pinus contorta Louden ssp. *murrayana* (Grev. & Balf.) Critchf.

Each subspecific epithet has its own type specimen. Not all species have infraspecific taxa.

Naming a Newly Discovered Plant

Although taxonomists have been naming plants for over two hundred years, new species are still being discovered. Occasionally taxonomists even discover previously unrecognized genera and families. Most of the new taxa now come from tropical areas that have been poorly explored by trained botanists. On occasion, however, new taxa are discovered even in well-botanized temperate regions like California. Sometimes new species are described when the taxonomy of a well-known genus is reevaluated. The plants may have been known all along, but they may not have been recognized as distinct until a detailed study has been carried out. In other cases something truly new and unknown comes to the attention of taxonomists. In the late 1980's, for instance, a previously unknown member of the Rosaceae was discovered near Lake Shasta in the foothills of the Cascade Range. This plant, since named *Neviusia cliftonii*, was found to be most closely related to a rare shrub from the southeastern U.S.

There are several steps that a taxonomist must follow to **validly publish** the name of a new plant (i.e., publish it in an acceptable format).

(1) The name must be properly constructed and it should not be a name anyone has ever used before.

(2) The taxonomist must clearly indicate the rank of the taxon being described; if it is being described as a subspecies, for instance, that must be so indicated.

(3) A specimen must be designated as the **holotype** to serve as a permanent reference point for the name. This specimen is deposited in a designated herbarium [plant museum] where it may be consulted by other botanists. The holotype is the tangible expression of the description and thus becomes the basis of comparison. Duplicate specimens of the holotype

(specimens of the same plant collected at the same time in the same place) are called **isotypes**. Type specimens are often specially curated in herbaria and are particularly important to a taxonomist who is attempting to determine the correct application of a name.

(4) The taxonomist must publish a description or diagnosis written in Latin. A diagnosis is a brief statement that indicates the ways in which the newly described taxon differs from other plants. A description of the plant in the language of the taxonomist and an illustration are often prepared but these are not requirements.

The requirement that a description or diagnosis written in Latin be part of the publication of a new taxon may seem archaic, but it actually is a perpetuation of the use of Latin as an international scientific language. Although the remainder of the publication may be in English, German, Russian, Japanese, or some other language, the essential description or diagnosis is in Latin. A taxonomist does not have to learn all the languages of the world to be able to understand the publication of a new species. Because so much of the early taxonomic work was written in Latin, a professional taxonomist must be able to work with this language. Some botanists who are more skilled than others in the use of Latin often are asked by their colleagues to prepare the required Latin descriptions or diagnoses. A very useful resource for writing and reading botanical Latin is a book entitled *Botanical Latin* (Stearn, 1983).

(5) The name and the accompanying information must be **effectively published** (i.e., printed in a publication that would be generally available to other botanists). New taxa are usually described in botanical journals or books such as manuals or floras. (No new species were described in *The Jepson Manual.*) Publishing the name of a new plant in a seed catalog, a newspaper, or some other ephemeral publication is not acceptable (although in earlier times such publications were allowed).

Why do Botanists Change Plant Names?

For many years the standard reference for California plants was *A California Flora* by Philip A. Munz and David D. Keck. When it was published in 1959 it was the most up to date compilation of information about the state's flora. Within a few years, however, so much new information had accumulated that Munz published a lengthy supplement. Over the years since the publication of the supplement, an accumulation of new information and corrections of old information made a new flora more and more necessary. When *The Jepson Manual* was finally published in 1993, many professional and amateur botanists pored over the pages to find out what names had been changed. Many were dismayed to learn that the long-familiar names of some plants had been replaced by new, unfamiliar names. Many species of owl's-clover, for instance, had been transferred from *Orthocarpus* to *Castilleja* where they joined the Indian paintbrushes. Others were now placed into *Triphysaria*, a genus unfamiliar to almost everyone; only a few remained in *Orthocarpus*. The familiar and widespread genera *Haplopappus* and *Stipa* had disappeared entirely and their species had been dispersed to various other genera. Changes had occurred in many other genera as well. Why were there so many name changes?

The answer to the question is not simple. Some changes are a result of the application of the rules of nomenclature. Earlier California botanists in some cases had made errors in the names they accepted and the rules require other names to be used. In other cases it was the advance of our knowledge of California's plants that brought about reinterpretations of the previous taxonomic treatments for particular genera and species. New methods of gathering or interpreting data sometimes resulted in evidence that relationships were not what earlier botanists had thought. Frequently these realignments result in name changes. To professional botanists many of these changes were old news. Over the 34 years since Munz and Keck published *A California Flora* many studies of California plants had been published in botanical journals. It was only when the changes that resulted from these studies were gathered together in *The Jepson Manual* that they seemed so numerous.

Some of the changes resulted from an application of the rules of botanical nomenclature. One of the most important principles is priority of publication. In its simplest terms, this means that the first correctly published name for a taxon is the one to use. The publication date of Linnaeus' *Species Plantarum* (May 1, 1753) effectively marks the beginning point for binomial nomenclature, and the publication of this book is considered the earliest date to which priority of publication applies. Any name published before this date (even if it was a binomial) has no standing as far as priority is concerned. Any binomial correctly published after that date must be considered for priority purposes.

Priority is very important when two or more names are discovered to apply to the same taxon. It is not hard to see how a plant could be named more than once. As early taxonomists attempted to sort out and name the many plants that were being sent to them or that they were collecting, it was inevitable that sometimes two or more taxonomists would independently name the same plant. Early botanists who visited California included Russians, Spanish, English, and Americans. It is not at all unusual for a species to have been discovered and named by several different botanists. Additionally some species are sufficiently variable that their extremes look quite different. Early taxonomists often worked from very limited samples and were not aware of the natural variation encountered in the wild. The importance of priority is that it allows us to decide which of the competing names to use—the earliest one published.

Ultimately, application of the principle of priority should serve to stabilize nomenclature. Even today, however, we are still discovering instances where the name that has long been in use for a plant is predated by another name. Many more names have been published for California plants than there are species in the state. Some early names that were published in obscure books or serials have only recently come to the attention of California botanists. In other cases modern studies indicate that two or more kinds of plants that had previously been treated as members of different species are actually so closely related that they belong in the same species. This sets up the conditions for application of the principle of priority. When species are merged, sometimes it turns out that the name that has been applied to a widespread, conspicuous species is not the oldest available name and the combined species end up with the name that had once been applied to a much less widespread plant.

Priority has been limited in certain specific cases by application of the concept of **conserved names** (nomina conservanda). In some cases botanists have discovered that a very widely used generic or family name is actually predated by an obscure, largely unknown name. Replacing all the names in the large genus with a duplicate set in the obscure genus would be likely to cause much confusion. In such cases the more commonly used name may be conserved or retained as the valid name, but this takes a special action of an International Botanical Congress. In this exception to the principle of priority, the name with the earlier publication date becomes a rejected name. The *International Code of Botanical Nomenclature* contains a list of conserved and rejected names. Until recently only generic and family names could be conserved. Under the provisions of the most recent edition of the *Code*, conservation of species names is possible, but only under very limited conditions.

The nomenclature of a group of plants is tightly tied to its classification. Sometimes botanists disagree as to the placement of a particular taxon or group of taxa. An early taxonomist may have considered a plant to be a member of one genus whereas a later botanist has evidence that it belongs in a different genus. When this happens, the later botanist may transfer the species from one genus to another. Some taxonomists (splitters) tend to see differences as more important than similarities and divide large groups into smaller taxa. Others (lumpers) may see similarities as more important than differences and merge small taxa into larger groups. As a result of differing taxonomic treatments a species may have correct names in several genera. A taxonomist authoring a particular treatment has to decide which classification is best supported and therefore which name to use.

Interior goldenbush, for example, is a common, spring-flowering shrub of southern California that has been classified in several different genera. David Keck (in *A California Flora*) followed a classification that treated it as *Haplopappus linearifolius*. Keck accepted *Haplopappus* as a broadly defined and highly variable genus. Evidence has accumulated from various sources that *Haplopappus* as treated by Keck consists of several lineages that are not each other's closest relatives. The original species of *Haplopappus* is a South American plant that is probably not closely related to any California species. Robert Hoover (in *The Vascular Plants of San Luis Obispo County, California*) followed a different system that placed species that Keck had treated as *Haplopappus* into several genera. He considered the interior goldenbush to be in a genus of its own, treating it as *Stenotopsis linearifolia*. Gregory Brown and David Keil (in *The Jepson Manual*) had access to published evidence that was unavailable to Hoover that indicates that the interior goldenbush is so closely related to members of the genus *Ericameria* that it readily forms hybrids with *E. cooperi*. Brown and Keil therefore treated the plant as *Ericameria linearifolia*. Each of the three names of the interior goldenbush is correct for the genus in which it was used.

Taxonomists may disagree as to the rank of a taxon as well, one treating a plant as a subspecies or variety and another treating the same plant as a distinct species. Once again in these cases there is a correct name for the plant in each of the alternative taxonomic placements. Alternate taxonomies for the Sierran lodgepole pine, for instance, treat it as a distinct species, *Pinus murrayana*, or as either a subspecies or a variety of *Pinus contorta*.

Alternate names for a plant are considered to be **synonyms**. There are two types of synonyms. **Taxonomic synonyms** are synonyms in the *opinion* of a taxonomist. They are names based on different type specimens and are considered to be synonyms because a taxonomist who has studied the plants has concluded that the specimens on which they are based are all members of the same taxon. **Nomenclatural synonyms** are based on the same type specimen (and almost always have the same specific or infraspecific epithet). Synonyms of either type can differ in rank.

Certain names cannot be used because they do not conform to the rules of nomenclature. These are called **illegitimate names**. For instance it is against the rules to use a name that has already been validly published for a different kind of plant. Such a name is a **later homonym** and it must be rejected. An example is the genus *Fremontia*. For many years this name was applied to flannelbush, one of California's most conspicuous shrubs. However, the name *Fremontia*, had previously been used for a completely different plant. The correct generic name for flannelbush is *Fremontodendron*, a name coined to replace the disallowed *Fremontia*. Later homonyms sometimes have resulted when a specific epithet that was correct in one genus is transferred to another genus where it is predated by an earlier use of the same epithet. Also against the rules are **tautonyms** (binomial names in which the same word is used for the generic name and for the specific epithet).

Pronouncing Scientific Names

Latin is no longer a spoken language and we do not know precisely how it was spoken in the Roman world. Many scientific names are words that were not a part of ancient Latin and would sound as foreign to the Romans as Latin does to us. Many English-speaking botanists pronounce Latin names as if the words were written in English. This is known as the Traditional English system. There are many variations and these are often passed on from teacher to student. Individual botanists are not always consistent in pronunciation, often pronouncing names as they first learned them, even if words of similar construction end up with differing pronunciations.

On the other hand, most classicists and many European botanists prefer Reformed Academic Latin in which strict rules govern the pronunciation of particular letters or combinations of letters. Phonetically the latter undoubtedly comes closer to the Roman pronunciation than does the English system. As an example, the family name, Rosaceae, is usually pronounced Ro-záy-see-ee by most English-speaking taxonomists and Ro-sáh-seh-ah by continental botanists. Differing pronunciations can hinder communication in some cases. In our increasingly internationalized world it is likely that American botanists will come in contact with scientists from other regions. A good source of information on pronunciation in the Reformed System is *Botanical Latin* by W. T. Stearn (1993). Weber (1986) suggested a set of pronunciation guidelines for American botanists who wish to communicate with botanists educated in other countries.

Although there is no consensus among botanists of the world regarding the pronunciation of vowel sounds, there are some general guidelines. Look at the word carefully and pronounce the word phonetically. All vowels in Latin are sounded. In some cases two vowels are pronounced together as a diphthong, making a single syllable. The most common of these are *ae* and *œ*. In most other two-letter combinations, both are pronounced. In an English word such as *trace* that ends in *e*, this terminal letter is usually silent, but in Latin a terminal *e* is always pronounced, as in the word *arvense* (ar-vén-se).

The editors of *The Jepson Manual* presented the following recommendations regarding pronunciation:

1. Divide the word carefully into syllables (it is safest to assume that every vowel belongs to a different syllable).
2. Pronounce each syllable (e.g., "co-to-ne-as-ter", not "cot-on-east-er).
3. Listen to others and practice what sounds good to your ear; conviction is important.
4. Attempt to accent all syllables equally; this is likely to show you where accents fall naturally (some manuals, but not this one, specify accents with stress marks).
5. Develop your own standards for pronouncing common endings like "aceae", "-iae", "-ensis", etc.
6. Retain pronunciation of proper names used in scientific names ("jones-eeee", not "jo-nes-ee-eye").
7. When someone presumes to correct your pronunciation, a knowing smile is an appropriate response.

Common names

Common names have a larger potential audience than do scientific names. Few people speak or write Latin today, and fewer still are learning classical languages. Far more people would recognize oak or maple than would recognize *Quercus* or *Acer*. Common names have the appeal of the familiar.

There are some disadvantages to the use of common names, however. Common names develop in the language of a given people and may not be useful to people with a different language or dialect. Wide-ranging plants often have several different common names, some of them widely used and others unique to a given locality or language. Because of California's Spanish language heritage, some of the state's plants have very dissimilar common names in English and Spanish. For example, alternate common names for *Encelia farinosa*, a common shrub of southern California, are brittlebush and incienso. Even in English,

Would you suspect from the names that California bay-laurel, Oregon-myrtle, and pepperwood are the same tree? Sometimes more than one kind of plant share the same common name. Three unrelated leafless shrubs of the southwestern deserts, for instance, share the common name, crucifixion thorn. On a local basis such common names may be very useful, but over a broader geographical area they may lead to problems in communication.

Common names can be misleading. Since there is no possible way to regulate the formation of common names or to legislate which ones gain acceptance, there is no way to be sure the names used are accurate. For example, seep-willow, arroyo willow, and willow-herb are unrelated plants even though they have similar names. Box-elder and elderberry are unrelated. Some botanists use hyphenation to indicate which common names show false relationships. Black oak (not hyphenated) is a true oak; poison-oak (hyphenated) is not an oak at all, but instead is a member of the cashew family. If everyone followed this practice, at least some of the confusion caused by common names would be eliminated. Unfortunately not all botanists use hyphenation in this fashion, and very few members of the general public are even aware of this convention. In this book we are using hyphenation of common names as described above.

People develop words only for those objects they notice. If a plant is inconspicuous or rare, it may not have a common name. Sometimes common names represent what a botanist would consider to be a single species and sometimes they represent a genus or still larger grouping. It is not always easy to determine which is the case in a particular instance. Often the common or conspicuous members of a genus have their own common names. The less common or less conspicuous species of the same genus may be known only by the common name of the genus (if it has one).

California has a remarkable diversity of plants that have no true common names. Some common and conspicuous genera and species lack widely accepted common names. Genera without accepted generic common names include *Clarkia*, *Cryptantha*, *Gilia*, *Monardella*, *Phacelia*, and many others. In some cases, the generic name has become the accepted common name, but this is by no means the rule.

Many "common names" have been coined by botanists to satisfy a perceived need for non-technical names for communication with members of the general public. Abrams and Ferris (1923–1960) coined thousands of "common names" for western plants, many of them direct translations of the scientific names. For example, the common name, "Hoover's eriastrum" is merely a translation of the scientific name, *Eriastrum hooveri*. However, laymen probably use these "common names" about as often as they use the scientific names. The editors of the California Native Plant Society's *Inventory of Rare and Endangered Vascular Plants of California*, have coined common names for some of the most uncommon, inconspicuous, and obscure plants in California because these are the names more likely to have meaning to policy makers, politicians, and the general public. Some translated common names have gained wide acceptance, though, particularly if a plant is conspicuous and there are no competing common names. For example, Jeffrey pine is the widely accepted common name for *Pinus jeffreyi*.

Problems sometimes occur when taxonomic realignments or nomenclatural problems require that a plant be given a new scientific name. A translated common name may seem inappropriate when the species to which it has been assigned is transferred to a different genus. Abrams used the common name "Blochman's hasseanthus" for *Hasseanthus blochmaniae*. However, the genus *Hasseanthus* has been merged with *Dudleya* in recent treatments of the Crassulaceae and "Blochman's hasseanthus" has been replaced by "Blochman's dudleya.

There have been some attempts to standardize common names and how they are constructed. As discussed, above, however, the realm of common names is often outside the botanical community. Members of the general public may ultimately determine acceptance and usage of common names. Attempts to standardize names sometimes cause problems. One botanist who was attempting to standardize common names for plants listed the common name of *Lupinus ludovicianus* as Louisiana lupine under the mistaken impression that the word, "*ludovicianus*", was derived from Louisiana. Actually, the plant is the county flower of San Luis Obispo County, California, the only place in the world where it grows. The word *ludovicianus* was derived from Luis instead of Louisiana, and the plant is usually called the San Luis Obispo lupine. Kartesz and Thieret (1991) have suggested a set of guidelines for botanists to use in standardizing the construction of common names.

Common names often carry unintended meanings. Common names that end in "weed" may be perceived negatively by members of the public. Many California natives that do not have the attributes of weediness described in Chapter 22 have common names that indicate otherwise: tarweeds, locoweeds, duckweeds, etc. Because of the negative connotation of these "weed" names among policy makers and many members of the general public, the editors of the California Native Plant Society's *Inventory of Rare and Endangered Vascular Plants of California* took the step of coining new "common names" for several kinds of rare plants.

The editors felt that politicians would be less likely to scoff at a rare plant as "just a weed" if it had a neutral or positive name.

Special mention should be given to one change of common name that we have adopted in this book. One of the most common trees in the foothills of cismontane California is *Pinus sabiniana*. Until recently the widely used common name for this tree has been digger pine. To most people this has been a neutral name of no more significance than blue oak or coast redwood. Some people might have wondered about the word, digger, but most were completely ignorant of the meaning. One group of California residents, however, consider this common name to be anything but neutral. During the 1800's the word "digger" was used in a pejorative sense to describe some of the native Americans of the state who dug into the ground to obtain edible bulbs and tubers. These people also harvested the large edible seeds of *Pinus sabiniana*. The tree gained the common name, digger pine, from this association. The descendants of these Indians regard the continued use of the common name as an insult. Two alternative common names have been proposed as replacements. Gray pine has been suggested because of the gray-green color of the wispy foliage of this pine. Foothill pine has been suggested because of the characteristic habitat of the tree. The latter name has been adopted in the *Jepson Manual*, and we have chosen to use it as well.

Nomenclatural Conventions used in *California Vegetation*

In *California Vegetation* we have adopted the scientific names used in *The Jepson Manual*. Because many users of *California Vegetation* may be more familiar with the nomenclature of *A California Flora*, we have included synonyms from Munz and Keck in brackets the first time a name is used in a chapter. In the index we have cross-referenced the names from *A California Flora* to their *Jepson Manual* equivalents.

We have tried to provide common names for all the plants discussed in this book. We have done so because of the wide audience for the book which includes laymen as well as serious students of the California flora. We want users to be able to communicate outside their area of expertise. A student using this book as a field botany text may need to communicate in the future with untrained laymen. A layman may need to look up additional information about a plant that can only be found through use of the scientific names.

In many cases the common names we have used are widely accepted. In other cases our personal biases and local usages have persuaded us to accept one of several available common names for a plant. Users in other areas may not always agree with the names we have accepted. For instance, we use Great Basin sagebrush as the common name for *Artemisia tridentata*. Other botanists or range managers may prefer to call this plant big sage. Should the name for *Chrysothamnus* be rabbitbrush or rabbitbush? Both are used in the literature. We prefer the latter. Some plants have no accepted common names. In some cases we have adopted names, probably of local origin, that we have seen used for the plants in question. We have coined a few common names ourselves over the years and taught them to enough students that they now have a common usage, at least in some circles. In other cases we have used the translation method or simply adopted the generic name as the common name when no "real" common name was available. We have attempted to be consistent in the names we have adopted. In the index the common names are referenced to the species to which they apply unless the name is a direct translation or mere use of the generic name as a common name.

References on Plant Names

Abrams, L., and R. S. Ferris. 1923–1960. Illustrated flora of the Pacific states. Stanford Univ. Press, Stanford. 4 vols.

Bailey, L. H. 1933. How plants get their names. Dover Publications, Inc. (1963 reprint).

Berlin, B. 1973. Folk Systematics in Relation to Biological Classification and Nomenclature. Annual Review of Ecology and Systematics 4:259-271.

Brown, R. W. 1956. Composition of scientific words. Smithsonian Institution Press, Washington, D.C. 882 pp. (1991 reprint).

Brummitt, R. K., and C. F. Powell (eds.). Authors of plant names. Royal Botanic Gardens, Kew, England. 732 pp.

Greuter, W., J. McNeill, et al. (eds.). 1994. International code of botanical nomenclature as adopted by the Fifteenth International Botanical Congress, Yokohama, August–September 1993. Regnum Vegetabile 131.

Hickman, J. C., ed. 1993. The Jepson manual. Higher plants of California. University of California Press, Berkeley, Los Angeles, and London. xvii + 1400 pp.

Hoover, R. F. 1970. The vascular plants of San Luis Obispo County, California. Univ. Calif. Press, Berkeley. 350 pp.

Jeffrey, C. 1977. Biological Nomenclature. The Systematics Association, London.

Kartesz, J. T., and J. W. Thieret. 1991. Common names for vascular plants. Guidelines for use and application. Sida 14:421–434.

Munz, P. A. 1968. Supplement to A California Flora. Univ. California Press, Berkeley. 224 pp.

——— and D. D. Keck. 1959. A California flora. Univ. California Press, Berkeley. 1681 pp.

Nicolson, D. H. 1974. Orthography of Names and Epithets: Latinization of Personal Names. Taxon 23:549-561.

———. 1986. Species Epithets and Gender Information. Taxon 35:323-328.

Stearn, W. T. 1993. Botanical Latin. History, grammar, syntax, terminology and vocabulary, 4th ed. David & Charles, Newton Abbot, England. xiv + 546 pp.

Stebbins, G. L. 1993. Concepts of species and genera. Pp. 229–246 *in* Flora of North America Editorial Committee, Flora of North America, Vol. 1. Introduction. Oxford University Pres, New York.

Walters, D. R., and D. J. Keil. 1995. Botanical nomenclature. Pp. 13–28 *in* Vascular plant taxonomy, 4th ed. Kendall-Hunt, Publ. Co., Dubuque.

Weber, W. A. 1986. Pronunciation of scientific names. Madroño 33:234-235.

Chapter 7

Marine Aquatic Communities

Marine aquatic plant communities form a narrow band along the continental shelf and immediate coastal areas. They occupy approximately one percent of the state's area (including the continental shelf). The plants comprising the communities are influenced by marine waters of the Pacific Ocean, either directly, or in protected bays and estuaries, by tidal action. Three fairly different types of communities occur in these maritime habitats: **subtidal and intertidal communities**, **coastal estuarine communities**, and **coastal salt marsh communities**. These are treated below in terms of decreasing influence of salt water. All are dominated by species of halophytic (salt-tolerant) plants.

1. Subtidal and Intertidal Communities

Subtidal and intertidal communities are dominated by a mixture of macrophytic algae and marine aquatic angiosperms. These communities occur in the photic (light) zone of coastal waters. There is considerable vertical and horizontal variation in these communities. The lower zones (subtidal communities) are never exposed to the atmosphere, even at low tides. The upper zones (intertidal communities) are exposed to the air on a regular or occasional basis. The uppermost of these are exposed at most low tides, and the plants living there must be tolerant of extreme exposure. The lowest intertidal zones are exposed to the air only at times of extreme low tides but otherwise are submersed. Between these extremes are communities that are regularly submersed at high tides and regularly exposed at low tides.

The diversity of the algae in marine aquatic communities is much greater than that of flowering plants. The algae range in size from giant kelps (up to about 50 m in length) to microscopic single-celled epiphytes. The conspicuous algae include red, green and brown algae. The only angiosperms in the subtidal and intertidal communities of California are sea-grasses—species of *Phyllospadix* (surf-grass) and *Zostera marina* (eel-grass).

Fig. 7-1. Colonies of sea palms (*Postelsia palmiformis*) on rocks exposed at high tide on Sonoma County coast.

Community composition is affected by several factors. Some algal species are intolerant of exposure to the drying effects of the atmosphere and exposure to direct sunlight. Others can withstand some desiccation without being harmed. Those intolerant of exposure are restricted to subtidal or to the lowermost intertidal zones or occur in tide pools that remain filled

Fig. 7-2. Coastal intertidal zone in northern San Luis Obispo County with exposed rocks and floating strands of kelp. Rocks and the algae and surf-grasses attached to them are alternately exposed to the air and submerged by the tides and by wave action. Photo by David Keil.

with sea water even at low tide. Only those algae very tolerant of exposure can occupy the upper intertidal areas (Fig. 7-1). Community composition is further influenced by variation in water temperature, salinity gradients, and circulation patterns. Turbidity can restrict the penetration of light into the water, preventing the establishment and growth of deep-water species.

Fig. 7-3. Surf-grass (*Phyllospadix scouleri*) and various algae on rocks exposed at low tide. Photo by David Keil.

Another variable is the nature of the substrate. Rocky sea bottoms support different communities than do sandy areas or other ocean sediments. Surf-grasses and those algae occurring in the upper subtidal and intertidal zones must be well anchored because of the strength of wave action and ocean currents, particularly during storms. Most have holdfasts that firmly attach the plants to submersed rocks. Because sand offers little in the way of anchorage, large algae are usually absent from sandy sea bottoms.

Biotic factors are also important. As in terrestrial communities there often is competition for space. Rocks in the intertidal zone may be completely covered by algae with no space left for additional plants to become established. Large algae may shade out smaller plants. Some species, however, grow as epiphytes, attached to the thalli of the larger species. Others are understory species, tolerant of shade. The selective grazing of marine animals, such as limpets and sea urchins, can affect community composition by reducing the size or numbers of the palatable algal species.

The kelps, the largest of the brown algae, form undersea "forests" in the subtidal zone along some areas of the California coast. These include species of such genera as *Egregia*, *Nereocystis*, *Laminaria*, and *Macrocystis*. These are often associated with smaller algae and with the only flowering plants of the subtidal zone, *Zostera marina* and *Phyllospadix torreyi*.

Fig. 7-4. *Fucus distichus* and *Pelvetia fastigiata* on rocks exposed at low tide. Both are brown algae. Photo by David Keil.

Phyllospadix torreyi is generally restricted to rocky coastal areas with high wave action. *Zostera*, on the other hand occurs in protected embayments. The forest-like underwater communities only occur on rocky substrates. The large algae have holdfasts that anchor them to the sea bottom. The kelps are unable to establish themselves on sandy or muddy substrates. Pollution and unusually high water temperatures have extirpated many of the kelps in southern California waters in historic times.

Plants growing in the intertidal zone must be tolerant of both underwater conditions and exposure to the air (Fig. 7-2). Thesc plants are pounded by surf and pulled by waves and tidal flows. When exposed to the air they may be subjected to desiccation, particularly when low tides occur on sunny days. There is a strong zonation among the members of the intertidal communities relative to their tolerance of exposure to the atmosphere and to the effects of grazing by marine herbivores. Some algae are restricted to tide pools that retain water or are repeatedly refilled by wave action. These are the least tolerant of the intertidal species to atmospheric exposure. *Phyllospadix scouleri* (surf-grass; Fig. 7-3) is usually the only angiosperm in the intertidal zone though *P. torreyi* also occasionally occurs here. Surf-beaten rocks that are fully exposed at low tide are often covered by a dense growth of green, red and brown algae (Fig. 7-4). These include such genera as *Ulva*, *Iridaea*, *Corallina*, *Pelvetia*, *Fucus*, and various others. In central and northern California the sea-palm (*Postelsia palmiformis*; Fig. 7-1) can be found standing above the wave-beaten rocks at low tide.

Fig. 7-5. Map illustrating the discontinuous distribution of coastal salt marshes and coastal estuarine communities along California's coastline from the San Diego area north to Eureka.

A detailed survey of the algal composition of the subtidal and intertidal communities is outside the scope of this publication. However, the bibliography at the end of this chapter provides several sources that cover this subject in detail.

2. Coastal Estuarine Communities

Estuarine communities occur where freshwater from streams mixes with sea water in a protected embayment. The result is a zonation of brackish waters of varying degrees of salinity. Several environmental factors distinguish these communities from those of the open shore and open waters. In estuarine areas there is protection from wave action and winds. These areas have strong variation in salinity. Additionally the bottom substrate is usually a thick layer of sediments, often a thick mud. These communities are located in widely scattered areas along the California coast (Fig. 7-5) in bays and estuaries such as Humboldt Bay, San Francisco Bay, Tomales Bay, Morro Bay (Fig. 7-6), Santa Barbara, Los Angeles, and San Diego.

Estuarine plants are subjected to much variation in salinity as a result of seasonal and daily fluctuations. During the rainy winter months fresh water draining from the land tends to dilute the salt in the estuaries. In summer months stream flow decreases and salinity increases. On a daily basis water level and salinity fluctuate with the tides. Tidal rivers and streams reverse their flow on a daily basis with fresh water flowing downstream into the channel at low tide and brackish water flowing upstream at high tide (Fig. 7-7). As a result of these fluctuations the plants of estuarine communitics must be tolerant of varying salinity. At extreme low tide estuarine areas may be exposed to the air with the plants stranded on mud flats.

Estuarine communities border and often interfinger with coastal salt marshes (Figs. 7-6, 7-7). Around the salt marsh fringes various stiff-bodied plants are firmly anchored on wet soils that are subjected to periodic inundation. (These salt marsh communities are discussed in the following section.) The estuarine communities are characterized by prolonged inundation and the plants are generally soft-bodied and rather flexible. These include two flowering plants, *Zostera marina* (eel-grass) and *Ruppia maritima* (ditch-grass) and various algae. Eel-grass usually occurs in areas of greater salinity than does ditch-grass. Eel-grass often forms dense stands in the still or gently flowing water of embayments. Ditch-grass usually is restricted to the brackish streams that penetrate the salt marshes. The most conspicuous algae are green algae such as *Ulva*, *Cladophora* and *Enteromorpha*.

Fig. 7-6. Dunes and coastal salt marsh at Shark Inlet in the lee of the Morro Bay Sand Spit in San Luis Obispo County. Eelgrass (*Zostera marina*) and various algae occupy the waters of the inlet. Tidal waters flood the areas bordering the inlet at high tide, regularly inundating the lower salt marsh. Sedges bordering the salt marsh in the center receive fresh water from springs that empty into the inlet. Taller vegetation in the foreground is above the level of the tides. Photo by Jesse Martin.

Fig. 7-7. Coastal tidal stream. At high tide salt water flows into the lower channel of Los Osos Creek in San Luis Obispo County. At low tide the stream flow reverses and fresh water flows into Morro Bay. Along the creek are areas of coastal salt marsh. **A.** View downstream toward Morro Bay. **B.** View upstream toward riparian woodland. Photos by Elizabeth Bergen.

3. Coastal Salt Marsh Communities

Coastal salt marshes occur in scattered locations along the California coastlines in bays and other areas that are protected from the wave action of the open ocean. Inlets and bays with a barrier, such as a sand spit or an island, are often locations for salt marshes (Fig. 7-6). These communities develop where there is a mixing of fresh water flowing from streams and springs with salt water from the ocean (Fig. 7-7). The result is a zonation of brackish water of varying degrees of salinity.

Salt marshes are scattered along the entire length of the California coastline from the mouth of the Smith River in Del Norte County to the Tijuana Slough in San Diego County (Fig. 7-5). The overall area of these communities is rather small. Only about 370 square kilometers (145 square miles) of salt marsh remain in California, about 90 percent of which is in the San Francisco Bay region. Coastal salt marshes often occur adjacent to the shallow waters occupied by coastal estuarine communities. The overall distribution of these communities is similar although small areas of salt marsh sometimes occur where coastal estuarine communities are absent (e.g., bordering lagoons at the mouths of small coastal streams).

As a result of human impact, coastal salt marshes now occupy a much smaller area than they did in pre-colonial times. Unfortunately most of the larger coastal salt marshes are located near large metropolitan areas. Many have been drained or filled for urban development whereas others have been transformed into marinas. Conservation efforts, which have met with some success, have concentrated on the importance of these communities as feeding and nesting sites for resident and migratory birds and as the habitats for some rare plants (e.g., *Cordylanthus maritimus* ssp. *maritimus*, salt marsh bird's beak). Salt marshes are very productive communities and are extremely important wildlife habitats. These communities are also very important parts of the Pacific flyway used by migratory waterfowl. A major threat to coastal salt marshes and their resident wildlife is the potential of oil spills from off-shore drilling and shipwrecks. Additionally, pollution of streams by various substances may cause damage to salt marsh ecosystems.

Salt marsh communities are unique in that they are exposed to continuously changing amounts of submergence and salinity. At times of low tide and winter storm action when there is much runoff from the land, water flowing through a salt marsh has low salinity. During inundation by high tides the water covering the marsh may be much saltier. In summer when stream-flow is low, salinity is at its highest. The most saline areas in a salt marsh are the "salt pans" where tidal water collects in small pools at

Fig. 7-8. Coastal salt marshes. Low tide at Morro Bay salt marsh (San Luis Obispo County). Dominants are *Salicornia virginica* (pickleweed), *Jaumea carnosa* (jaumea), and *Distichlis spicata* (salt grass). The large-leafed plants are *Limonium californicum* (sea-lavender). Photo by David Keil.

highest tides and then dries out. These hypersaline environments are so salty that they are often unvegetated.

Coastal salt marshes are ecologically somewhat similar to alkali sink communities (Chapter 19). These coastal and desert communities share some species, and many of the plants have similar adaptations to life in saline habitats. Coastal climates are, however, much milder than desert climates because of the proximity to the ocean, and plants of coastal salt marshes seldom are exposed to high temperatures.

Species diversity tends to be low relative to other communities, as few species can tolerate a salt marsh environment. The halo-

Fig. 7-9. Salt marsh in Orange County at high tide. Dominants are *Spartina foliosa* (cordgrass) and *Salicornia virginica* (pickleweed) with patches of the introduced *Cotula coronopifolia* (brass buttons). Photo by David Keil.

phytes that occur in these communities must be able to withstand waterlogged, saline, often clayey soils with low oxygen concentration. They must be able to tolerate being periodically covered by water of variable salinity and then being exposed as the tide moves out. This usually results in a characteristic zonation pattern in the coastal salt marsh communities. Those plants able to withstand total immersion grow in the lower zones whereas those that are intolerant of prolonged immersion occupy the upper zones. In a coastal salt marsh a difference in elevation of only a few centimeters can be very significant.

Vegetation of coastal salt marsh communities consists of halophytic plants that are mostly low-growing herbaceous perennials (Figs. 7-8, 7-9). Most species have reduced leaves and several are succulents. Many have aerenchyma (tissues with many air cavities) that allow the plants to respire in low-oxygen environments. Oxygen absorbed by stems or leaves can diffuse through the internal air chambers to the roots which are anchored in nearly anaerobic soils. The dominant plants all have features that allow them to live in saline soils and to absorb water despite its dissolved salts. Some, such as *Distichlis spicata* (saltgrass), *Monanthochloe littoralis* (shoregrass), *Limonium californicum* (sea-lavender), and *Spartina foliosa* (California cordgrass) have salt glands through which excess salts are excreted. Others have cell sap with such a high concentration of dissolved solutes that they can absorb water without suffering any osmotic imbalances. These plants are mostly succulents and include *Salicornia* spp. (pickleweeds), *Suaeda* spp. (sea-blite), and *Jaumea carnosa* (fleshy jaumea). The flowers of several of the most common species are highly reduced, and wind pollination is common. Vegetative reproduction by rhizomes is very common.

Ecotones between coastal salt marshes and adjacent communities may be broad or quite abrupt. Coastal salt marshes grade into coastal estuarine communities along an elevational and moisture gradient and into freshwater marsh and riparian communities along a salinity gradient. In areas with an abrupt elevational gradient, a coastal salt marsh community may be bordered by such communities as coastal dune scrub, coastal live oak woodland, coastal grasslands, etc.

Some of common species of coastal salt marshes are listed below:

Atriplex californica	California saltbush
Atriplex patula	spear oracle
Atriplex watsonii	Watson's saltbush
Batis maritima	saltwort
Cuscuta salina	saltmarsh dodder
Distichlis spicata	salt grass
Frankenia salina [*F. grandifolia*]	frankenia
Jaumea carnosa	fleshy jaumea
Juncus acutus	giant rush
Juncus lesueurii	rush
Limonium californicum	sea-lavender
Monanthochloe littoralis	monanthochloe
Potentilla anserina ssp. *pacifica* [*P. egedii*]	coastal silverweed
Salicornia bigelovii	pickleweed
Salicornia subterminalis	pickleweed
Salicornia virginica	pickleweed
Scirpus americanus [*S. olneyi*]	Olney's three-square
Spartina foliosa	cordgrass
Suaeda spp.	sea-blite
Triglochin spp.	arrow-grass

Locally other species occur in salt marshes including some rare endemics. Some of the dominant salt marsh plants are very widespread and are not limited to California. *Salicornia virginica*, for instance, is widely distributed on both the Atlantic and Pacific coasts of North America.

For the most part, salt marshes have not been as affected by the influx of alien plants that has accompanied modern civilization as have many other California plant communities. There are a few exceptions, however. Several alien species of *Spartina* (cordgrass) have become established in California, and *S. densiflora*, a South American species is now dominant in much of the Humboldt Bay salt marsh and established elsewhere along the California coast. *Cotula coronopifolia* (brass buttons) occurs in some salt marshes where it may be locally common (Fig. 7-9). Several halophytic members of the Chenopodiaceae have also established footholds in the California salt marshes.

Marine Aquatic Plant Communities—References

Abbot, I. A., and G. J. Hollenberg. 1976. Marine algae of California. Stanford Univ. Press, Stanford, California.

Argent, G., ed. 1991. Delta-Estuary. California's inland coast. A public trust report. Prepared for the California State Lands Commission. 208 pp.

Armstrong, W. P., and R. F. Thorne. 1989. California seagrasses. Fremontia 16(4):15–21.

Asen, P. A. 1976. Upper intertidal algal zonation on Bodega Head, Sonoma County, California. Madroño 23:257–263.

Atwater, B. F., and C. W. Hedel. 1976. Distribution of seed plants with respect to tide levels and water salinity in the natural tidal marshes of northern San Francisco Bay estuary, California. U.S. Geol. Surv. Open File Rept. 76-389. 41 pp.

Barbour, M. G., R. B. Craig, F. R. Drysdale and M. T. Ghiselin. 1973. Coastal ecology. Bodega Head. Univ. Calif. Press, Berkeley, Los Angeles and London.

Barnes, R. S. K. 1977. The coastline. Cambridge Univ. Press, Cambridge.

Cain, D. J., and H. T. Harvey. 1983. Evidence of salinity-induced ecophenic variation in cordgrass (*Spartina foliosa* Trin.). Madroño 30:50–62.

Caplan, R. I., and R. A. Boolootian. 1967. Intertidal ecology of San Nicolas Island. Pp. 203–217 *in* R. N. Philbrick (ed.). Proceedings of the Symposium on the Biology of the California Islands. Santa Barbara Bot. Gard., Santa Barbara.

Carefoot, T. 1977. Pacific seashores. A guide to intertidal ecology. Univ. Washington Press, Seattle and London.

Chapman, V. J. 1960. Salt marshes and salt deserts of the world. Interscience, N.Y. 392 pp.

———. 1974. Salt marshes and salt deserts of the world. 2nd ed. J. Cramer, Lehre.

———. 1976. Coastal vegetation, 2nd ed. Pergamon Press, Oxford.

Cooper, A. W. 1974. Salt marshes. Pp. 55–98 *in* H. T. Odum, B. J. Copeland, and E. A. McMahan (eds.). Coastal ecological systems of the United States. vol. 2.

Cowardin, L.M., V. Carter, F. C. Golet, and E. T. LaRoe. 1979. Classification of wetland and deepwater habitats of the United States. U.S.D. I. Fish and Wildlife Service, Office of Biological Services Publ. OBS-79/31. Washington, D.C. 103 pp.

Dawes, C. J. 1981. Marine botany. John Wiley & Sons, N.Y. 628 pp.

Dawson, E. Y. 1959. A preliminary report on the benthic marine flora of southern California. Pp. 169–264 *in* Oceanographic survey of the continental shelf area of southern California. Publ. Calif. State Water Poll. Contr. Bd. 20.

———. 1965. Intertidal algae. Pp. 220–231, 351–438 *in* An oceanographic and biological survey of the southern California mainland shelf. Publ. Calif. State Water Qual. Contr. Bd. 27.

———, and M. S. Foster. 1982. Seashore plants of California. Univ. Calif. Press, Berkeley, Los Angeles and London.

Dayton, P. K., V. Currie, T. Gerrodette, B. D. Keller, R. Rosenthal and D. Ven Treska. 1984. Patch dynamics and stability of California kelp communities. Ecol. Monogr. 54:253–289.

Doty, M. S. 1946. Critical tide factors that are correlated with the vertical distribution of marine algae and other organisms along the Pacific Coast. Ecology 27:315–328.

———. 1957. Rocky intertidal surfaces. Pp. 535–585 *in* J. W. Hedgpeth (ed.). Treatise on marine ecology and paleoecology. Memoir 67, Geol. Soc. Amer., N.Y.

Ellison, J. P. 1984. A revised classification of native aquatic communities in California. Calif. State Dept. Fish and Game Planning Br. Admin. Rept. 84-1. 30 pp.

Emery, K. O. 1960. The sea off southern California. John Wiley & Sons, N.Y.

Ferren, W. R., Jr. 1985. Carpinteria salt marsh. Environment, history, and botanical resources of a southern California estuary. Herbarium Publ. No. 4, Univ. Calif., Santa Barbara. 300 pp.

———. 1989. A preliminary and partial classification of wetlands in southern and central California with emphasis on the Santa Barbara Region. Prepared for Wetland plants and vegetation of coastal southern California, a workshop organized for the California Department of Fish and Game and the United States Fish and Wildlife Service. University of California , Santa Barbara. 54 pp.

———. 1990. Recent research on and new management issues for southern California estuarine wetlands. Pp. 55–79 in A. A. Schoenherr (ed.), Endangered plant communities of southern California. Proceedings of the 15th Annual Symposium. Southern California Botanists spec. publ. 3.

———, and P. L. Fiedler. 1993. Rare and threatened wetlands in central and southern California. Pp. 119–131 *in* J. E. Keeley (ed.), Interface between ecology and land development in California. Southern California Academy of Sciences, Los Angeles.

Filice, F. P. 1954. An ecological survey of the Castro Creek area in San Pablo Bay. Wassman J. Biol. 12:1–24.

Foster, M., M. Neushul, and R. Zingmark. 1971. The Santa Barbara oil spill, part 2: initial effects on intertidal and kelp bed organisms. Environ. Pollut. 2:115–134.

Gale, J., and Poljakoff, M. (eds.). 1975. Plants in saline environments. Springer-Verlag, N.Y.

Gessner, F. 1970. Temperature; Plants. Pp. 363–406 *in* O. Kinne (ed.). Marine Ecology: a comprehensive integrated treatise on life in oceans and waters. Vol. 1. Environmental factors. Part 1. Wiley-Interscience, N.Y.

Glynn, N. L. 1965. Community composition, structure and interrelationship in the marine intertidal *Endocladia muricata - Balanus glandula* association in Monterey, California. Beaufortia 12:1–198.

Henrickson, J. 1976. Ecology of southern California coastal salt marshes. Pp. 49–64 *in* J. Latting (ed.). Plant communities of southern California. Calif. Native Pl. Soc. Spec. Publ. 2.

Hinde, H. P. 1954. Vertical distribution of salt marsh phanerogams in relation to tide levels. Ecol. Monogr. 24:209–225.

Hopkins, D. R., and V. T. Parker. 1984. A study of the seed bank of a salt marsh in northern San Francisco Bay. Amer. J. Bot. 71:348–355.

Jensen, J. B., and S. J. Tanner. 1973. A preliminary checklist of the marine algae of the Moss Landing Jetty: an annotated floristic compilation. Contrib. Moss Landing Marine Lab. 38 (Tech. Publ. 73–7).

Kemp, P. R., and G. L. Cunningham. 1981. Light, temperature and salinity effects on growth, leaf anatomy and photosynthesis of *Distichlis spicata* (L.) Greene. Amer. J. Bot. 68:507–516.

Kraeuter, J. N., and P. L. Wolf. 1974. The relationship of marine macro-invertebrates to salt-marsh plants. Pp. 449–462 *in* R. J. Reimold and W. H. Queen (eds.). Ecology of halophytes. Academic Press, N.Y.

Lewis, J. R. 1964. The ecology of rocky shores. English Universities Press, London.

Littler, M. M. (ed.). 1978. The annual and seasonal ecology of Southern California subtidal, rocky intertidal and tidepool biotas. Bur. of Land Managem., U.S. Dept. Interior, Washington, D.C.

——— (ed.). 1979. The distribution, abundance and community structure of rocky intertidal and tidepool biotas in the Southern California Bight. Bur. Land Managem., U.S. Dept. Interior, Washington, D.C.

———. 1980. Overview of the rocky intertidal systems of southern California. Pp. 265–306 *in* D. M. Power (ed.). The California Islands: proceedings of a multidiciplinary symposium. Santa Barbara Mus. Nat. Hist., Santa Barbara.

———. 1980. Southern California rocky intertidal ecosystems: methods, community structure and variability. Pp. 565–608 *in* J. H. Price, D. E. G. Irvine, and W. H. Farnham (eds.). The shore environment: methods and ecosystems. Acad. Press, London.

MacDonald, K. B. 1976. Plant and animal communities of Pacific North American salt marshes. Pp. 167–191 *in* R. J. Reimold and W. H. Queen (eds.). Ecology of halophytes. Academic Press, N.Y.

———. 1977. Coastal salt marsh. Pp. 263–294 *in* M. G. Barbour and J. Major (eds.). Terrestrial vegetation of California. John Wiley and Sons, N.Y.

———, and M. G. Barbour. 1974. Beach and salt marsh vegetation of the North American Pacific coast. Pp. 175–234 *in* R. J. Reimold and W. H. Queen (eds.). Ecology of halophytes. Academic Press, N.Y.

Mahall, F. E. 1976. A vegetation survey of San Pablo Bay salt marshes. J. Ecol. 64:81–88.

Mahall, B. E., and R. B. Park. 1976. The ecotone between *Spartina foliosa* Trin. and *Salicornia virginica* L. in salt marshes of northern San Francisco Bay I. Biomass and production. J. Ecol. 64:421–433.

———, and R. B. Park. 1976. The ecotone between *Spartina foliosa* Trin. and *Salicornia virginica* L. in salt marshes of northern San Francisco Bay II. Soil water and salinity. J. Ecol. 64:793–809.

Mall, R. E. 1969. Soil - water - salt relationships of waterfowl food plants in the Suisun Marsh of California. The Resources Agency Dept. of Fish and Game. Wildlife Bull. No. 1.

McLusky, D. S. 1971. Ecology of estuaries. Heinemann Educational Books, London.

Minckley, W. L., and D. E. Brown. 1982. Californian maritime and interior marshlands. Pp. 257–262 *in* D. E. Brown (ed.). Biotic communities of the American Southwest—United States and Mexico. Desert Plants 4:1–341.

Murray, S. N. 1974. Benthic algae and grasses. Pp. 9.1–9.61 *in* M. D. Dailey, B. Hill, and N. Lansing (eds.). A summary of knowledge of the southern California coastal zone and offshore areas. Vol. II, Biological environment. U. S. D. I., Bureau of Land Management, Washington, D.C.

———, and M. M. Littler. 1981. Biogeographical analysis of intertidal macrophyte floras of southern California. J. Biogeogr. 8:339–351.

———, ———, and I. A. Abbott. 1980. Biogeography of the California marine algae with emphasis on the Southern California Islands. Pp. 235–339 *in* D. M. Power (ed.). The California Islands: proceedings of a multidiciplinary symposium. Santa Barbara Mus. Nat. Hist., Santa Barbara.

Neushul, M., W. D. Clark, and D. W. Brown. 1967. Subtidal plant and animal communities of the Southern California Islands. Pp. 37–55 *in* R. N. Philbrick (ed.). Proceedings of the symposium on the biology of the California Islands. Santa Barbara Bot. Gard., Santa Barbara.

Odum, E. P. 1961. The role of tidal marshes in estuarine production. Conservationist 15:12–15.

Phelger, F. 1977. Soils of marine marshes. Pp. 69–77 *in* V. J. Chapman (ed.). Salt marshes and salt deserts of the world. 2nd ed. J. Cramer, Lehre.

Pestrong, R. 1972. San Francisco Bay tidelands. Calif. Geol. 25:3–9.

Phillips, R. C. 1979. Ecological notes on *Phyllospadix* (Potamogetonaceae) in the northeast Pacific. Aquat. Bot. 6:159–170.

Poinssot, B. 1977. The Stinson Beach salt marsh. Stinson Beach Press, Stinson Beach, Calif.

Purer, E. A. 1942. Plant ecology of the coastal salt marshlands of San Diego County, California. Ecol. Monogr. 12:81–111.

Queen, W. H. 1974. Physiology of coastal halophytes. Pp. 345–353 *in* R. J. Reimold and W. H. Queen (eds.). Ecology of halophytes. Academic Pr, N.Y.

———. 1977. Human uses of salt marshes. Pp. 363–368 *in* V. J. Chapman (ed.). Salt marshes and salt deserts of the world. 2nd ed. J. Cramer, Lehre.

Ranwell, D. S. 1972. Ecology of salt marshes and sand dunes. Chapman and Hall, London.

Reimold, R. J., and W. H. Queen (eds.). 1974. Ecology of halophytes. Academic Press, N.Y.

Ricketts, E. F., J. Calvin, and J. W. Hedgpeth. Between Pacific tides, fourth ed. Stanford Univ. Press, Stanford.

Rigg, G. B., and R. C. Miller. 1949. Intertidal plant and animal zonation in the vicinity of Neah Bay, Washington. Proc. Calif. Acad. Sci. 26:323–351.

Scagel, R. F. 1963. Distribution of attached marine algae in relation to oceanographic conditions in the northeast Pacific. Pp. 37–50 *in* M. I. Dunbar (ed.). Marine Distributions. Spec. Publ. Roy. Soc. Canada 5.

Sculthorpe, C. D. 1967. The biology of aquatic vascular plants. St. Martin's Press, N.Y. 610 pp.

Setchell, W. A. 1935. Geographic elements of the marine flora of the north Pacific Ocean. Amer. Nat. 69:560–577.

Smith, G. M. 1944. Marine algae of the Monterey Peninsula, California. Stanford Univ. Press, Stanford.

Stephenson, T. A., and A. Stephenson. 1972. Life between tidemarks of rocky shores. Freeman Publ. Co., San Francisco.

Thom, R. 1980. A gradient in benthic intertidal algal assemblages along the southern California coast. J. Phycol. 16:102–108.

Vogl, R. J. 1966. Salt marsh vegetation of upper Newport Bay, California. Ecology 47:80–87.

Waisel, Y. 1972. The biology of halophytes. Acad. Press, N.Y.

Warme, J. E. 1971. Paleoecological aspects of a modern coastal lagoon. Univ. Calif. Publ. Geol. Sci. 87:1–131. J. Chapman, Salt marshes and salt deserts of the world. 2nd ed. J. Cramer, Lehre.

Zedler, J. B. 1982. The ecology of southern California coastal salt marshes: a community profile. San Diego State Univ. FWS Rept. OBS-81/54.

Chapter 8

Coastal Sand Dune Communities

Coastal sand dune communities are the first terrestrial plant communities above the high tide line where sandy beaches and/or sand dunes occur. These communities, like the sand dunes they occupy, are discontinuous along the California coastline from San Diego to the Oregon border (Fig. 8-1). Between coastal sand dune areas are various non-sandy substrates ranging from steep, rocky coasts with poorly developed soils to marine terraces with deep alluvial soils. Narrow strips of sandy beach and small pockets of sand dunes often occur in these otherwise non-sandy coastal areas. The coastal dune habitat is a restricted natural community along the Pacific Coast of North America. Only 23% of California's coastline is occupied by sandy beaches and considerably less by sand dunes. Beaches and coastal dunes occupy much less than one percent of the state's total land area.

Sandy beaches and coastal sand dunes form where ocean currents deposit large amounts of sand in shallow coastal waters. Much of the sand originates on land as a result of stream erosion and is swept out to sea at times of high water flow. Ocean currents can move sand considerable distances from the original source areas. Local coastal profiles determine where sand-bearing currents can flow. Coastal dunes form when sea breezes blow sand grains inland from beaches where the sand has been cast ashore by the surf.

The fluctuations of sea level that accompanied the glaciations and deglaciations of the land during the Quaternary Period greatly affected the formation and persistence of coastal dunes. During the last ice age of the Pleistocene, sea level was over 100 m (330 ft) lower than it is at present and the shoreline was west of its current location. Dunes that formed during the time of lower sea level were later submerged and their sand redistributed by waves and ocean currents. Most dunes located along the present-day shoreline are less than 10,000 years old. The Guadalupe

Fig. 8-1. Discontinuous and scattered distribution of coastal sand dune communities along the California coastline from Mexico to the Oregon border.

dunes complex of San Luis Obispo County (Fig. 8-2) is a good example. California also has some ancient dunes, now located away from the immediate coast. The Nipomo Mesa (San Luis Obispo Co.) is many thousands of years old, having formed at a time in the Pleistocene Epoch when sea level was higher relative to the local landscape than it is today. Old, long-stabilized dunes may support such communities as chaparral, coast live oak woodland, or coastal coniferous forests.

Climatic conditions on the coastal dunes are relatively mild. Because the nearby ocean serves as a vast heat reservoir, diurnal and seasonal temperature fluctuations are comparatively small. Coastal fogs are frequent. Frost is uncommon and the growing season may exceed 350 days. Annual rainfall ranges from about 25 cm (10 in) in the south to as much as 175 cm (70 in) in the north.

Despite the moderate climate, the environment of plants growing on the coastal dunes may be quite harsh. The persistent coastal winds carry with them abrasive sand particles and salt. The salt tends to accumulate on the plants and in the soil. The sandy soil is unstable, has low fertility, very little humus, and consequently, a low water-holding capacity. When exposed to direct sunlight the sand can become very hot (54–71 °C [130–160 °F] at the sand surface). Insolation of the plants may be intense because of the direct light from the sun and the high reflectivity of the sand.

1. Pioneer Dune Communities

Once sand has been blown above the high tide level, colonization of the sand surfaces by terrestrial plants can occur (Fig. 8-3). Active dunes occur in areas where the rate of sand movement exceeds the rate of colonization and establishment of plants. Active dunes may occupy a zone immediately adjacent to the beach, or they may gradually be blown inland, sometimes engulfing and burying entire plant communities in their path. Plants growing on active dunes tend to collect sand and form vegetated hummocks. If enough plants become established to significantly reduce the rate of air movement along the sand surface, dune movement will slow or stop. Species of plants intolerant of rapid burial can become established once a dune has stopped moving. A dune that has ceased movement as a result of vegetative cover is a stabilized dune

Pioneer dune communities occur on beaches and active dunes close to the ocean (Fig. 8-3). This zone of unstable sand is often described as the coastal strand. Because of the harsh environmental conditions, pioneer dune communities usually have low species diversity. The plants are mostly prostrate herbs with

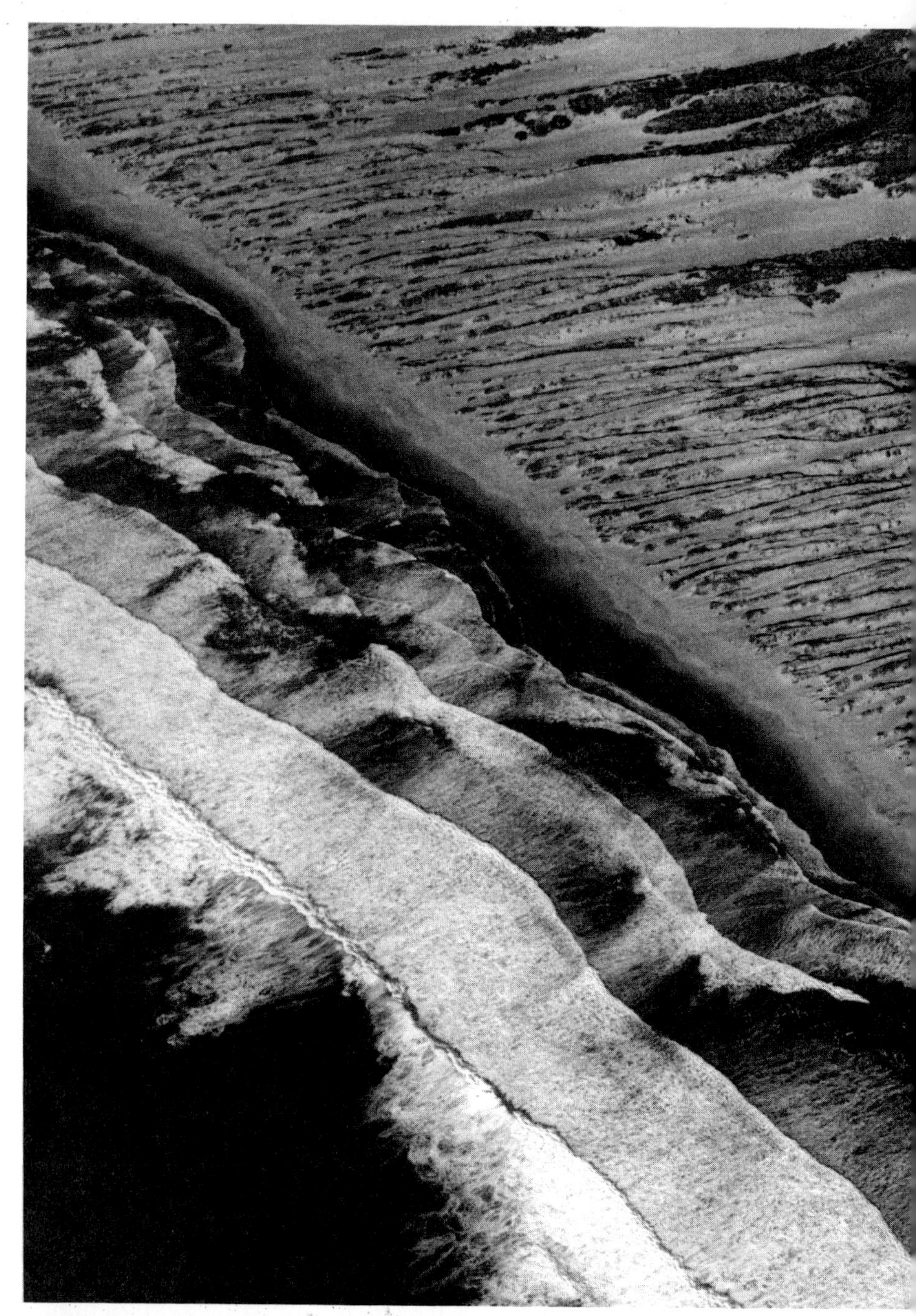

Fig. 8-2. Aerial photo of Nipomo Dunes, San Luis Obispo County. From the beach inland the habitats include active and partially stabilized foredunes, dune scrub communities, depressions with dune wetlands,

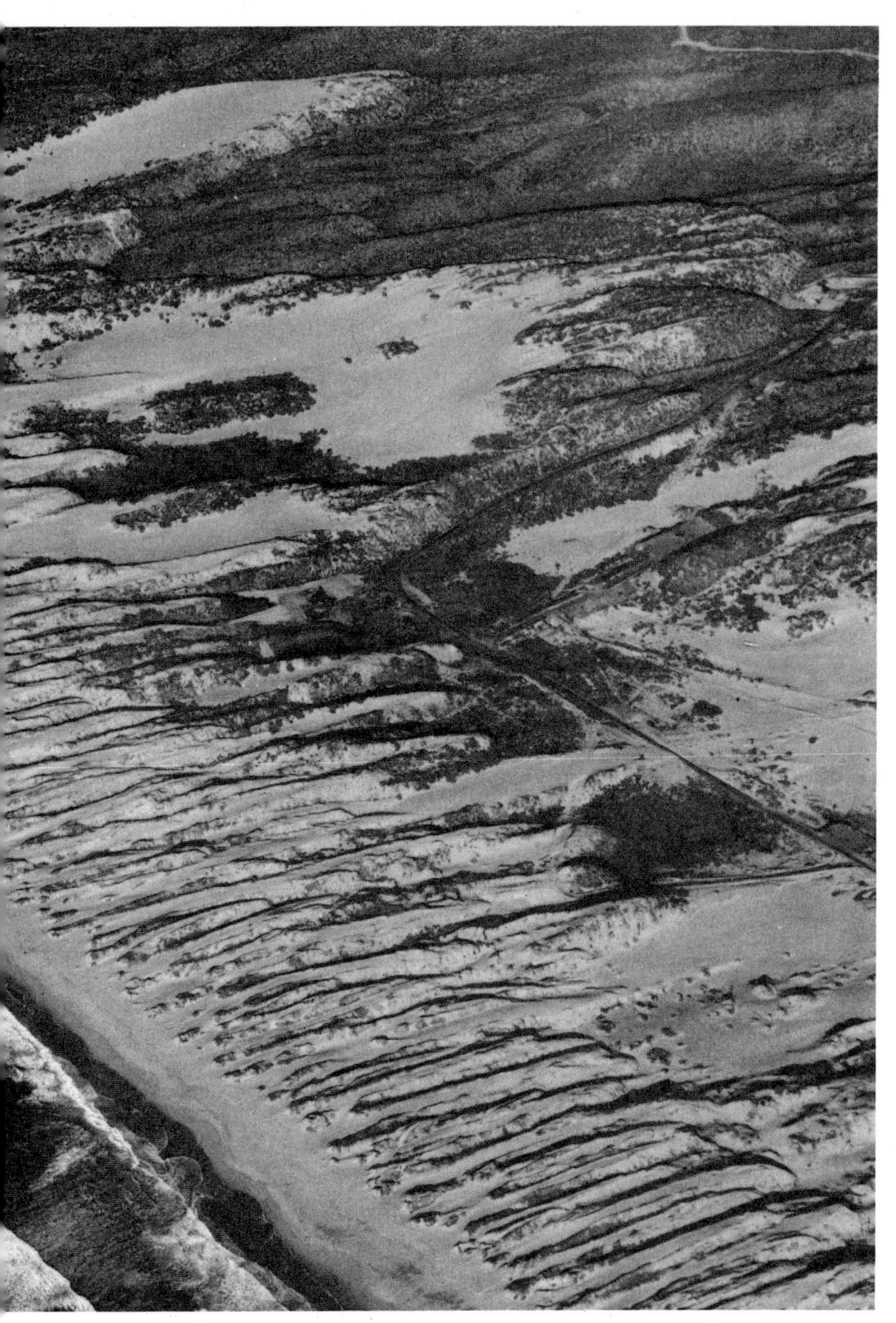

and advancing active dunes. The dunes move from the northwest to the southeast as they blow inland. Photo by Robert Rodin.

Fig. 8-3. Beach and low foredunes at Morro Bay Sand Spit. This view, looking to the north shows effects of the prevailing winds actions in moving sand. Small vegetated hummocks provide wind shadows allowing foredunes to grow. Photo by Marlin Harms.

creeping stems (Fig. 8-4). Most are able to root at the nodes, and vegetative reproduction often results in large colonies developing from a single original plant. These plants are tolerant of repeated burial by shifting sands. The root systems are usually extensive, and long tap-roots may exploit deep-seated water supplies. These growth features result in an interlaced system of buried roots and stems that bind the sand and reduce wind erosion. Leaves of the members of pioneer dune communities are small and often succulent. Their small surface to volume ratio is similar to that found in many desert plants. Leaf and stem surfaces are often densely grayish pubescent. The light color of such leaves increases their reflectivity and reduces the drying effects of sunlight.

There is often a marked zonation in pioneer dune communities. Only the most salt-tolerant species are able to grow immediately adjacent to the beach and the high tide line. Species such as *Abronia maritima* (beach sand-verbena), *A. latifolia* (yellow sand-verbena), *Atriplex leucophylla* (beach saltbush), and *Cakile maritima* (sea rocket) often occur in this zone. In areas with well established foredunes (Fig. 8-5), pioneer dune communities grade into coastal dune scrub communities. With increasing distance from the beach, dune vegetation is less influenced by salt spray. Species composition changes and diversity increases in mid-dune areas. Common native dune species include:

Fig. 8-4. Small dune hummock with beach bur (*Ambrosia chamissonis*) and dunedelion (*Malacothrix incana*). Sea rocket (*Cakile maritima*) grows in the foreground. Photo by David Keil.

Abronia latifolia	yellow sand-verbena
Abronia maritima	beach sand-verbena
Achillea millefolium	yarrow
Ambrosia chamissonis [*Franseria c.; F. bipinnatifida*]	beach-bur
Atriplex californica	California saltbush
Atriplex leucophylla	beach saltbush
Calystegia soldanella [*Convolvulus s.*]	dune morning-glory
Croton californicus	croton
Camissonia cheiranthifolia [*Oenothera c.*]	dune evening primrose
Lathyrus littoralis	beach-pea
Leymus mollis [*Elymus m.*]	American dune grass
Lupinus chamissonis	coastal silver lupine
Malacothrix incana	dune-dandelion, dunedelion

Several introduced species are widespread as members of pioneer dune communities including *Cakile maritima Ammophila arenaria* and *Carpobrotus chilensis* and *C. edulis* (ice plants). Other native or introduced species occur in these communities in some parts of the state.

Vegetation on the windward side of the dunes is often quite different in appearance from that on the leeward side. Whereas

Fig. 8-5. Foredunes at Morro Bay Sand Spit. The principal dune-stabilizing plants are *Ambrosia chamissonis*, *Carpobrotus chilensis*, and *Croton californicus*. Photo by Marlin Harms.

one or two species might grow together on an exposed foredune, the leeward side of a mid-dune may be vegetated by half a dozen or more species. Protection from continuous winds and salt spray also allows pioneer species to achieve greater stature on the leeward side.

Once the dune is stabilized by the pioneer dune community, and the sand surface becomes more stable and protected from the direct winds, soil development starts to take place (Fig. 8-5). The sandy soils gradually accumulate more organic matter, retain more water, become more fertile, and decrease in salt content. Nutrients are added to the soil by decomposition of leaves, stems, and roots of established or formerly established plants and by breakdown of sea wrack (remains of marine plants and animals, driftwood, etc., cast ashore by the waves). Networks of fungal hyphae play a role in stabilizing the sand, holding moisture, breaking down organic remains, and forming mycorrhizal associations with dune plants.

It is important to note that succession is complex and will vary from place to place within a dune system. The plant community gradient from west to east (shoreline to inland) may or may not represent a successional progression. Because of the harsh conditions along the immediate shoreline, succession will not likely lead to a coastal dune community with the same species composition of the more protected mid-dune or back dune areas. The final stage of succession along the shoreline and in the exposed

Fig. 8-6. A blowout removes sand from the roots of the plants that have been stabilizing the dune. Most of the plants affected like this soon die and a once-stabilized dune becomes active again. Photo by Marlin Harms.

foredunes just inland from the ocean may be the pioneer plants. Some of the plants common in sections of the back dunes will never be able to invade and become established in the foredune areas because of the environmental stresses of the shoreline habitat. Plant community succession in these dune areas may be more a pattern of community zonation in response to environmental conditions over space rather than changes in community composition through time

Stabilization of dunes is a process that occupies considerable lengths of time and may be reversed by changing environmental conditions. Because of the instability of sandy soils, disturbances that remove the vegetative cover can cause a previously stabilized dune to become active. These changes may come about as a result of natural disturbances such as storms or fire, or as a result of human activities. Wind erosion of previously stabilized dunes results in the formation of blowouts (Fig. 8-6) where the sand is stripped away from roots and buried stems of dune-stabilizing plants. This may cause the death of the plants and the degradation of established plant communities. Once a dune has become destabilized, wind may cause it to shift its position and to engulf plants growing in its path. On such an active dune only pioneer species or plants that have survived the effects of shifting sands will grow.

Fig. 8-7. Old dunes vegetated by coastal dune scrub in lee of younger dunes of the Morro Bay Sand Spit. Shrubs in the foreground are mostly 0.5–1 m tall. Most of the dunes in the background are well-stabilized, but the dune to the left is actively advancing across previously vegetated areas. Photo by Jesse Martin.

2. Dune Scrub Communities

Dune scrub communities (Figs. 8-7, 8-8) are generally located inland from the pioneer dune communities and are successionally older and more integrated communities. Because they are usually located in the wind-shadow of the foredunes or in areas away from the immediate coast, and because they have a well-developed vegetative cover, dune scrub communities have soils which are considerably more stable than those of pioneer dune communities. The soils of dune scrub communities have more organic matter, retain more water, are more fertile, and have a lower salt content than soils of pioneer dune communities. Shade and litter from the vegetation greatly reduce the reflectivity and temperature fluctuation of the soil. Sometimes a thin, fragile surface layer of mosses and lichens binds the sand particles together.

Dune scrub communities have greater species diversity than do pioneer dune communities, and the vegetation is usually denser and taller as well. Subshrubs and shrubs predominate and include such species as:

Artemisia californica	California sagebrush
Baccharis pilularis	coyote bush
Ericameria ericoides [*Haplopappus e.*]	mock-heather
Eriogonum parvifolium	coastal buckwheat
Lotus scoparius	deerweed
Lupinus arboreus	tree lupine
Lupinus chamissonis	coastal silver lupine
Salvia mellifera	black sage

Because coastal dune-scrub vegetation shares several species with coastal scrub (Chapter 9) some authors consider dune scrub vegetation to be an edaphically controlled component of the complex of coastal scrub communities. We have chosen here to classify dune scrub vegetation together with pioneer dune communities because of their close relationships and shared species.

In areas where dune scrub and chaparral communities intergrade, a variety of additional species may become members of the resulting intermediate communities. Herbaceous species are also common components of dune scrub communities and these include *Abronia umbellata* (purple sand-verbena), *Croton californicus* (croton), *Lessingia filaginifolia* [*Corethrogyne f.*] (California-aster), *Carpobrotus chilensis*, *Eschscholzia californica* (California poppy), and assorted others. A diversity of annual herbs occur in dune scrub communities, whereas sea rocket is often the only annual in pioneer dune communities. In addition to the indigenous taxa, an assortment of introduced annuals and perennials are now widespread in dune scrub communities.

Fig. 8-8. Coastal dune scrub community at Los Osos in San Luis Obispo County. Dominant shrubs include mock heather (*Ericameria ericoides*), black sage (*Salvia mellifera*), and California sagebrush (*Artemisia californica*). The trees in the background are blue gums (*Eucalyptus globulus*), an exotic species that has displaced dune communities in some areas. Photo by Jesse Martin.

Fig. 8-9. Dune Lakes on the Nipomo Dunes, San Luis Obispo County. This series of natural ponds and lakes support a mixture of dune wetlands from freshwater marshes to dune swales. The trees in the foreground are *Eucalyptus globulus*, planted as a windbreak. Photo by Dennis Johansen.

Dune scrub communities do not develop in all parts of California and are sometimes seral in areas where they do occur. In much of northern California old dunes are ultimately invaded by coniferous trees (Chapters 12, 13). *Picea sitchensis* (sitka spruce) and *Pinus contorta* ssp. *contorta* (shore pine) are common climax forest species on old dunes of northern California. In central California old dunes may ultimately support chaparral (Chapter 10), coast live oak woodlands (Chapter 15), or Monterey pine forests (Chapter 12). In southern California coastal dunes are somewhat desert-like in species composition.

Even though the sandy soils underlying coastal dune scrub vegetation are generally well-stabilized, they are still vulnerable to disturbance and wind erosion. Long-stabilized dunes can be reactivated and dunes moving inland can bury existing vegetation. Unstabilized back dunes undergo a process of revegetation similar to that in foredune areas, but often with different pioneer species.

3. Dune Wetland Communities

In areas where the surface of a dune is at or near the water table, communities may develop that are dominated by

phreatophytes (plants with a high moisture requirement with roots extending to the fringe of the water table). These areas range from shallow lakes or ponds or mere depressions where water accumulates during a part of the year (Fig. 8-9). Some of these depressions are created by wind erosion of the dunes and others are the result of other factors such as the partial filling by sand of abandoned stream channels. Some dune wetlands are marsh-like communities that occupy these depressions. These communities are commonly dominated by sedges and rushes or by an assemblage of other marsh plants. The marsh plants are often mixed with species of the adjacent drier dune areas. In well-developed dune wetlands, woody species such as *Salix lasiolepis* (arroyo willow), *Myrica californica* (wax-myrtle), *Toxicodendron diversilobum* [*Rhus d.*] (poison-oak), and *Rubus ursinus* (blackberry) may become established. When subsequent sand movement fills in a dune depression, some phreatophytic species that are tolerant of burial may persist.

Dune wetland communities share many species with riparian communities and freshwater marsh communities. Those dune wetlands in which typical dune species are a major component of the vegetation are treated here. Those wetlands in which the species composition is essentially the same as in riparian or freshwater marsh communities on non-sandy substrates are grouped with other riparian and freshwater marsh communities (Chapters 19 and 20).

Coastal dune swales are found where the water table is close to the soil surface, and the sand is slightly to moderately moister than the surrounding dune slopes. These communities are at the dry end of the moisture gradient for wetland communities and often are composed of a mixture of semi-aquatic and coastal dune scrub plants. Swale communities usually have large numbers of semi-aquatic phreatophytes, plants with high moisture requirements and the ability to tap the fringe of the water table.

Coastal dune swales are intermediate between the "true" wetlands and the surrounding foredunes or dune scrub communities. With time and continued sand deposits from adjacent dunes, these areas may become transformed into typical dune communities. At the wetter end of the gradient, coastal dune swale communities overlap with freshwater marsh and riparian communities. Near dune slopes, coastal dune swales grade into typical dune scrub communities and there are often large areas of overlap. Many of the species of the coastal dune swale occur as understory where riparian woodland communities occur in dune areas.

Fig. 8-10. Effects of off-road vehicular activity on dune vegetation. A sand roads pioneered by a single vehicle soon becomes a sand thoroughfare and a blowout ensues. A year after this picture was taken, almost all the vegetation had been destroyed. Photo by Robert Rodin.

4. Human Impacts on Coastal Dunes

Man's impact upon coastal sand dune communities has been considerable. Many dunes have been used as building sites and in the more populous areas of California, little remains of the natural plant communities that originally occupied the dunes. During World War II coastal fortifications were established in some dune areas and others were used as sites for amphibious landing exercises. Roads have been built across dunes for oil field exploration and development. In many areas large amounts of sand has been mined for use in construction.

Recreational use of dunes, particularly by off-road vehicles has resulted in degradation of many dune communities, removal of natural plant cover from stabilized dunes, and the development of extensive blowouts (Fig. 8-10). Heavy foot traffic in some areas has also seriously damaged the plant cover.

In an effort to control dune movement, various private and governmental agencies have planted such introduced species as *Ammophila arenaria* (European beach grass) and *Carpobrotus edulis* (ice plant). These aggressive species have often succeeded in stabilizing the dunes but in the process have sometimes crowded out the indigenous species (Fig. 8-11). Other introduced species that have become widespread on coastal dunes include *Cakile maritima* (sea rocket) and *Ehrharta calycina* (veldt grass).

During the late 1800's extensive groves of *Eucalyptus globulus* (blue gum and other *Eucalyptus* species were planted on dunes in areas such as the Nipomo Mesa. These plantations have essen-

Fig. 8-11. Introduced plants have modified the vegetation of many coastal sand dunes. Both ice plant (*Carpobrotus edulis*) and European beach grass (*Ammophila arenaria*) can dominate the dunes to the extent that native species are excluded. Photo by David Keil.

tially no understory development and have completely replaced the original dune scrub vegetation.

Human-caused changes in shoreline profiles have greatly modified the patterns of sand deposition and erosion. Changes in direction and strength of ocean currents affect the ability of the water to move sand grains. When water movement is retarded, less sand can be carried. On the other hand, rapidly moving currents can cause previously established sand deposits to be eroded and new deposits to develop in other areas. Construction of breakwaters, jetties, piers, etc. in recent times has prevented sand from reaching the shore in some locations and has resulted in rapid wave erosion of beaches and dunes. On land the damming and channelization of streams and other flood-control measures have greatly reduced the amount of sand reaching the ocean. Without an annual replenishment of sand, the beaches often erode more rapidly than new sand deposits can replace them. The result is a transformation of the California coastline.

Coastal Sand Dune Communities—References

Axelrod, D. I. 1978. Origin of coastal sage vegetation, Alta and Baja California. Amer. J. Bot. 65:1117–1131.

Barbour, M. G. 1976. Synecology of beach vegetation along the Pacific Coast of the United States: a first approximation. J. Biogeogr. 3:55–79.

———. 1978. Salt spray as a microenvironmental factor in the distribution of beach plants at Point Reyes, California. Oecologia 32: 213–224.

———. 1990. The coastal beach plant syndrome. Pp. 197-214 *in* Proceedings. Canadian Symposium on coastal sand dunes.

———. 1992. Life at the leading edge: the beach plant syndrome. Pp. 291–307 *in* U. Seeliger (ed.), Coastal plant communities of Latin America. Academic Press.

———, R. B. Craig, F. R. Drysdale, and M. T. Ghiselin. 1973. Bodega Head: coastal ecology. Univ. Calif. Press, Berkeley. 338 pp.

———, and A. F. Johnson. 1977. Beach and dune. Pp. 223–261 *in* M. G. Barbour and J. Major (eds.). Terrestrial vegetation of California. John Wiley and Sons, N.Y.

———, and T. M. DeJong. 1977. Response of West Coast beach taxa to salt spray, seawater inundation, and soil salinity. Bull. Torrey Bot. Club 104:29–34.

———, and J. E. Rodman. 1970. Saga of the west coast sea rockets, *Cakile edulenta* ssp. *californica* and *C. maritima*. Rhodora 72:370–386.

Bluestone, V. 1981. Strand and dune vegetation at Salinas River State Beach, California. Madroño 28:49–60.

Boorman, L. A. 1977. Sand dunes. Chapter 9 *in* R. S. K. Barnes (ed.). The coastline. John Wiley and Sons, N.Y.

Boyd, R. S. 1992. Influence of *Ammophila arenaria* on foredune plant microdistribution at Point Reyes National Seashore, California. Madroño 39:67–76.

Breckon, G. J., M. G. Barbour. 1974. Review of North American Pacific Coast beach vegetation. Madroño 22:333–360.

Chapman, V. J. 1976. Coastal vegetation, 2nd ed. Pergamon Press, N.Y.

Cooper, W. S. 1936. The strand and dune flora of the Pacific Coast of North America: a geographic study. Pp. 141–187 *in* T. H. Goodspeed (ed.). Essays in geobotany. Univ. Calif. Press, Berkeley.

———. 1967. Coastal dunes of California. Mem. Geol. Soc. Amer. 104:1–131.

Davidson, E. D., and M. G. Barbour. 1977. Germination, establishment and demography of coastal bush lupine (*Lupinus arboreus*) at Bodega Head, California. Ecology 58:592–600.

de Jong, T. M. 1979. Water and salinity relations of California beach species. J. Ecol. 67:647–663.

Fink, B. H., and J. B. Zedler. 1990. Maritime stress tolerance studies of California dune perennials. Madro˜ö 37:200–213.

Holton, B., and A. F. Johnson. 1979. Dune scrub communities and their correlation with environmental factors at Point Reyes National Seashore, California. J. Biogeogr. 6:317–328.

Jenny, H., R. J. Arkley, and A. M. Schultz. 1969. The pygmy forest-podsol ecosystem and its dune associates of the Mendocino coast. Madroño 20:60–74.

Jones, K. G. 1984. The Nipomo dunes. Fremontia 11(4):3–10.

MacDonald, K. B., and M. G. Barbour. 1974. Beach and salt marsh vegetation along the Pacific Coast. Pp. 175–234 *in* R. J. Reimold and W. H. Queen (eds.). Ecology of halophytes. Academic Press, N.Y.

McBride, J. R., and E. C. Stone. 1976. Plant succession on the sand dunes of the Monterey Peninsula, California. Amer. Midl. Naturalist 96:118–132.

McCoy, R. J. 1980. Analysis of off-road vehicle damage: distribution and status of endangered native vascular plants. Nipomo Dunes, California, January 1977–March 1980. Senior Project, Calif. Polytechnic State Univ., San Luis Obispo.

Minckley, W. L., and D. E. Brown. 1982. Californian maritime strands. Pp. 263–264 *in* D. E. Brown (ed.). Biotic communities of the American Southwest—United States and Mexico. Desert Plants 4:1–341.

Purer, E. A. 1936. Studies of certain coastal and sand dune plants of southern California. Ecol. Monogr. 6:1–87.

Ranwell, D. 1972. Ecology of salt marshes and sand dunes. Chapman and Hall, London. 258 pp.

Salisbury, E. 1952. Downs and dunes. G. Bell. Sons, London.

Wiedemann, A. M. 1984. The ecology of Pacific Northwest coastal sand dunes: a community profile. U.S.D.I., Fish and Wildlife Service Rept. FWS/OBS-84/04. 130 pp.

Williams, W. T. 1974. Species dynamism in the coastal strand plant community at Morro Bay, California. Bull. Torr. Bot. Club 101:83–89.

———, and J. R. Potter. 1972. The coastal strand community at Morro Bay State Park, California. Bull. Torr. Bot. Club 99:163–171.

———, and J. A. Williams. 1984. Ten years of vegetation change on the coastal strand at Morro Bay, California. Bull. Torr. Bot. Club 111:145–152.

Wilson, R. C. 1972. *Abronia*. I. Distribution, ecology and habit of 9 species of *Abronia* found in California. Aliso 7:421–437.

Chapter 9

Coastal Scrub Communities

Coastal scrub communities occupy a narrow discontinuous band extending along almost the entire coast of California (Fig. 9-1). They cover about 2.5 percent of the state's land surface. Although these communities are largely restricted to the immediate coastal zone in northern California, south of San Francisco Bay they extend in localized patches as far inland as the flanks of the inner Coast Ranges. In southern California, communities are located both in the coastal zone and the interior where they occur in the cismontane valleys and on the flanks of the Transverse and Peninsular Ranges. They approach the fringes of the desert in the mountain passes that separate cismontane from transmontane California (Fig. 9–1). The elevational distribution of these communities ranges from near sea level to over 700 m (2100 ft). In northern California they mostly occur at low elevations. The higher elevational stands are restricted to the interior areas on mountain slopes in southern California.

The climate in areas where coastal scrub communities occur varies from moist to rather dry. In northern California, coastal scrub communities receive as much as 200 cm (80 in) of rainfall per year whereas the most southerly stands in San Diego County receive as little as 22.5 cm (9 in) of annual rainfall. These communities extend beyond the California boundary into still drier areas of northern Baja California where they intergrade with, and are ultimately replaced by, desert communities.

Species composition of coastal scrub changes drastically from cool, moist areas of the north to warm dry areas of the south. Because of this variability, ecologists often treat the northern and southern portions of coastal scrub separately. "Typical" examples of these communities may differ markedly in species composition and community structure, but there is no clear separation and intermediate stands are common. We have chosen to recognize northern and southern phases of coastal scrub in this treatment.

Fig. 9-1. Map illustrating the distribution of coastal scrub communities in California. These communities occupy a diversity of habitats from the sea bluffs immediately above the ocean to drier hillsides miles from the ocean. However, they are most characteristic of coastal hillsides near the ocean.

Fig. 9-2. Northern coastal scrub in Marin County. Dominants include *Baccharis pilularis* (coyote bush), *Eriophyllum staechadifolium* (coastal golden-yarrow), *Scrophularia californica* (figwort) and *Artemisia californica* (California sagebrush). Also present were various ferns, perennial forbs and grasses. Photo by David Keil.

1. Northern Coastal Scrub Communities

Northern coastal scrub communities occur in a discontinuous strip from northern San Luis Obispo County north to southern Oregon in the immediate coastal zone and on adjacent slopes of the coast ranges. Most stands are located below 500 m (1500 ft) elevation. These communities occupy the coolest and most mesic habitats of any of the coastal scrub types. Precipitation varies from about 200 cm (80 in) in the north to about 100 cm (40 in) in the south.

Soils and parent materials vary widely in the areas where these communities occur. Included are well-weathered clay soils and shallow, coarse soils. Some northern coastal scrub communities occur on old, long-stabilized sand dunes. The distribution and variation in community structure in relation to soils and parent materials have not been adequately studied to date.

Northern coastal scrub communities are dominated mostly by evergreen shrub species one to two meters tall and usually have a well-developed herbaceous or low-woody understory. In most

Fig. 9-3. Shrubs common in coastal scrub communities. A. Coyote bush (*Baccharis pilularis*) is a common component of northern coastal scrub communities and some stands of southern coastal scrub communities. B. Poison-oak (*Toxicodendron diversilobum*) occurs in many cismontane plant communities. Photos by Paula Kleintjes.

stands one of two species is dominant: *Baccharis pilularis* (coyote bush; Fig. 9-2A), or *Lupinus arboreus* (tree lupine). Coyote bush is one of relatively few species that occurs throughout the entire range of the northern coastal scrub and also extends into southern California. Under the shrub canopy the understory is nearly continuous. Common shrubby members of the understory include *Eriophyllum staechadifolium* (coastal golden-yarrow), *Gaultheria shallon* (salal), *Mimulus aurantiacus* [*Diplacus a.*] (bush monkey-flower), *Rubus ursinus* (blackberry), and *Toxicodendron diversilobum* [*Rhus d.*] (poison-oak; Fig. 9-2B). Herbaceous members of the community include grasses and numerous forbs. Some of the more common species are:

Achillea millefolium	yarrow
Anaphalis margaritacea	pearly everlasting
Artemisia suksdorfii	sagewort
Heracleum lanatum	cow-parsnip
Polystichum munitum	sword fern
Pteridium aquilinum	bracken fern
Scrophularia californica	figwort

Northern coastal scrub communities are ecotonally related to several other coastal communities. They commonly occur in close association with coastal prairies and sometimes invade these grassland communities. Some understory species of northern coastal scrub grow under the canopy of mixed evergreen forests, coastal closed-cone coniferous forests, coast redwood forests and

Fig. 9-4. Hillsides in northern Santa Barbara County with southern coastal scrub dominated by *Artemisia californica* and *Salvia leucophylla*. Oaks (*Quercus agrifolia* and *Q. douglasii*) occupy canyons and lower slopes. Photo by David Keil.

other coastal coniferous forests. In some areas northern coastal scrub intergrades with coastal forms of chaparral. Often ecotones between northern coastal scrub and adjacent communities are indistinct. In some areas the coastal scrub communities are seral and are successionally replaced by forests.

Human impact upon northern coastal scrub has taken several forms. Man-caused fires have been a periodic disturbance since prehistoric times. Fire-prevention in modern times has modified successional pathways. Construction activities, urbanization, and agriculture have destroyed many stands. Overall human impact is difficult to measure in some areas because northern coastal scrub communities are among the least studied types in California.

2. Southern Coastal Scrub Communities

South of the San Francisco Bay area, drought-tolerant species assume a progressively greater role in the composition of coastal scrub communities. From Santa Cruz to northern San Luis Obispo Counties both northern and southern phases of coastal scrub occur along with intermediate assemblages of species. Stands of northern coastal scrub occupy mesic sites and those of southern coastal scrub occur in more xeric areas. Southern

Fig. 9-5. Shrubs of southern coastal scrub. A. California sagebrush (*Artemisia californica*) and purple sage (*Salvia leucophylla*) are often dominants in coastal scrub communities. Purple sage occurs from southern San Luis Obispo southward. California sagebrush occurs throughout the range of southern coastal scrub communities. Photo by David Keil. B. Black sage (*Salvia mellifera*) is a common component of southern coastal scrub communities. Photo by Robert Ederer.

coastal scrub occurs in patches in the outer and inner south coast ranges and in scattered areas along the immediate coast. There are additional inland occurrences in the Cuyama River valley and in valleys and lower mountain slopes of southern California.

Southern coastal scrub communities occur on a variety of substrates including such infertile parent material as sandstone, diatomite and serpentine. In some areas these communities occupy very coarse and shallow soils, and in other areas southern coastal scrub occurs on fine-grained clays. Generally southern coastal scrub communities occur on soils with moisture available in the upper horizons only during the winter-spring growing season. Subsequently the upper horizons dry out and little or no deep subsurface moisture persists during the summer and autumn. The coarse soils on which these communities occur are usually shallow and overlie parent material that retains little moisture. The fine grained clay soils impede deep penetration of moisture and promote runoff.

Plants of southern coastal scrub are admirably suited for these environmental conditions. Most have shallow root systems that permeate upper soil horizons but seldom penetrate to deeper soil horizons. The dominant plants are mostly soft-stemmed shrubs or suffrutescent herbs that have thin, often deciduous leaves. Few are sclerophyllous. Active growth in coastal scrub communities occurs mostly during winter and spring when soil moisture is available. During summer when soils dry out, the dominant plants lose some or all of their leaves, and often terminal portions of the stems die back. Drought-deciduous shrubs are more prevalent in southern coastal scrub than in northern coastal scrub. Photosynthesis occurs at high rates during winter and spring months and drops off markedly during the summer.

Southern coastal scrub communities are a mixture of herbaceous, suffrutescent, and shrubby species that usually average two meters or less in height. The herbage of the dominant plants is often glutinous or resinous and may be pungently scented with volatile oils. Such shrubs as *Artemisia californica* (California sagebrush; Fig. 9-4A), *Mimulus aurantiacus* (bush monkey-flower), and the various species of *Salvia* (sage; Fig. 9-4A, B) exhibit these features. Although most dominant species are fully or partially drought-deciduous or have terminal dieback, some species of are evergreen, including *Baccharis pilularis* (coyote bush), *Rhus integrifolia* (lemonade berry), *Malosma laurina* [*Rhus l.*] (laurel sumac), and *Rhamnus californica* (coffeeberry), and some are winter-deciduous, such as *Toxicodendron diversilobum* (poison oak).

Southern coastal scrub communities are often called coastal sage scrub. This is an appropriate name for many stands in which species such as *Artemisia californica* and *Salvia* spp. are dominant. These strongly scented plants are not always dominant in coastal scrub communities, however, and in some they are entirely absent. Using the name, coastal sage scrub, for communities in which sagebrush and sages are absent can be confusing. We prefer the name, southern coastal scrub, which groups physiognomically similar communities, whether or not sagebrush or sages are present.

Southern coastal scrub communities are often grouped together with chaparral (Chapter 10) communities as brushlands. Because of the relatively soft texture of stems and leaves southern coastal scrub communities are sometimes called "soft chaparral". "Typical" chaparral communities by contrast are sometimes called "hard chaparral" because the dominants have stiff branches and sclerophyllous leaves. Chaparral plants, in general, have deeply penetrating root systems that can exploit moisture in subsoil and in cracks in bedrock. Chaparral species are mostly evergreen plants that carry on photosynthesis and often reproduce during summer months. In winter and spring, plants of coastal scrub photosynthesize more rapidly than the sclerophyllous chaparral plants. However, overall annual productivity in chaparral is

Fig. 9-6. Stand of southern coastal scrub in northern Santa Barbara county with dense population of *Yucca whipplei*. California sagebrush (*Artemisia californica*), purple sage (*Salvia leucophylla*), and other shrubs are codominants. Photo by David Keil.

greater than in coastal scrub. Stands of southern coastal scrub in which the evergreen sclerophyllous species, *Malosma laurina*, *Rhus integrifolia*, and *Rhus ovata* are common are intermediate in character between coastal scrub and chaparral.

There is considerable north-south differentiation in southern coastal scrub along moisture and temperature gradients. The wide variety of substrates upon which coastal scrub communities occur contributes selective factors favoring a diversity of community composition. Coastal scrub communities grow in close proximity to various other plant communities and intergrade with them. Topographic variation also influences coastal scrub community composition and structure. As a consequence of this assortment of factors, only generalized trends can be discerned and much site-specific variation in community appearance and species composition makes characterization of these communities difficult.

Species that are common or dominant in at least some southern coastal scrub communities are listed below.

Artemisia californica	California sagebrush
Baccharis pilularis	coyote bush
Brickellia desertorum	brickelbush

Clematis lasiantha	old man's beard
Encelia californica	encelia
Encelia farinosa	brittlebush
Eriodictyon crassifolium	yerba santa
Eriodictyon trichocalyx	yerba santa
Eriogonum fasciculatum	California buckwheat
Eriogonum parvifolium	coastal buckwheat
Eriophyllum confertiflorum	golden yarrow
Lepidospartum squamatum	scalebroom
Leymus condensatus [*Elymus c.*]	giant ryegrass
Lotus argophyllus	lotus
Lotus scoparius	deerweed
Malosma laurina	laurel sumac
Mirabilis laevis	four-O'clock
Opuntia littoralis	prickly pear
Rhamnus californica	coffeeberry
Rhus integrifolia	lemonade berry
Rhus ovata	sugarbush, sugar sumac
Salvia apiana	white sage
Salvia leucophylla	purple sage
Salvia mellifera	black sage
Scrophularia californica	figwort
Toxicodendron diversilobum	poison-oak
Yucca whipplei	yucca

Some of these species, such as *Artemisia californica*, *Salvia mellifera*, *Salvia leucophylla*, and *Eriogonum fasciculatum* are rather widespread. Others are restricted as dominants ecologically or geographically. *Baccharis pilularis*, for instance, is primarily a coastal species and does not extend very far inland. The ubiquitous poison-oak is most common in relatively mesic areas. Succulents such as *Opuntia littoralis* and *Yucca whipplei* are sometimes dominant in areas with coarse, rocky, infertile soils (Fig. 9-6). Communities dominated by *Yucca* are sometimes treated separately as coastal yucca scrub.

Species composition of southern coastal scrub changes gradually along a north-south gradient. North of Santa Barbara County coastal scrub communities most commonly consist of mixtures of one or more of the following:

Artemisia californica	California sagebrush
Baccharis pilularis	coyote bush
Eriogonum fasciculatum	California buckwheat
Lotus scoparius	deerweed
Mimulus aurantiacus	bush monkey-flower
Salvia mellifera	black sage
Toxicodendron diversilobum	poison-oak

Several additional species occur from Santa Barbara County southward but are absent or rare farther north:

Malosma laurina	laurel sumac
Rhus integrifolia	lemonade berry
Rhus ovata	sugarbush, sugar sumac
Salvia apiana	white sage

In addition to a north-south differentiation, there are also east-west changes in community composition. These changes are representative of climatic changes along a gradient extending from the coast inland. Toward the Central Valley in the inner Coast Ranges central California and toward the desert fringes in the valleys of southern California, there is a decrease of oceanic influence with an increase in summer temperatures and a decrease of precipitation. This is reflected in changes in species composition toward an assemblage of the most drought-tolerant species, (e.g., *Artemisia californica*, *Eriogonum fasciculatum*, *Salvia* spp.) in association with desert or semidesert taxa such as:

Atriplex spp.	saltbushes
Chrysothamnus nauseosus	rabbitbush
Encelia farinosa	brittlebush (only in S. Calif.)
Ericameria linearifolia [*Haplopappus l.*]	interior goldenbush
Gutierrezia californica	snakeweed

Several communities form ecotones with southern coastal scrub. Forest and woodland communities that border southern coastal scrub include coastal live oak woodland, coastal closed cone coniferous forests, coastal redwood forests and mixed evergreen forests. In ecotonal areas where the tree canopy is relatively open, plants of coastal scrub may grow as an understory layer. Most coastal scrub species are not shade-tolerant and consequently are absent from areas with dense tree cover. The coastal scrub species are most common in xeric sites with shallow, relatively infertile soils.

In much of southern California, coastal scrub communities are a part of a complicated vegetational mosaic involving grasslands, coastal scrub, chaparral, coastal live oak woodlands, and riparian woodlands. Ecotones among these communities vary from extremely sharp to rather indistinct. Coastal dune scrub (Chapter 8) intergrades very extensively with coastal scrub and represents a transition between beach and foredune vegetation and the coastal scrub of non-sandy substrates. On the other hand, the border between coastal scrub and grassland areas is often sharp and frequently is accompanied by a bare unvegetated zone. This

Fig. 9-7. Burned stand of southern coastal scrub on Figueroa Mountain in Santa Barbara County. The above-ground parts of most of the shrubs were consumed in the fire. *Yucca whipplei* survived the fire because its fleshy leaves protected the stem and terminal bud from the heat of the fast-moving fire. Lupines (*Lupinus succulentus*) and whispering bells (*Emmenanthe penduliflora*) grew from long-dormant seeds buried in the soil. Photo by David Keil.

situation has been the subject of considerable ecological study and controversy. Both allelopathy, the release of phytotoxic substances by certain coastal scrub species, and foraging by rodents and birds have been invoked to explain the bare zones. The actual explanation apparently involves aspects of both factors.

Chaparral and southern coastal scrub intergrade both spatially and temporally. The communities share some species, such as *Rhamnus californica*, *Toxicodendron diversilobum* and species of *Ceanothus*. In ecotonal areas the dominant plants may intermix or the communities may change abruptly. In at least some areas of southern California, coastal scrub species are seral to a chaparral climax. In many chaparral burn sites such coastal scrub plants as *Lotus scoparius*, *Eriogonum fasciculatum*, *Salvia mellifera*, and *Artemisia californica* for a time form a part of the canopy until the longer-lived chaparral species reassert dominance. Some authors consider most or all coastal scrub stands to be a degraded form of chaparral that eventually would change to chaparral if periodic or long-term disturbances were removed.

Fire is important in southern coastal scrub communities. Brush fires often sweep through these communities during summer and fall. The foliage of many of the dominant plants contains highly flammable volatile oils and resins. Above-ground portions of the plants may be completely consumed during a fire (Fig. 9-7). As in chaparral communities some of the dominant species sprout vigorously from root crowns after a fire. Other coastal scrub dominants are killed by fire and recolonize burned sites from seed. Recolonization may occur from unburned seed in the soil or from seeds dispersed from unburned sites. Seed germination of some species may be stimulated by heat from the fire. Enhanced soil fertility in the growing season immediately following a fire may allow rapid regrowth of the community. Leguminous species, such as *Lotus scoparius*, have nitrogen-fixing root-nodules and are often first-year colonists on burned sites. Frequent brush fires may modify the species composition. Some species, such as *Eriogonum fasciculatum* are killed by fire and require several years to grow to reproductive maturity from seed. Frequent fires can eliminate these species by breaking their reproductive cycles. Other species, such as *Lotus scoparius*, are killed by fire but grow back from seed very rapidly, sometimes reaching reproductive maturity during one growing season.

Human impact in southern coastal scrub has most frequently been destructive. In southern California where urbanization has proceeded at an unprecedented rate during the past 25 years, many stands of southern coastal scrub have been cleared for housing or other construction. In other areas these communities have been cleared for agriculture or through "type conversion" have been changed to annual grassland for livestock grazing. Some coastal scrub species are good colonizers in disturbed areas. These often invade waste lots and colonize roadcuts or other sites from which natural cover has been removed. These species are also often seral in burned or otherwise disturbed chaparral communities.

3. Southern Semidesert Coastal Scrub Communities

In extreme southern California and adjacent areas of Baja California, desert species form a part of coastal scrub communities. These communities are very poorly represented in California where they occur only in southern San Diego County, but they are more extensive in northwestern Baja California. The existing stands are located on arid, semidesert sea cliffs. Precipitation is lower in these areas than in any other coastal scrub community in California, but often there is a small but significant amount of summer rainfall. The regions in which these communities occur are frost-free.

Fig. 9-8. Seabluff coastal scrub in San Luis Obispo County. Shrubs on the bluffs include *Eriogonum parvifolium* (coastal buckwheat), *Eriophyllum staechadifolium* (coastal golden-yarrow), *Isocoma menziesii* (coastal goldenbush), *Lupinus arboreus* (tree lupine) and *Baccharis pilularis* (coyote bush). Photo by David Keil.

The vegetation of these communities is characterized by a mixture of shrubs and succulents including a variety of cacti. Shrubs include *Lycium californicum* (wolfberry), *Rhus integrifolia*, and *Encelia californica*. Succulents such as *Bergerocactus emoryi* (cunado), *Dudleya* spp. (dudleya), *Euphorbia misera* (cliff spurge), *Opuntia littoralis*, and *Opuntia prolifera* (prickly pear) form a conspicuous part of the community. In Baja California other desert species, such as *Simmondsia chinensis* (jojoba) and *Ephedra* spp. (Mormon-tea), grow among the coastal scrub plants. In addition several species endemic to Baja California contribute to these communities south of the Mexican border.

4. Sea-bluff Coastal Scrub Communities

In areas with sea bluffs or rocky headlands, a coastal scrub community is often well developed just above the high tide level or at the margins of the erosion face of the bluff (Fig. 9-8). Plants of these communities sometimes cling to nearly vertical rock faces just above the pounding surf. Seabluff coastal scrub communities are widespread along the California coastline as a very narrow band, often not extending more than a few meters inland. These communities are particularly well developed on the Channel Islands where the influence of the maritime climate extends further inland than it usually does on the mainland.

Plants that grow on sea bluffs must be able to tolerate constant exposure to salt spray. Often the soils on which these plants grow contain considerable amounts of salt. The soils, particularly on rocky headlands, may dry out during the summer drought.

Most plants that comprise these communities are low-growing shrubs, herbs or succulents. The plants that grow atop bluffs may be very similar to those on the sides or may be quite different. Some coastal terraces, where deep alluvial soils have formed, support a grassland vegetation. Where the terrace ends abruptly above the ocean, often with a steep, nearly vertical slope, there is an abrupt transition to a narrow band of coastal sea bluff scrub. As the terrace face erodes, coastal scrub plants disperse inland with their rate of successful establishment approximately keeping pace with the rate of erosion.

Plants of steep bluff faces include such succulents as *Dudleya* spp. and *Carpobrotus chilensis* [*Mesembryanthemum c.*] (ice plant), *Carpobrotus edulis* (ice plant), and various shrubs and subshrubs such as:

Artemisia pycnocephala	beach sagewort
Baccharis pilularis	*coyote bush*
Eriogonum parvifolium	coastal buckwheat
Eriophyllum staechadifolium	coastal golden-yarrow
Hazardia squarrosa [*Haplopappus s.*]	sawtooth goldenbush
Isocoma menziesii [*Haplopappus venetus*]	coastal goldenbush
Lessingia filaginifolia [*Corethrogyne f.*]	California-aster

Common herbs include *Armeria maritima* (thrift), *Erigeron glaucus* (seaside daisy), *Eriogonum latifolium* (beach buckwheat), and assorted others. In the zone closest to the waves, plants otherwise characteristic of coastal salt marshes such as *Distichlis spicata*(salt grass), and *Jaumea carnosa* (fleshy jaumea) may cling to the bluff.

Vegetation on top of a bluff or terrace may grade smoothly into northern or southern coastal scrub communities or into a coastal prairie, or there may be a distinct assemblage of plants growing next to the cliff. In the northern half of the state this assemblage may be dominated by one or more species of *Lupinus*, particularly *L. variicolor* close to the edge and *L. arboreus* a few meters inland.

Coastal Scrub Communities —References

Alberts, A. C., A. D. Richman, D. Tran, R. Sauvajot, C. McCalvin, and D. T. Bolger. 1993. Effects of habitat fragmentation on native and exotic plants in southern California coastal scrub. Pp. 103–110 *in* J. E. Keeley (ed.), Interface between ecology and land development in California. Southern California Academy of Sciences, Los Angeles.

Atwood, J. L. 1993. California gnatcatchers and coastal sage scrub: the biological basis for endangered species listing. Pp. 149–169 *in* J. E. Keeley (ed.), Interface between ecology and land development in California. Southern California Academy of Sciences, Los Angeles.

Axelrod, D. I. 1978. Origin of coastal sage vegetation, Alta and Baja California. Amer. J. Bot. 65:1117–1131.

Barbour, M. G., R. B. Craig, F. R. Drysdale, and M. T. Ghiselin. 1973. Bodega Head: coastal ecology. Univ. Calif. Press, Berkeley. 338 pp.

Bartholomew, B. 1970. Bare zone between California shrub and grassland communities, the role of animals. Science 170:1210–1212.

Cole, K. 1980. Geological control of vegetation in the Purisima Hills, California. Madroño 27:79–89.

Cooper, W. S. 1922. The broad sclerophyll vegetation of California. An ecological study of the chaparral and its related communities. Carnegie Inst. Wash. Publ. 319. 124 pp.

Davidson, E. D., and M. G. Barbour. 1977. Germination, establishment and demography of coastal bush lupine (*Lupinus arboreus*) at Bodega head, California. Ecology 58:592–600.

Desimone, S. A., and J. H. Burk. 1992. Local variation in floristics and distributional factors in Californian coastal sage scrub. Madroño 39:170–188.

Epling, C., and H. Lewis. 1942. The centers of distribution of the chaparral and coastal sage associations. Amer. Midl. Naturalist 27:445–462.

Grams, H. J., K. R. McPherson, V. V. King, S. A. MacLeod, and M. G. Barbour. 1977. Northern coastal scrub on Point Reyes peninsula, California. Madroño 24:18–24.

Gray, J. T. 1982. Community structure and productivity in *Ceanothus* chaparral and coastal sage scrub of southern California. Ecol. Monogr. 52:415–435.

———. 1983. Competition for light and a dynamic boundary between chaparral and coastal sage scrub. Madroño 30:43–49.

———. 1983. Nutrient use by evergreen and deciduous shrubs in southern California. J. Ecol. 71:21–41.

———, and W. H. Schlesinger. 1981. Biomass, production, and litterfall in the coastal sage scrub of southern California. Amer. J. Bot. 68:24–33.

———, and ———. 1983. Nutrient use by evergreen and deciduous shrubs in southern California. J. Ecol. 71:43–56.

Halligan, J. 1973. Bare areas associated with shrub stands in grassland: the case of *Artemisia californica*. Bioscience 23:429–432.

———. 1974. Relationship between animal activity and bare areas associated with California sagebrush in annual grassland. J. Range Manage. 27:358–362.

———. 1976. Toxicity of *Artemisia californica* to four associated herb species. Amer. Midl. Naturalist 95:406–421.

Hanes, T. L. 1971. Succession after fire in the chaparral of southern California. Ecol. Monogr. 41:27–52.

———. 1976. Vegetation types of the San Gabriel Mountains. Pp. 65–76 *in* J. Latting (ed.). Plant communities of southern California. Calif. Native Pl. Soc. Spec. Publ. 2.

———, and H. Jones. 1967. Postfire chaparral succession in southern California. Ecology 48:259–264.

Harrison, A., E. Small, and H. Mooney. 1971. Drought relationships and distribution of two Mediterranean climate Californian plant communities. Ecology 52:869–875.

Heady, H. F., T. C. Foin, M. M. Hektner, D. W. Taylor, M. G. Barbour, and W. J. Barry. 1977. Coastal prairie and northern coastal scrub. Pp. 733–760 *in* M. G. Barbour and J. Major (eds.). Terrestrial vegetation of California. John Wiley and Sons, N.Y.

Holland, V. L. 1977. Major plant communities of California. Pp. 3–41 *in* D. R. Walters, M. McLeod, A. G. Meyer, D. Rible, R. O. Baker, and L. Farwell (eds.). Native plants: a viable option. Calif. Native Pl. Soc. Spec Publ 3, Berkeley.

Howell, J. T. 1929. The flora of the Santa Ana Canyon region. Madroño 1:243–253.

Keeley. 1988. Bibliographies on chaparral and the fire ecology of other Mediterranean systems, 2nd ed. Calif. Water Resource Cntr., No. 69. Univ. Calif. Davis, Rep. 28 pp.

Kirkpatrick, J. B., and C. F. Hutchinson. 1977. The community composition of Californian coastal sage scrub. Vegetatio 35:21–33.

———, and ———. 1980. The environmental relationships of Californian coastal sage scrub and some of its component communities and species. J. Biogeogr. 7:23–38.

Minnich, R. A. 1976. Vegetation of the San Bernardino Mountains. Pp. 99–124 *in* J. Latting (ed.). Plant communities of southern California. Calif. Native Plant Soc. Spec. Publ. 2.

———. 1983. Fire mosaics in southern California and northern Baja California. Science 219:1287–1294.

Mooney, H. A. 1977. Southern coastal scrub. Pp. 471–489 *in* M. G. Barbour and J. Major (eds.). Terrestrial vegetation of California. John Wiley and Sons, N.Y.

Muller, C. H., W. H. Muller, and B. L. Haines. 1964. Volatile growth inhibitors produced by aromatic shrubs. Science 143:471–473.

Nilsen, E. T., and W. H. Muller. 1981. The influence of low plant water potential on the growth and nitrogen metabolism of the native California shrub *Lotus scoparius* (Nutt. in T. & G.) Ottley. Amer. J. Bot. 68:402–407.

———, and ———. 1982. Phenology of the drought-deciduous shrub *Lotus scoparius*: climatic controls and adaptive significance. Ecol. Monogr. 51:323–341.

O'Leary, J. F. 1988. Habitat differentiation among herbs in postburn California chaparral and coastal sage scrub. Amer. Midl. Naturalist 120:41–49.

———, and W. E. Westman. 1988. Regional disturbance effects on herb succession patterns in coastal sage scrub. J. Biogeogr. 15:775–786.

———. 1990. Californian coastal sage scrub: general characteristics and considerations for biological conservation. Pp. 24–41 *in* A. A. Schoenherr (ed.), Endangered plant communities of southern California. Proceedings of the 15th Annual Symposium. Southern California Botanists spec. publ. 3.

Pase, C. P., and D. E. Brown. 1982. Californian coastal scrub. Pp. 86–90 *in* D. E. Brown (ed.). Biotic communities of the American Southwest—United States and Mexico. Desert Plants 4:1–341.

Philbrick, R. N., and J. R. Haller. 1977. The southern California islands. Pp. 893–906 *in* M. G. Barbour and J. Major (eds.). Terrestrial vegetation of California. John Wiley and Sons, N.Y.

Poole, D. K., and P. C. Miller. 1975. Water relations of selected species of chaparral and coastal sage communities. Ecology 56: 1118–1128.

Shreve, F. 1936. The vegetation of a coastal mountain range. Ecology 8:27–44.

———. 1936. The transition from desert to chaparral in Baja California. Madroño 3:257–264.

Smith, R. L. 1980. Alluvial scrub vegetation of the San Gabriel River floodplain, California. Madroño 27:126–138.

Specht, R. L. (ed.). 1981. Mediterranean-type shrublands. Ecosystems of the World, 11. Elsevier Sci. Publ. Co., N.Y.

Vogl, R. J. 1982. Chaparral succession. Pp. 81–85 *in* U.S.D.A. Forest Service Gen. Tech. Rep. PSW-58.

Wells, P. V. 1962. Vegetation in relation to geological substratum and fire in the San Luis Obispo quadrangle, California. Ecol. Monogr. 32:79–103.

Westman, W. E. 1979. A potential role of coastal sage scrub understories in the recovery of chaparral after fire. Madroño 26:64–68.

———. 1981. Factors influencing the distribution of species of California coastal sage scrub. Ecology 62:170–184.

———. 1981. Diversity relations and succession in California coastal sage scrub. Ecology 62:170–184.

———. 1982. Seasonal dimorphism in foliage of California coastal scrub. Oecologia 51:385–388.

———. 1991. Measuring realized niche spaces: climatic response of chaparral and coastal sage scrub. Ecology 72:1678–1684.

———, J. F. O'Leary, and G. P. Malanson. 1981. The effect of fire intensity, aspect and substrate on post-fire growth of California coastal sage scrub. Pp. 151–180 *in* N. S. Margaris and H. A. Mooney (eds.). Components of productivity of Mediterranean regions—basic and applied aspects. Dr. W. Junk, The Hague, Netherlands.

Wieslander, A. E., and C. H. Gleason. 1954. Major brushland areas of the Coast Ranges and Sierra-Cascade foothills in California. U.S.D.A. Forest Service, Calif. Forest and Range Exp. Stn., Misc. Paper 15. 9 pp.

Zedler, P. H., C. R. Gantier and G. S. McMaster. 1983. Vegetation changes in response to extreme events; the effect of a short interval between fires in California chaparral and coastal scrub. Ecology 64:809–818.

Chapter 10

Chaparral Communities

Few plant communities are more widespread in California or more characteristic of the state than chaparral (Fig. 10-1). The total area of the state covered by these communities is about 9 percent. Chaparral communities also occur in southern Oregon, northern Baja California and in the mountains of Arizona, and similar communities occur in the mountains of northern Mexico. However, it is in California that these communities reach their greatest development. In the Coast Ranges, the Sierra Nevada, the southern part of the Cascade Range and in the Klamath Mountains, chaparral communities are mostly restricted to steep, dry slopes (Fig. 10-1). In southern California, however, chaparral may form a continuous plant cover over both slopes and valleys. In the coast ranges and in southern California, chaparral often occurs in close association with southern coastal scrub communities (Chapter 9). Away from the immediate coast, chaparral often occupies hotter, drier slopes, but the soil and slope characteristics are often very similar to those described for the coastal scrub communities. Chaparral communities often experience greater seasonal temperature fluctuation than do coastal scrub communities, with higher summer temperatures and lower winter temperatures.

Chaparral soils are often shallow and rocky, but chaparral can occur on a variety of substrates including stabilized sand dunes. The effective soil layer may be quite deep because of penetration of roots of chaparral plants into subsoil or into cracks in bedrock. Soil fertility is often low. The depth of accumulated soil litter is variable and depends both on species composition of the chaparral stand and steepness of the terrain. Chaparral dominated by microphyllous species such as *Adenostoma fasciculatum* (chamise) may have very little soil litter, whereas stands dominated by broadleafed sclerophylls such as *Quercus berberidifolia* (scrub oak) or *Arctostaphylos* (manzanita) species may have an accumulation of litter several centimeters thick. On steep slopes, surface runoff may remove large amounts of litter during winter storms. The waxes and resins released from decomposing litter may coat the surfaces of soil particles, making

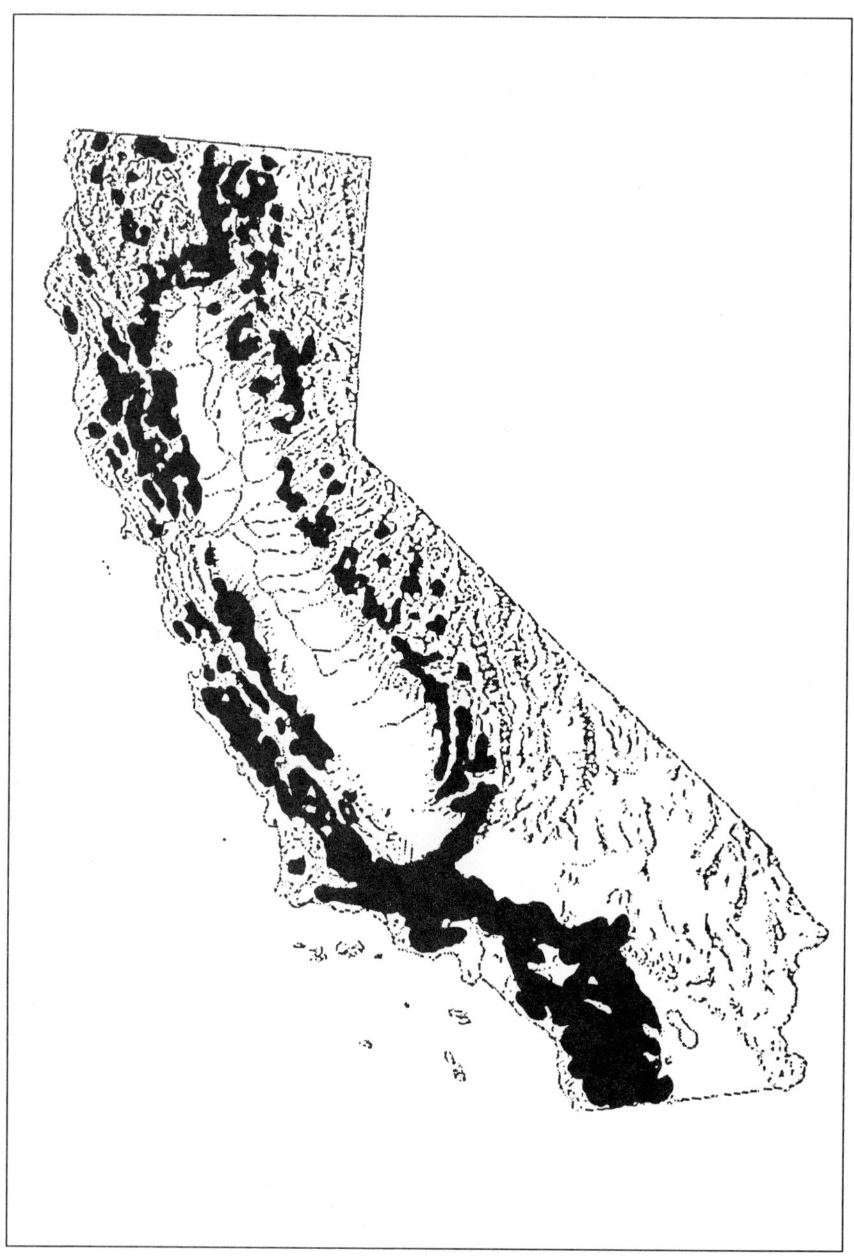

Fig. 10-1. Map illustrating the distribution of chaparral throughout all the mountain ranges of California. Chaparral is best developed and most widespread in southern California. However, it is also a common and characteristic community in the central and northern California mountain ranges as well.

Fig. 10-2. Mixed chaparral stand in La Panza Mountains of eastern San Luis Obispo County dominated by *Arctostaphylos glauca* (bigberry manzanita) and *Adenostoma fasciculatum* (chamise). Photo by David Keil.

the soil rather hydrophobic and reducing the ability of rainfall to penetrate into the soil. The soil may also contain chemicals released from living plants or from decomposing foliage that reduce seed germination and seedling development.

Some chaparral stands are true shrublands and others could be characterized as dwarf woodlands, with the dominant plants growing as gnarled dwarf trees (Fig. 10-2). The dominant plants range in height from a few centimeters in some maritime chaparral areas to between one and three meters in the shrublands, and to four meters or more in dwarf woodlands. In most chaparral communities branches of adjacent plants grow so closely together that the canopies are interlaced, forming dense impenetrable thickets. Chaparral vegetation is characterized by needle-leafed (Fig. 10-3) or broad-leafed (Fig. 10-4) sclerophyllous (hard-leafed) shrubs and dwarf trees. The branches of these plants are often very stiff and woody, and the dominant shrubs are usually long-lived plants. The chaparral communities are often referred to as "hard chaparral" in contrast to the coastal scrub or "soft chaparral" (Chapter 9).

Herbaceous undergrowth is often nonexistent in mature chaparral stands. A combination of factors contributes to this condition. Accumulations of undecomposed litter may make

Fig. 10-3. Needle-like leaves of *Adenostoma*. **A.** *A. fasciculatum* (chamise). **B.** *A. sparsifolium* (red shanks). Photos by David Keil.

seedling establishment difficult. Germination inhibitors present in the soil may prevent seeds from sprouting. These compounds may be leached from the shrub canopy or litter during rainy periods or volatilized into the air during warm weather. In the cover provided by shrubs, foraging herbivores may eat any seedlings that do become established. The interlaced canopies of adjacent shrubs may cast dense shade on the soil surface. Some chaparral stands are sufficiently open that individual canopies do not overlap and under these conditions herbaceous species characteristic of grasslands or woodlands may form a sparse understory. The shrubs may, however, be surrounded by bare areas similar to those around coastal scrub (Chapter 9).

Chaparral communities are very closely associated with fire. The resinous foliage, numerous woody stems and accumulated litter, and standing dead branches combine to make chaparral shrubs highly flammable, particularly during the long dry season. Wildfires in chaparral are a recurrent event and have played a significant ecological role in the development and perpetuation of these communities. Most chaparral species have one or more features that allow them either to survive fires or to become reestablished after fires. In fact, chaparral communities have increased in total area occupied at the expense of other communities as a result of these features. In historic times wildfires that originated in chaparral areas have burned into less fire-adapted forest communities, opening the forest canopies and

Fig. 10-4. Sclerophyllous foliage characteristic of many chaparral shrubs. **A.** Buckbrush (*Ceanothus cuneatus*). **B.** Mountain-mahogany (*Cercocarpus betuloides*). Photos by David Keil.

allowing chaparral species to become established in place of the trees. Frequent fires of this sort may result in complete replacement of forests by chaparral. Chaparral fires tend to recur with a frequency generally averaging from ten to fifty years, sufficient time for chaparral species to regenerate but inadequate for most forest regeneration.

The impact of fire in chaparral is sudden and devastating. The entire environment is changed in a matter of minutes. Soil surface temperatures as high as 700 °C (1200 °F) can be reached in a chaparral fire, killing non-sprouting shrub species and burning off the litter. Populations of herbivorous animals are

Fig. 10-5. Chaparral shrubs burned in fast-moving brush fire. Foliage, litter, and many of the smaller branches were consumed by the fire and the remaining above-ground parts were charred. The partially buried stem bases and the roots were insulated from the heat by the soil. In many species sprouts develop from the stem bases. Photo by David Keil.

drastically reduced as the flames sweep through the brush. When driven by high winds, such as the dry Santa Ana winds of southern California, chaparral fires can become all-consuming firestorms, leaping from ridgetop to ridgetop as windblown embers spread the conflagration. In a matter of minutes a dense chaparral stand can be reduced to a barren hillside denuded except for a few charred snags (Fig. 10-5).

Fires in chaparral have numerous secondary effects. The removal of vegetative cover exposes the soil surface to direct insolation and to direct impact of raindrops. Soil temperatures fluctuate over a greater diurnal range when no plant cover is present. The first winter rains following a chaparral fire may result in rapid erosion of topsoil, gullying, and landslides. The ashing of litter and above-ground plant parts releases non-volatile nutrients that had been bound up in plant tissues. These nutrients may temporarily enhance soil fertility, but many of the nutrients may ultimately be lost to runoff. In addition to the release of nutrients into the soil, the heat of a fire generally breaks down most of the germination inhibitors present in the soil and heat-treats any seeds buried near the soil surface.

Fig. 10-6. Resprouting of burned shrubs from buried stem bases. Photo by David Keil.

Most dominant chaparral species have "adaptations" to fire. Some are able to survive fires because they have fire-resistant underground structures (Fig. 10-6, 10-7A). These structures are below the soil surface and are protected from the fire and extreme temperatures by the insulating properties of the soil. Some have thick, woody, underground structures, variously described as root-crowns, lignotubers, and basal burls. Plants with these structures include *Adenostoma fasciculatum*, *Lonicera interrupta* (chaparral honeysuckle), and several species of *Arctostaphylos*. Some species have extensive underground rhizome systems, such as *Toxicodendron diversilobum* [*Rhus d.*] (poison oak), and *Pickeringia montana* (chaparral pea). The herbaceous vines of the genus *Marah* (manroot, wild-cucumber) have large, deeply buried, fleshy roots.

Not all species have features that allow them to survive fires. Although the adult plants of these species are killed outright by fires, seeds survive the fire if the upper layers of the soil are not heated too much. The seeds of both sprouting and non-sprouting species often retain viability for many years, and some have germination rates greatly enhanced by heat scarification. Although seed germination may be negligible in a mature chaparral stand, profuse seed germination follows fire.

In addition to the germination of seeds of the dominant shrubs, numerous other species also make an appearance following a

Fig. 10-7. Fire succession in chaparral. **A.** Regeneration of chamise (*Adenostoma fasciculatum*) after fire. Sprouts arise from the basal burl of this chamise bush. Chamise seedlings grow together with a mixture of fire annuals. **B.** *Phacelia brachyloba* is a common fire-following ephemeral in the La Panza Mountains of San Luis Obispo County. Photos by Robert Ederer.

chaparral fire. Some represent seeds dispersed from nearby areas. Others come from seeds that have lain dormant in the soil for many years. During the first wet season following a fire, great quantities of herbs may dominate the burn site (Fig. 10-7B). A distinctive feature of many of these plants is that they are "pyrophyte endemics" and only occur after fires. They may appear on a burn site for several years after a fire has occurred, but soon diminish in quantity and disappear entirely as the chaparral stand regenerates. These species then exist on the site only as a buried seed load. The seeds remain viable for many years and germinate only after they have been heat-treated by a subsequent fire. Plants such as *Papaver californicum* (fire poppy) and *Emmenanthe penduliflora* (whispering bells) are examples of typical fire annuals.

Fire succession in chaparral may involve merely a rapid regeneration of the previously existing cover, or it may include the temporary ascendancy to dominance of some species that are later successionally replaced as the stand returns to maturity. The former condition, sometimes termed "autosuccession" occurs when a species such as *Adenostoma fasciculatum* regrows from its lignotubers without significant development of other taxa. The latter is a common situation and often involves short-lived chaparral shrubs or species more characteristic of coastal scrub as seral components of the chaparral. Such species as *Artemisia californica* (California sagebrush), *Baccharis pilularis* (coyote bush), *Lotus scoparius* (deerweed), *Eriogonum fasciculatum* (California buckwheat), and *Salvia mellifera* (black sage), are principally coastal scrub species but are often seral in chaparral. Other taxa, such as *Dendromecon rigida* (bush poppy) and *Leptodactylon californicum* (prickly phlox) are often common shrubs in chaparral after fires, but decrease in representation as the community approaches maturity.

Human activities in chaparral communities have had several effects. In prehistoric and early historic times, Indians periodically set fire to chaparral areas to increase the forage available for game and to reduce the density of the brush. Some of the later Causasian residents also carried out the same practices. In more recent times, development of residential tracts in chaparral areas has brought a large segment of the population of southern California into close contact with this fire-prone vegetation. Suppression of brush-fires in inhabited areas has allowed the fuel load to increase to a level that often results in extremely hot fires when the chaparral finally does burn.

Many chaparral fires result from human activities. Some are accidental and others are the result of arson. Management of fires and fire-prone areas is a major concern of governmental agencies in California. In attempts to prevent or slow the spread of fires, wide firebreaks or fuel breaks have been cleared in many areas. Governmental agencies such as the United States Forest Service have experimented with the use of controlled burning to break continuous chaparral stands into blocks in various stages of succession. Nevertheless, chaparral fires annually cause extensive damage in California with millions of dollars in losses. Following fires, burned areas are often seeded with annual grasses such as *Lolium multiflorum* (ryegrass) in an attempt to stabilize the slopes against erosion. The effectiveness and long-term ecological effects of this seeding is a matter of much controversy.

Chaparral plant communities are a diverse assemblage with much variation in species composition. Nearly 900 species of vascular plants occur in chaparral. About 240 of these are woody plants (mostly shrubs) from a diversity of families such as Asteraceae, Ericaceae, Fabaceae, Fagaceae, Rhamnaceae, Rosaceae and Scrophulariaceae. The remainder of the species are annual or perennial herbs from a wide array of families. These communities are recognizable more from their physiognomy than from their species composition. Physiognomically similar com-

Fig. 10-8. Common chaparral shrubs. **A.** Foliage and fruits of *Arctostaphylos glauca* (bigberry manzanita). Photo by David Keil. **B.** Foliage of scrub oak (*Quercus berberidifolia*). Photo by Paula Kleintjes.

munities occur in other parts of the world with Mediterranean climates. Chaparral-like communities occur in the area around the Mediterranean Sea, along the coast of Chile, in southern Africa, and in southern Australia. These communities have essentially no species in common but are remarkably similar in structure and adaptations to fire.

It is useful for the study of chaparral communities to group them into "phases" or "series". These phases intergrade on a broad scale, and not all chaparral stands can be so classified. Some chaparral phases can be recognized by the dominance of one or several species of shrubs. Other chaparral phases are taxonomically diverse but occupy particular ecological zones. The latter are perhaps defined too broadly, but insufficient information is presently available to categorize all of them by species composition.

1. Mixed Chaparral Communities

Mixed chaparral communities are characterized by codominance of several shrubby species (Fig. 10-2). Common species of mixed chaparral stands include:

Adenostoma fasciculatum (Fig. 10-3A)	chamise
Arctostaphylos spp. (Fig. 10-8A)	manzanita
Ceanothus spp. (Fig. 10-4A)	ceanothus
Cercocarpus betuloides (Fig. 10-4B)	mountain-mahogany
Eriodictyon spp.	yerba santa
Garrya spp.	silk-tassel bush
Fremontodendron californicum	flannelbush, fremontia
Heteromeles arbutifolia	toyon
Pickeringia montana	chaparral-pea
Prunus ilicifolia	holly-leafed cherry
Quercus berberidifolia (Fig. 10-8B)	scrub oak
Quercus wislizenii var. *frutescens*	shrub form of interior live oak
Rhamnus californica	coffeeberry
Rhamnus crocea	redberry
Rhus ovata	sugarbush
Ribes spp.	currants, gooseberries
Toxicodendron diversilobum	poison-oak

Not all of these species necessarily occur together in a given stand, and other taxa may be locally present or even dominant. However, the mixed nature of the community is generally evident. The California Native Plant Society's new classification system list several series that we have included in mixed chaparral. Usually the shrubs grow to a height of from one to three meters in mixed chaparral, but sometimes these communities take the form of a dwarf woodland with a canopy four meters or more in height.

Mixed chaparral stands are usually dense, and understory development is negligible in mature stands.

2. Chamisal Chaparral Communities

Chamise (*Adenostoma fasciculatum*) is a very common component of stands and frequently is the dominant species. The term chamisal is applied when 80% or more of a stand is composed of this species. In some areas, particularly in southern California, chamise forms almost pure stands. Chamisal chaparral occurs on some of the most xeric sites that are occupied by chaparral. On more mesic sites, these communities grade into mixed chaparral. Litter accumulation is slight in chamisal chaparral and bare ground is exposed between the shrubs. Herb growth is essentially nil in mature chamisal chaparral.

3. Red-Shanks Or Ribbon-Bush Chaparral Communities

Red-shanks or ribbon-bush (*Adenostoma sparsifolium*) is restricted to southern California in widely separated, rather small areas from San Luis Obispo County to San Diego County. Chaparral stands dominated by *A. sparsifolium* are usually more open than those of mixed chaparral or chamisal chaparral. Red-shanks grows to large shrub or small tree size (Fig. 10-9) and consequently the stands that it dominates often have the appearance of a dwarf woodland. Red-shanks can grow in pure stands, but often it is mixed with *Adenostoma fasciculatum*, *Ceanothus* spp., and *Rhus ovata*. These communities sometimes have a well-developed herb layer, in contrast to most other chaparral associations. Red-shanks chaparral occurs from 600 to 1800 meters elevation and on both coastal and desert slope exposures. These communities are largely restricted to soils derived from granitic parent material.

4. Manzanita Chaparral Communities

In some areas chaparral consists almost entirely of manzanita (*Arctostaphylos* spp.). Manzanita chaparral is not as widespread as are communities dominated by chamise or by *Ceanothus*. These communities occur mostly on relatively deep, fertile soils and can be found from near sea level to relatively high elevations. The site requirements are more-or-less intermediate between those of chamisal chaparral and forest communities. Manzanita chaparral often forms stiff, almost impenetrable stands and is usually dominated by one or two species of *Arctostaphylos*. Man-

Fig. 10-9. A large individual of *Adenostoma sparsifolium* (red-shanks). Photo by David Keil.

zanitas range from prostrate shrubs to small trees four meters or more tall. There is generally no understory in these stands. The appearance of a manzanita chaparral stand from a distance is often that of a light green velvety mantle.

5. Ceanothus Chaparral Communities

Numerous species of *Ceanothus* occur in chaparral communities in California and in some areas *Ceanothus* is dominant. Ceanothus chaparral stands are commonly dominated by a single species of *Ceanothus* which may occur as a pure stand. Ceanothus chaparral in southern California is often a seral stage, and the relatively short-lived *Ceanothus* bushes are successionally replaced as the stand reaches maturity. In northern California, however, *Ceanothus* species often are the climax species. Ceanothus chaparral usually reaches a height of one to three meters and forms a dense cover. Understory development is usually absent and litter accumulation is low. These communities rarely occur above 1200 m elevation although some *Ceanothus* species occur at much higher elevations. Ceanothus chaparral usually develops on more mesic sites than does chamisal chaparral. Various other shrubs occur as subordinate members of Ceanothus chaparral stands and these often include *Adenostoma fasciculatum*, *Quercus berberidifolia*, *Heteromeles arbutifolia*, and *Rhus ovata*.

6. Scrub Oak Chaparral Communities

The name chaparral has its origin from Spain. The Spanish term "chaparro" refers to scrub oak, and the suffix -al means "the place of". Thus, the term chaparral literally translates to "the place of scrub oak". While scrub oaks are often a common component of chaparral, they are not common in all chaparral stands. Many phases of chaparral lack scrub oaks entirely. However, scrub oaks occur in mixed chaparral and may completely dominate mesic chaparral sites where soils are deep. When scrub oaks comprises over 60% of the chaparral shrub cover, we refer to the community as scrub oak chaparral. Scrub oak chaparral communities occur at elevations above chamisal chaparral and often interfinger with forest communities in both northern and southern California. They form one of the most interesting and variable phases of chaparral.

The most common scrub oaks in lower elevation chaparral communities are *Quercus berberidifolia* (scrub oak), *Q. wislizenii* var. *frutescens* (shrub form of interior live oak), *Q. johntuckeri* (desert scrub oak), *Q. agrifolia* (shrub forms of coast live oak), *Q. durata* (leather oak) and *Q. garryana* (shrub form of Oregon oak). Occasionally *Q. chrysolepis* (canyon live oak) occurs as a shrub form as well. At higher elevations, *Q. vaccinifolia* (huckleberry oak) dominates many montane chaparral stands and often occurs in monospecific populations (please refer to section on montane chaparral).

Perhaps the most common and widespread of the scrub oaks is *Quercus berberidifolia* (scrub oak) which is characteristic of many California chaparral stands. However, the others are also common and dominate localized sites. For example, desert scrub oak dominates many chaparral areas on hot, dry slopes that border desert and semi-desert areas of the interior mountains such as the Temblor Range and the Transverse Range. Shrubby interior live oaks are locally dominant on mesic sites of the coast ranges as well as the interior mountains of both northern and southern California, often on poor soils. On old, stabilized coastal dunes along the Central Coast, coast live oaks form dense monospecific populations, and on serpentine soils, leather oak may be dominant in serpentine chaparral stands.

An unusual form of scrub oak chaparral communities occurs in scattered locations in the Sierra-Cascade axis from Kern County northward. This chaparral is dominated by *Quercus garryana* var. *breweri* (shin oak). These communities are usually associated with upper foothill woodland and lower zones of montane coniferous forests. They differ from other chaparral communities in that the dominant shrubs are winter-deciduous and fire-resistant. Although shin oak often occurs in pure stands, it also grows with other typical higher elevation chaparral species. This unique phase of chaparral resembles communities of the southern Rocky Mountains that are dominated by a similar deciduous oak, *Quercus gambelii* (Gambel oak).

Scrub oak chaparral is 2–4 meters tall and often forms a tall chaparral that resembles a dwarf or pygmy woodland. Although the trunks on the trees may be large, the canopy is low, usually a short distance above the ground. Unlike most chaparral communities, it is often possible to crawl (sometime walk) under the canopy of the shrub oaks. This form of chaparral is often referred to as a dwarf or pygmy woodland. For example, Coast live oaks form a picturesque pygmy woodland along the Central Coast in areas of stabilized coastal dunes . Locals refer to this community as the elfin forest (refer to Chapter 15). However, scrub oak, desert scrub oak, shrubby interior live oak, and Oregon oak form similar pygmy woodlands on interior mountain slopes in California.

Because scrub oak chaparral communities occupy such a diversity of habitats, associated species are quite variable from stand to stand. However, the most common associated species usually include:

Arctostaphylos spp. (Fig. 10-8A)	manzanita
Ceanothus spp. (Fig. 10-4A)	ceanothus
Cercocarpus betuloides (Fig. 10-4B)	mountain-mahogany
Fraxinus dipetala	flowering ash
Fremontodendron californicum	fremontia
Garrya spp.	silk-tassel bush
Heteromeles arbutifolia	toyon
Prunus ilicifolia	holly-leafed
Rhamnus californica	coffeeberry

Rhamnus crocea	redberry
Rhus ovata	sugarbush
Ribes spp.	currants, gooseberries
Toxicodendron diversilobum	poison-oak

Understory vegetation in scrub oak chaparral, like other chaparral communities, is sparse. The litter layer is usually quite thick, 20 cm or more, which makes it difficult for plants to become established. Thus, there are usually only a few scattered shade tolerant herbs and ferns.

Like other chaparral communities, scrub oak chaparral is highly adapted to fire. Scrub oaks generally sprout vigorously following fires, and because they occupy mesic chaparral sites, these oaks grow rapidly to re-establish the chaparral community within a relatively short time (80% cover in 25 years). Unlike many chaparral species, however, oaks generally do not reestablish themselves by seed after a fire. The seeds within the acorns are killed by the heat of a fire, and it may be several years after a fire before new acorns are produced.

7. Maritime Chaparral Communities

Maritime chaparral communities occur in windswept coastal areas of central and northern California, such around Monterey Bay in Monterey and Santa Cruz Counties, north of San Simeon and around the southern end of Morro Bay in San Luis Obispo County, and on the Burton Mesa in Santa Barbara County. In several of these areas maritime chaparral communities are best developed on sandy soils of old stabilized sand dunes, but in northwestern San Luis Obispo County they occur on fine grained clays and serpentine-derived soils. These communities form a mosaic with closed-cone coniferous forests, coastal live oak woodlands, and coastal scrub in the northern areas and with coastal prairie, coastal scrub, coastal dune scrub, and coastal live oak woodland in the southern area. Maritime chaparral communities are often dominated by a mixture of *Arctostaphylos* and *Ceanothus* species, some of which have very localized distributions. The dominant plants range in height from prostrate shrubs a few centimeters tall to large bushes 2–3 meters tall (Fig. 10-10). Associated with the *Arctostaphylos* and *Ceanothus* species are such species as *Adenostoma fasciculatum*, *Heteromeles arbutifolia*, *Rhamnus californica*, *Rhamnus crocea*, *Salvia mellifera*, *Toxicodendron diversilobum*, and other shrubs. On slopes exposed to strong coastal winds, prostrate, wind-pruned shrub forms of *Quercus agrifolia* (coast live oak) may form a part of the community. The herb flora varies from a well-developed grass and herb zone in sites where the shrub cover is incomplete to nearly non-existent in dense patches of chaparral.

Fig. 10-10. Maritime chaparral in the Arroyo de la Cruz area of San Luis Obispo County. Shrubs on this windswept ridge include prostrate *Arctostaphylos hookeri* var. *hearstiorum* (Hearst manzanita), knee-high *Adenostoma fasciculatum* (chamise), mounds of *Arctostaphylos cruzensis* (la Cruz Manzanita), and wind-pruned *Quercus agrifolia* (coast live oak). Photo by David Keil

8. Island Chaparral Communities

The chaparral of the California Channel Islands is dominated by species familiar on the mainland. However, the physiognomy differs in terms of openness of cover, stand maturity and diversity of growth form. There are some endemic species in the island chaparral but these are usually not dominant. The most common species in island chaparral are:

Adenostoma fasciculatum	chamise
Arctostaphylos spp.	manzanita
Ceanothus spp.	ceanothus
Quercus berberidifolia	scrub oak
Rhus integrifolia	lemonade berry

Other associates in lesser densities are:,

Cercocarpus betuloides	mountain-mahogany
Comarostaphylis diversifolia	summer-holly
Crossosoma californicum	crossosoma
Dendromecon harfordii	island bush poppy

Fig. 10-11. Island chaparral on Santa Catalina Island. Dominant shrubs include *Arctostaphylos catalinae* (Santa Catalina Island Manzanita), *Quercus berberidifolia* (scrub oak), *Rhamnus pirifolia* (island redberry), and *Adenostoma fasciculatum* (chamise). Note the openness of the shrub canopy. Photo by David Keil.

Garrya elliptica	silk-tassel bush
Heteromeles arbutifolia	toyon
Prunus ilicifolia ssp. lyonii	Catalina cherry
	island scrub oak
Rhamnus pirifolia	island redberry
Rhus ovata	sugarbush, sugar sumac
Toxicodendron diversilobum	poison oak

Crossosoma californicum, *Dendromecon harfordii*, and *Rhamnus pirifolia* are island endemics. Although widespread on the mainland, *Adenostoma fasciculatum* is less common on the islands. Unlike most mainland chaparral stands, island chaparral is usually rather open (Fig. 10-11). Individual plants may be considerably larger than in mainland chaparral stands. Shrub density is less than in mainland areas and often the island stands resemble an open dwarf woodland with a grassy understory and scattered shrubs common to coastal scrub. This community physiognomy may at least in part be a result of the introduction of goats to the islands. These animals have established large feral populations in some areas. They have selectively eaten some species and have created browse lines on others. The overgrazing has removed much of the potential fuel for brush-fires and consequently has modified the fire cycle of the community. In the absence of fire individual plants of some species are able to grow to a larger stature than in other areas. In some areas there have been no fires for 100 years or more and short-lived plants that reproduce in mass after a fire have been eliminated or greatly reduced in numbers.

9. Serpentine Chaparral Communities

Chaparral communities that occur on serpentine soils are generally composed of a mixture of broadly tolerant species such as *Adenostoma fasciculatum*, *Cercocarpus betuloides*, *Pickeringia montana*, and *Heteromeles arbutifolia*, widespread species that are largely restricted to serpentine, such as *Quercus durata* (leather oak) and *Garrya congdonii*, and localized serpentine endemics such as *Arctostaphylos obispoensis* (San Luis Obispo manzanita) and *Ceanothus jepsonii* (musk brush). Shrubs of the serpentine chaparral often exhibit dwarfing and may be more widely spaced than those of mixed chaparral. In openings between the shrubs a herb layer may or may not occur. Not all serpentine chaparral stands are open shrublands and some closely resemble typical chamisal or mixed chaparral in density. In some areas *Adenostoma fasciculatum* is sufficiently dominant that the stands are essentially indistinguishable from the chamisal chaparral that occurs on non-serpentine soils.

Serpentine chaparral communities intergrade in some areas with mixed evergreen forests and foothill woodland and are frequently found together with Sargent cypress forests. In these areas of intergradation, the chaparral shrubs often grow as an understory under the canopy of the dominant trees.

10. Montane Chaparral Communities

The highest elevational chaparral communities occur in association with montane coniferous forests (Fig. 10-12) and may be found as high as subalpine forest areas. The climate is cool to cold and much precipitation occurs as snow. The sites occupied by montane chaparral are usually exposed areas, often on steep, rocky, south-facing slopes with very shallow soils. Montane chaparral stands also occur on sites where the original forest cover has been removed by fire, logging, landslides or other disturbances. Species of montane chaparral often occur as understory elements of montane coniferous forests, especially where the forest canopy is open.

Montane chaparral communities have several genera in common with lower elevation chaparral communities, but these communities have few species in common. The dominant genera

Fig. 10-12. Montane chaparral and conifers in the Sierra Nevada. Photo by V. L. Holland.

are often species of *Arctostaphylos* or *Ceanothus*. Other common elements are: species of *Ribes* (gooseberries and currants) and *Prunus* (cherries), *Quercus vaccinifolia* (huckleberry oak), *Chrysolepis sempervirens* [*Castanopsis s.*] (mountain chinquapin), and *Symphoricarpos parishii* (snowberry). At intermediate elevations and in southern California, various species more characteristic of mixed chaparral communities may occur in montane chaparral. Chaparral areas dominated by Oregon oak (discussed previously) may also be included as a form of montane chaparral .

11. Semidesert Chaparral Communities

Semidesert chaparral communities occur in transmontane California on the mountains bordering the Mojave and Colorado deserts, in the Walker Pass area of the southern Sierra Nevada, and extend along the western Transverse Ranges and inner Coast Ranges into cismontane California. In the latter areas, these communities border valley grassland in some sites and cismontane desert areas in others. These communities are usually associated with piñon pine-juniper woodlands and with Joshua tree woodlands and occur in the lower parts of the montane coniferous forest as in the Mount Pinos region. They are the only chaparral communities that extend into transmontane regions.

Semidesert chaparral communities are often considerably more open than are most other types of chaparral. Spaces between the shrubs may be bare or may support a mixture of grasses and forbs. Species composition is often highly variable and includes elements derived from several different communities. Some stands of semidesert chaparral are floristic grab-bags. Species typical of these communities include:

Arctostaphylos pungens	manzanita
Ceanothus greggii	desert ceanothus
Eriogonum fasciculatum	California buckwheat
Quercus johntuckeri [*Q. turbinella* in part]	desert scrub oak
Rhus trilobata	skunkbush

Elements of mixed chaparral that sometimes grow in desert chaparral communities include:

Adenostoma fasciculatum	chamise
Arctostaphylos spp.	manzanita
Ceanothus spp.	ceanothus
Cercocarpus betuloides	mountain-mahogany
Dendromecon rigida	bush poppy
Eriodictyon spp.	yerba santa
Fremontodendron californicum	fremontia, flannelbush
Garrya flavescens	silk-tassel bush
Quercus wislizenii var. *frutescens*	interior live oak
Rhamnus tomentella	hoary coffeeberry
Rhamnus ilicifolia	holly-leafed redberry
Rhus ovata	sugarbush

Desert and desert woodland elements include:

Cercocarpus ledifolius	curl-leaf mountain-mahogany
Chrysothanmus nauseosus	rabbitbush
Ephedra spp.	Mormon-tea
Ericameria linearifolia [*Haplopappus l.*]	interior goldenbush
Juniperus californica	California juniper
Opuntia spp.	prickly-pear
Purshia mexicana [*Cowainia m.*]	cliff-rose
Purshia tridentata ssp. glandulosa [*P. glandulosa*]	antelope bush

Even coastal scrub plants such as *Salvia leucophylla* (purple sage) and *Artemisia californica* occasionally join these communities.

Chaparral Communities—References

Axelrod, D. I. 1989. Age and origin of chaparral. Pp. 7–19 *in* S. C. Keeley (ed.). The California chaparral. Paradigms reexamined. Natural History Museum of Los Angeles County, Science Series no. 34.

Baker, G. A., P. W. Rundel, and D. J. Parsons. 1983. Comparative phenology and growth in three chaparral shrubs. Bot. Gaz. 143:94–100.

Bauer, H. L. 1936. Moisture relations in the chaparral of the Santa Monica Mountains, California. Ecol. Monogr. 6:409–454.

Biswell, H. H. 1952. Factors affecting brush succession in the coast region and the Sierra Nevada foothills. Pp. 63–97 *in* Proceedings of the 4th annual California weed conference. San Luis Obispo.

———. 1974. Effects of fire on chaparral. Pp. 321–364 *in* T. T. Kozlowski and C. E. Ahlgren (eds.). Fire and ecosystems. Acad. Press, N.Y.

Bradbury, D. E. 1978. The evolution and persistence of a local sage/chamise community pattern in southern California. Assoc. Pacific Coast. Geogr. Yearbook 40:29–56.

Burcham, L. T. 1974. Fire and chaparral before European settlement. Pp. 101–120 *in* M. Rosenthal (ed.). Proceedings of the symposium on living with the chaparral. Sierra Club, San Francisco.

Chew, R. M., B. B. Butterworth, and R. Grecham. 1959. The effect of fire on the small mammal population of chaparral. J. Mammal. 40:253.

Christensen, N. L. 1973. Fire and the nitrogen cycle in California chaparral. Science 181:66–68.

———, and C. H. Muller. 1975. Effects of fire on factors controlling plant growth in *Adenostoma* chaparral. Ecol. Monogr. 45:29–55.

———, and C. H. Muller. 1975. Relative importance of factors controlling germination and seedling survival in *Adenostoma* chaparral. Amer. Midl. Naturalist 93:71–78.

Cole, K. 1980. Geological control of vegetation in the Purisima Hills, California. Madroño 27:79–89.

Cooper, W. S. 1922. The broad sclerophyll vegetation of California. An ecological study of the chaparral and its related communities. Carnegie Inst. Wash. Publ. 319. 124 pp.

Davis, F. W., D. E. Hickson, and D. C. Odion. 1988. Composition of maritime chaparral related to fire history and soil, Burton Mesa, Santa Barbara County, California. Madroño 35:169–195.

DeBano, L. F., and C. E. Conrad. 1978. The effect of fire on nutrients in a chaparral ecosystem. Ecology 59:489–497.

Detling, W. R. 1961. The chaparral formation of southwestern Oregon with considerations of its postglacial history. Ecology 42:348–357.

Dunne, J. A. Dennis, J. W. Bartolome, and R. H. Barrett. 1991. Chaparral response to a prescribed fire in the Mount Hamilton Range, Santa Clara County, California. Madroño 38:21–29.

Epling, C., and H. Lewis. 1942. The centers of distribution of the chaparral and coastal sage associations. Amer. Midl. Naturalist 27:445–462.

Frazer, J. M., and S. D. Davis. 1988. Differential survival of chaparral seedlings during the first summer drought after wildfire. Oecologia 76:215–221.

Fulton, R. E., and F. L. Carpenter. 1979. Pollination, reproduction and fire in California *Arctostaphylos*. Oecologia 38:147–157.

Gray, J. T. 1982. Community structure and productivity in *Ceanothus* chaparral and coastal sage scrub of southern California. Ecol. Monogr. 52:415–435.

———. 1983. Competition for light and a dynamic boundary between chaparral and coastal sage scrub. Madroño 30:43–49.

———, and W. H. Schlesinger. 1981. Nutrient cycling in Mediterranean type ecosystems. Pp. 259–286 *in* P. C. Miller (ed.). Resource use by chaparral and matorral. A comparison of vegetative function in two Mediterranean type ecosystems. Springer, N.Y.

Grieve, B. J. 1955. The physiology of sclerophyllous plants. J. R. Soc. West. Aust. 39:31–45.

Griffin, J. R. 1978. Maritime chaparral and endemic shrubs of the Monterey Bay region, California. Madroño 25:65–81.

Haidinger, T. L., and J. E. Keeley. 1993. Role of high fire frequency in destruction of mixed chaparral. Madro˜õ 40:141–147.

Hanes, T. L. 1965. Ecological studies on two closely related chaparral shrubs in southern California. Ecol. Monogr. 35:213–235.

———. 1971. Succession after fire in the chaparral of southern California. Ecol. Monogr. 41:27–52.

———. 1976. Vegetation types of the San Gabriel Mountains. Pp. 65–76 *in* J. Latting (ed.). Plant communities of southern California. Calif. Native Pl. Soc. Spec. Publ. 2.

———. 1977. Chaparral. Pp. 417–469 *in* M. G. Barbour and J. Major (eds.). Terrestrial vegetation of California. John Wiley and Sons, N.Y.

———, and H. Jones. 1967. Postfire chaparral succession in southern California. Ecology 48:259–264.

Harrison, A., E. Small, and H. Mooney. 1971. Drought relationships and distribution of two Mediterranean climate Californian plant communities. Ecology 52:869–875.

Harvey, R., and H. Mooney. 1964. Extended dormancy of chaparral shrubs during severe drought. Madroño 17:161–165.

Head, W. S. 1972. The California chaparral—an elfin forest. Naturegraph, Healdsburg, Calif.

Hellmers, H., J. Horton, G. Juhren, and J. O'Keefe. 1955. Root systems of some chaparral plants in southern California. Ecology 36:667–678.

Hochberg, M. C. 1980. Factors affecting leaf size of chaparral shrubs on the California Islands. Pp. 189–203 *in* D. M. Power (ed.). The California Islands: Proceedings of a Multidisciplinary Symposium. Santa Barbara Museum of Natural History, Santa Barbara.

Horton, J. S., and C. J. Kraebel. 1955. Development of vegetation after fire in chamise chaparral of southern California. Ecology 36:244–262.

Howell, J. T. 1929. The flora of the Santa Ana Canyon region. Madroño 1:243–253.

Jow, W. M., S. H. Bullock, and J. Kummerow. 1980. Leaf turnover rates of *Adenostoma fasciculatum* (Rosaceae). Amer. J. Bot. 67:256–261.

Keeley, J. E. 1975. Longevity of non-sprouting *Ceanothus*. Amer. Midl. Naturalist 93:505–507.

———. 1977. Seed production, seed populations in soil, and seedling production after fire for two congeneric pairs of sprouting and non-sprouting chaparral shrubs. Ecology 58:820–829.

———. 1984. Bibliographies on chaparral and the fire ecology of other Mediterranean systems. Calif. Water Resource Cntr., Univ. Calif. Davis, Rep. No. 58. 190 pp.

———. 1988. Bibliographies on chaparral and the fire ecology of other Mediterranean systems, 2nd ed. Calif. Water Resource Cntr., Univ. Calif. Davis, Rep. No. 69. 328 pp.

———, and S. C. Keeley. 1987. Role of fire in the germination of chaparral herbs and suffrutescents. Madroño 34:240–249.

———, and ———. 1988. Chaparral. Pp. 165–208 *in* M. G. Barbour and W. D. Billings (eds.), North American Terrestrial Vegetation. Cambridge Univ. Press, Cambridge.

———, and P. H. Zedler. 1978. Reproduction of chaparral shrubs after fire: a comparison of sprouting and seeding strategies. Amer. Midl. Naturalist 99:142–161.

Keeley, S. C. (ed.). 1989. The California chaparral. Paradigms reexamined. Natural History Museum of Los Angeles County, Science Series no. 34. 171 pp.

Kittredge, J. 1939. The annual accumulation and creep of litter and other surface materials in the chaparral of the San Gabriel Mountains, California. J. Agric. Res. 58:537–541.

———. 1939. The forest floor of the chaparral in San Gabriel Mountains, California. J. Agric. Res. 58:521–535.

———. 1955. Litter and forest floor of the chaparral in parts of the San Dimas Experimental Forest, California. Hilgardia 23:563–596.

Krause, D., and J. Kummerow. 1977. Xeromorphic structure and soil moisture in the chaparral. Oecol. Plant. 12:133–148.

Kummerow, J., D. Krause, and W. Jow. 1977. Root systems of chaparral shrubs. Oecologia 29:163–177.

———, ———, and W. Jow. 1978. Seasonal changes of fine root density in the southern Californian chaparral. Oecologia 37:201–212.

Lariguaderie, A., T. W. Hubbard, and J. Kummerow. 1990. Growth dynamics of two chaparral shrubs with time after fire. Madroño 37:225–236.

Lawrence, G. E. 1966. Ecology of vertebrate animals in relation to chaparral fire in the Sierra Nevada foothills. Ecology 47:278–291.

Macey, A., and J. Gilligan (eds.). 1961. Man, fire and chaparral—a conference on southern California wildland research problems. Univ. Calif. Agric. Res. Stn., Berkeley. 101 pp.

Marion, L. H. 1943. The distribution of *Adenostoma sparsifolium*. Amer. Midl. Naturalist 29:106–116.

McMaster, G. S. 1982. Response of *Adenostoma fasciculatum* and *Ceanothus greggii* chaparral to nutrient additions. J. Ecol. 70:745–756.

McPherson, J. K., and C. H. Muller. 1969. Allelopathic effects of *Adenostoma fasciculatum*, "chamise," in the California chaparral. Ecol. Monogr. 39:177–198.

Miller, E. H. 1947. Growth and environmental conditions in southern California chaparral. Amer. Midl. Naturalist 37:379–420.

Miller, P. C., and D. K. Poole. 1979. Patterns of water use by shrubs in southern California. For. Sci. 25:84–98.

———, and ———. 1980. Partitioning of solar and net irradiance in mixed and chamise chaparral in southern California. Oecologia 47:328–332.

———, ———, and P. M. Miller. 1983. The influence of annual precipitation, topography, and vegetative cover on soil moisture and summer drought in southern California. Oecologia 56:385–391.

Mills, J. N. 1986. Herbivores and early postfire succession in southern California chaparral. Ecology 67:1637–1649.

Minnich, R. A. 1976. Vegetation of the San Bernardino Mountains. Pp. 99–124 *in* J. Latting (ed.). Plant communities of southern California. Calif. Native Plant Soc. Spec. Publ. 2.

———. 1980. Vegetation of Santa Cruz and Santa Catalina Islands. Pp. 123–135 *in* D. M. Power (ed.). The California Islands: Proceedings of a Multidisciplinary Symposium. Santa Barbara Museum of Natural History, Santa Barbara.

———. 1983. Fire mosaics in southern California and northern Baja California. Science 219:1287–1294.

Mooney, H. A., and C. E. Conard (eds.). 1977. Proceedings of the symposium on the environmental consequences of fire and fuel management in Mediterranean ecosystems. U.S. Forest Serv. Gen. Tech. Rep. WO-3.

———, and E. L. Dunn. 1970. Convergent evolution of Mediterranean climate evergreen sclerophyll shrubs. Evolution 24:292–303.

———, S. L. Gulmon, D. J. Parsons, and A. T. Harrison. 1974. Morphological changes within the chaparral vegetation type as related to elevational gradients. Madroño 22:281–316.

———, and D. J. Parsons. 1973. Structure and function in the California chaparral—an example from San Dimas. Pp. 83–112 *in* F. di Castri and H. Mooney (eds.). Mediterranean type ecosystems, origin and structure. Springer-Verlag, N.Y.

Moreno, J. M., and W. C. Oechel. 1991. Fire intensity effects on germination of shrubs and herbs in southern California chaparral. Ecology 72:1993–2004.

Naveh, Z. 1960. The ecology of chamise (*Adenostoma fasciculatum*) as affected by its toxic leachates. Bull. Ecol. Soc. Amer. 41:56–57.

Ng, E., and P. C. Miller. 1980. Soil moisture relations in the southern California chaparral. Ecology 61:98–107.

O'Leary, J. F. 1988. Habitat differentiation among herbs in postburn California chaparral and coastal sage scrub. Amer. Midl. Naturalist 120:41–49.

Parker, V. T. 1984. Correlation of physiological divergence with reproductive mode in chaparral shrubs. Madroño 31:231–242.

Pase, C. P. 1982. Californian (coastal) chaparral. Pp. 91–94 *in* D. E. Brown (ed.). Biotic communities of the American Southwest—United States and Mexico. Desert Plants 4:1–341.

Parsons, D. J., P. W. Rundel, R. P. Hedlund, and G. A. Baker. 1981. Survival of severe drought by a non-sprouting chaparral shrub. Amer. J. Bot. 68:973–979.

Patric, J. H., and T. L. Hanes. 1964. Chaparral succession in a San Gabriel Mountain area of California. Ecology 45:353–360.

Pavlik, B. M., P. C. Muick, S. Johnson, and M. Popper. 1991. Oaks of California. Cachuma Press and the California Oak Foundation, Los Olivos, CA. 184 pp.

Perry, D., and S. Merschel. 1982. In the chaparral life ends and begins in a great blaze. Smithsonian 13(7):132–143.

Philbrick, R. N., and J. R. Haller. 1977. The southern California islands. Pp. 893–906 *in* M. G. Barbour and J. Major (eds.). Terrestrial vegetation of California. John Wiley and Sons, N.Y.

Poole, D. K., and P. C. Miller. 1975. Water relations of selected species of chaparral and coastal sage communities. Ecology 56:1118–1128.

———, and ———. 1981. The distribution of plant water stress and vegetation characteristics in southern California chaparral. Amer. Midl. Naturalist 105:32–43.

Radosevich, S. R., and S. G. Conard. 1980. Physiological control of chamise shoot growth after fire. Amer. J. Bot. 67:1442–1447.

Riggan, P. J., P. M. Jacks, and R. N. Lockwood. 1988. Interaction of fire and community development in chaparral of southern California. Ecol. Monogr. 58:155–176.

Rosenthal, M. (ed.). 1974. Symposium on living with the chaparral. Proceedings. Sierra Club, San Francisco. 225 pp.

Rundel, P. W. 1977. Water balance in Mediterranean sclerophyll ecosystems. Pp. 95–106 *in* H. A. Mooney and C. E. Conard (eds.). Proceedings of the symposium on the environmental consequences of fire and fuel management in Mediterranean ecosystems. U.S. Forest Serv. Gen. Tech. Rep. WO-3.

———, and D. J. Parsons. 1979. Structural changes in chamise (*Adenostoma fasciculatum*) along a fire-induced age gradient. J. Range Manage. 32:462–466.

———, and ———. 1980. Nutrient changes in two chaparral shrubs along a fire-induced age gradient. Amer. J. Bot. 6:51–58.

Sampson, A, W. 1944. Plant succession on burned chaparral lands in northern California. Univ. Calif. Agr. Exp. Stn. Bull. 685, Berkeley. 144 pp.

Schlesinger, W. H., and D. S. Gill. 1980. Biomass, production and changes in the availability of light, water, and nutrients during the development of pure stands of the chaparral shrub, *Ceanothus megacarpus*, after fire. Ecology 61:781–789.

———, J. T. Gray, D. S. Gill, and B. E. Mahall. 1982. *Ceanothus megacarpus* chaparral: a synthesis of ecosystem processes during development and annual growth. Bot. Rev. 48:71–117.

Sparks, S. R., and W. C. Oechel. 1993. Factors influencing postfire sprouting vigor in the chaparral shrub, *Adenostoma fasciculatum*. *Madroño 40:224–235.*

Specht, R. L. (ed.). 1981. Mediterranean-type shrublands. Ecosystems of the world, 11. Elsevier Sci. Publ. Co., N.Y.

Swank, S. E., and W. C. Oechel. 1991. Interactions among the effects of herbivory, competition, and resource limitation in chaparral herbs. Ecology 72:104–115.

Sweeney, J. R. 1956. Responses of vegetation to fire: a study of the herbaceous vegetation following chaparral fires. Univ. Calif. Publ. Bot. 28:143–150.

———. 1968. Ecology of some "fire type" vegetations in northern California. Pp. 111–125 *in* Proceedings of the Tall Timbers fire ecology conference. Tall Timbers Res. Stn., Tallahassee, Fla.

Thrower, N. J. W., and D. E. Bradbury (eds.). Chile-California Mediterranean scrub atlas. Dowden, Hutchinson & Ross, Inc. Stroudsburg, Pennsylvania.

Vogl, R. J. 1968. Fire adaptations of some southern California plants. Pp. 79–110 *in* Proceedings of the Tall Timbers fire ecology conference. Tall Timbers Res. Stn., Tallahassee, Fla.

———. 1976. An introduction to the plant communities of the Santa Ana and San Jacinto Mountains. Pp. 77–98 *in* J. Latting (ed.). Plant communities of southern California. Calif. Native Plant Soc. Spec. Publ. 2.

———. 1982. Chaparral succession. Pp. 81–85 *in* U.S.D.A. Forest Service Gen. Tech. Rep. PSW-58.

———, and P. K. Schorr. 1972. Fire and manzanita chaparral in the San Jacinto Mountains, California. Ecology 53:1179–1188.

Watkins, V. M., and H. de Forest. 1941. Growth in some chaparral shrubs in California. Ecology 22:79–83.

Wells, P. V. 1962. Vegetation in relation to geological substratum and fire in the San Luis Obispo quadrangle, California. Ecol. Monogr. 32:79–103.

Westman, W. E. 1979. A potential role of coastal sage scrub understories in the recovery of chaparral after fire. Madroño 26:64–68.

———. 1991. Measuring realized niche spaces: climatic response of chaparral and coastal sage scrub. Ecology 72:1678–1684.

Whitney, S. 1979. A Sierra Club naturalist's guide to the Sierra Nevada. Sierra Club Books, San Francisco.

Wieslander, A. E., and C. H. Gleason. 1954. Major brushland areas of the Coast Ranges and Sierra-Cascade foothills in California. U.S.D.A. Forest Service, Calif. Forest and Range Exp. Stn., Misc. Paper 15. 9 pp.

Wilson, R. C., and R. J. Vogl. 1965. Manzanita chaparral in the Santa Ana Mountains, California. Madroño 18:47–62.

Wright, R. D. 1966. Lower elevational limits of montane trees. I. Vegetational and environmental survey in the San Bernardino Mountains of California. Bot. Gaz. 127:184–193.

Chapter 11

Grassland Communities

Grassland communities originally covered about 13 percent of the land area of California. The most extensive areas of grassland communities are (or were) located in the San Joaquin, Sacramento and Salinas Valleys and similar smaller valleys that are too hot and dry to support woodland vegetation (Fig. 11-1). In addition there are scattered and discontinuous areas of grassland along the coast. In central and southern California, coastal grasslands are rather extensive on marine terraces. Grasslands occur from near sea level in coastal areas to approximately 1200 m (4000 ft) in dry mountain regions of southern and central California.

Several environmental factors combine to favor development of grassland communities. In some areas climatic factors seem to be of overriding importance. A combination of low (25–50 cm; 10–20 in), and often irregular precipitation, and high summer temperatures (often exceeding 100° F), plus the annual long summer drought, create conditions unsuitable for most woody plants. These features characterize the environment of most of the Central Valley region of California. Where moisture is available in somewhat greater and more dependable supplies, as in foothill regions or along broad drainages around major waterways, trees are often present, with grassland species forming the understory of the resulting woodland communities (Chapter 14).

Soil types vary greatly in California grasslands. Soils that support grassland communities range from coarse and gravelly to very fine clays. Parent materials are highly variable. Soil depth is variable but often exceeds 50 cm. The deepest soils occur in valley areas. These soils often are very fertile and are suitable for agriculture. Grassland soils generally have adequate winter and spring moisture to support grasses and other shallowly rooted plants but become too dry in the summer to support woody plants. Grassland soils may be underlain at a relatively shallow depth by a claypan or hardpan layer that prevents deep penetration of rainfall. In much of California the upper layers of grassland soils often dry out completely during the summer.

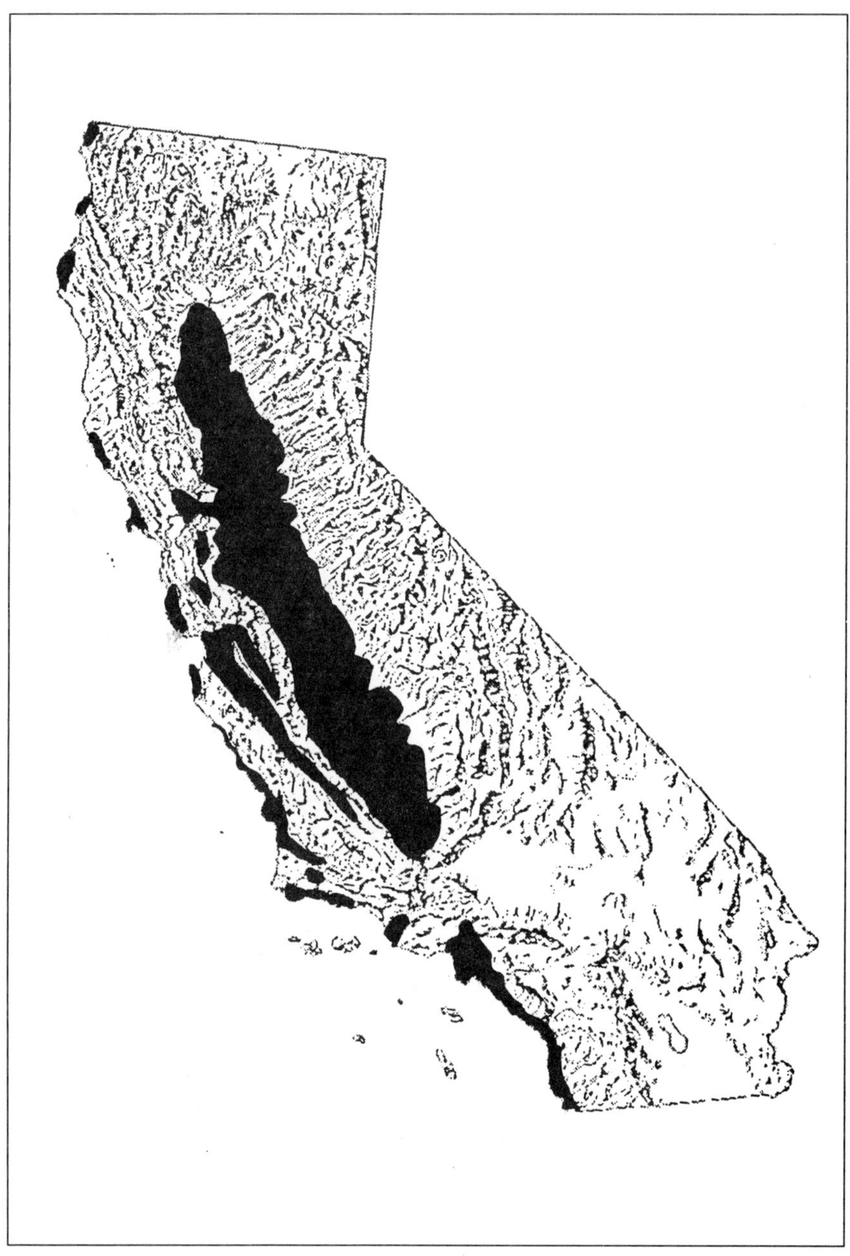

Fig. 11–1. Distribution of grasslands in California. Grassland communities are the most common vegetation of the interior valleys and coastal plains where they are found from the Mexican border to near the Oregon border. Variation in grassland communities from site to site is discussed in the text. Large areas of the grasslands mapped here have been converted to agricultural use or urban development.

In coastal regions grassland communities often form a mosaic with coastal scrub (Chapter 9), chaparral (Chapter 10), and coastal live oak woodland (Chapter 14). Coastal grasslands usually occur on deep, fine-grained soils. Coastal scrub communities often occur on shallow, well drained soils. However, coastal scrub and grasslands often occur together on heavy clay-rich soils, and one may successionally replace the other. The bare-zone ecotone between coastal scrub and grasslands is discussed in Chapter 9. Chaparral occurs mostly on steeper slopes with coarse, well drained soils. Coast live oaks occur on intermediate sites where moisture is available for longer periods into the late spring or summer, e.g., on north-facing slopes and in ravines.

The lack of woody plants in California grasslands may also be due to anthropogenic factors. In various areas of the state shrublands, woodlands and forests are known to have been converted to grassland through removal and suppression of the woody overstory. This is especially true in the valley and foothill regions where blue oak (*Quercus douglasii*) and valley oak (*Quercus lobata*) have been cleared either for firewood or for agriculture. Human disturbance of the grasslands did not begin with the Spanish settlers. Native American Indian tribes did not have domesticated livestock but instead hunted wild game. Most of the California Indian tribes practiced rangeland burning to improve the ability of the land to support game species. This burning tended to favor some species over other less fire-adapted plants. The widespread *Nassella pulchra*, for instance, is well adapted to periodic fires. Today various areas that once supported woody vegetation continue to be maintained as grasslands by man's direct activities, by grazing, or by fire.

To a casual observer, grassland communities may seem monotonous and uniform. Actually, however, these communities may be quite diverse. Hundreds of plant species occur in California's grasslands. In addition to the dominant grasses, California's grasslands often contain numerous annual and perennial **forbs** (non-grassy herbs). There is much variation in the species composition of grassland communities. In seasons with ample winter rainfall, annual and sometimes perennial wildflowers often create beautiful displays.

California's grasslands can be grouped into several phases. These phases intergrade to a considerable extent and vary with geographical location, elevation, exposure, climate, substrate and history of disturbance. Four of the phases are considered in this chapter. The first is in large part historical. It represents the native grassland communities that once occupied extensive areas of the state. California's native grasslands have been profoundly modified by human agency and largely replaced by communities

Fig. 11-2. *Nassella pulchra* (purple needle grass). This native bunch grass was one of the dominants in the California native grasslands prior to the Mission Period and is still common in some areas in grasslands dominated by introduced annuals. Photo by David Keil.

of altered composition. They are now reduced to scattered relict stands. The second phase, the valley and southern coastal grassland communities, comprises the annual-dominated grasslands that have replaced the native grassland communities in much of California. The third phase is the northern coastal grasslands. These too are communities of more or less altered composition. Desert grasslands are scattered in various locations in the Mojave and Great Basin deserts. Mountain meadows, which generally are at least seasonally wet, are discussed in chapter 20 with other wetland communities.

1. Native Bunchgrass Grassland Communities

Prior to the late 1800s native grasslands (California prairies) and related oak woodland communities with a grassland understory (Chapter 15) covered about 25% of California. The historical distribution of these communities was very similar to that of the state's present-day grasslands (Fig. 11-1). Although large areas of grasslands are still present in California, these exist mainly in variously modified forms, and large areas are now in cultivation. A large portion of the grasslands of California were in the past plowed for dry-land farming and then allowed to go fallow. Usually these areas now support a community highly modified from the original native grasslands, often with few of the original native grasses and forbs present. Other portions of the grassland have been used for urban construction, highways, oil-well drilling fields and other activities of modern civilization.

The grasslands of California have been modified to a greater extent than have any other California plant communities. We do not know exactly what the original grasslands of California looked like! There are no areas of the grassland communities that have gone untouched by man's activities. Because of this, there is considerable debate about the original nature and composition of the California grasslands. There are no existing scientifically accurate records of precolonial "pristine" grasslands. On the basis of fragmentary non-technical historical records, modern distributions and modern ecological studies, ecologists have attempted to determine the nature of these original grasslands. Here and there we can find examples of grasslands dominated by native perennials (Fig. 11-3), but whether these are like the original grasslands cannot be determined.

It is often stated that in precolonial times California grasslands were dominated by perennial bunch grasses, particularly *Nassella pulchra* [*Stipa p.*] (purple needlegrass; Fig. 11-2). Certainly in some regions, bunch grasses were dominant, but in much of the Central Valley the dominant plants may have always been annual herbs along with some drought-tolerant shrubs. According to a recent article on the subject, bunch grass-dominated areas in the San Joaquin Valley were located in areas with particularly favor-

Fig. 11-3. Relict native grassland community at Laguna Lake Park in San Luis Obispo County. The serpentine-derived soils on this hillside support a grassland community dominated by native bunch grasses. Most introduced annual grass species (except *Avena barbata*) are unable to grow on these serpentine soils. Photo by David Keil.

able soil moisture, and bunch grasses were not the widespread dominants that many have supposed. However, these grasslands were probably dominated by perennial bunch grasses with an variety of erect and stoloniferous perennial grasses and a large diversity of native annuals interspersed.

Nassella pulchra (purple needlegrass) and *Nassella cernua* (nodding needlegrass) were apparently the dominants in many areas. However, several other native perennial grasses were also common including:

Aristida spp.	three-awn
Danthonia californica	California oatgrass
Elymus elymoides [*Sitanion hystrix*]	squirreltail
Elymus glaucus	blue wild-rye
Elymus multisetus [*Sitanion jubatum*]	squirreltail
Festuca californica	fescue
Festuca idahoensis	fescue
Koeleria macrantha [*K. cristata*]	junegrass
Leymus triticoides [*Elymus t.*]	beardless wild-rye
Melica californica	melic grass
Melica imperfecta	melic grass

Muhlenbergia rigens	deergrass
Nassella lepida [*Stipa l.*]	slender needlegrass
Poa secunda [*P. scabrella*]	pine bluegrass

Most of these are bunch grasses, but *Leymus triticoides*, a sod-former rather than a bunch grass, was common in native grasslands along the coast as well as in the Central Valley. Native grassland communities on coastal terraces, valleys and foothills from the Central Coast to southern California were likely dominated by *Nassella lepida* and *Achnatherum coronatum* [*Stipa c.*] (large needlegrass) along with *Nassella pulchra* and *N. cernua.* In addition *Koeleria macrantha Melica imperfecta*, *Aristida* spp., and *Muhlenbergia rigens* were also common. A diversity of several hundred native annual and perennial wildflowers and some annual grasses occurred interspersed with the perennial grasses. The annual wildflowers provided spectacular color displays of blue, white, orange and yellow throughout the native grassland areas.

However, these are in large part historical communities. Why have California's grasslands changed so much? Several factors are involved. One of the most important reasons for the changes was introduction of cattle, sheep and horses by Spanish colonists which led to heavy overgrazing. From the end of the last ice age about 12,000 years ago to the coming of Spanish colonists in the 1700's, California's grasslands were subjected to rather low intensity grazing pressure. There were few species of large herbivores. Bison were absent from most of the state. The native grasslands were home to pronghorn antelope, tule elk and deer. The native herbivores moved through the grasslands, grazing them seasonally. They did not graze the bunch grasses down to their bases where their growing buds occur; consequently, the natives were able to withstand the light, seasonal grazing. Domesticated livestock placed a much heavier burden on California's grasslands. The California grasslands were exposed to intense livestock grazing throughout much of the 19th century.

At the same time that the grazing regime of California's grasslands was being altered, alien plant species were introduced in packing, ballast, grain shipments, and hay from Spain, or were carried into California by livestock. Many of these plants were native to the Mediterranean region which is climatically very similar to California. In the Mediterranean region cultivation of crops and grazing of domesticated animals has been carried out for thousands of years. The introduced species, including many annual grasses, were much more tolerant of grazing than the native perennials. These "pre-adapted" annuals spread rapidly into the California grasslands, ecologically replacing native bunch grasses. This introduction of alien annuals has profoundly

Fig. 11-4. Grassland communities at Laguna Lake Park in San Luis Obispo County. The vegetation in the foreground is a valley grassland community dominated by annual grasses native to the Mediterranean. The serpentine hillside in the background is dominated by native perennial bunch grasses (see Fig. 11-3). Photo by David Keil.

changed California's grasslands. Many of these species are now fully integrated dominant members of California's flora. The resulting communities are sometimes called the "California annual type" or even "new natives". It is difficult to imagine what the California landscape looked like without these plants.

The aggressive aliens were not only highly adapted to the new grazing intensity and patterns but were also very successful in competition with natives for soil nutrients, water and space. There is evidence that indicates ungrazed native grasslands were still overtaken by aliens because many of these introduced species, such as *Avena fatua* (wild oats), were able to out-compete and gradually replace the native perennials on favorable sites that were not grazed.

As a result of the land conversion to agriculture, new grazing pressures and competition with introduced annual grasses, California native grasslands were reduced to a few small scattered relict stands up and down the state. These stands are in areas with light grazing histories and often occur under somewhat harsh soil conditions in which aliens are not as competitive. For example, many relict stands occur on rocky hillsides or on unusual soil types (Fig. 11-3). Because local populations of many of the native grasses are serpentine tolerant, they are successful in competition on serpentine soils where introduced grasses do not grow as well. The relict native grassland stands that occurs on the Cal Poly, San Luis Obispo campus are found on rocky, serpentine hillsides and often associate with *Yucca whipplei* (chaparral yucca). Native grasslands also remain on other such serpentine areas in the state such as Mt. Tamalpias in Marin County. Many of the larger stands have been inventoried and mapped by the California Department of Parks and Recreation (Barry, 1972). However, numerous small, patchy stands exist that are not part of the inventory. For example, native bunch grasses can still be found sporadically in the present-day grasslands where grazing pressures have not been great. Efforts to protect the existing stands of native grasslands have been strongly encouraged and should continue. These relict stands represent what is left of an important part of California's heritage and should be conserved and protected.

2. Valley and Southern Coastal Grassland Communities

Valley and southern coastal grassland communities (Figs. 11-4, 11-5) occur in the Central Valley, interior valleys of the Coast Ranges, and along the coast of central and southern California as well as some of the off-shore islands (Fig. 11-1). They also occur in semiarid areas such as the Temblor Range and Diablo Range, and some slopes above the Mojave Desert. They are normally located at low elevations but may extend up to 1200 meters or more in southern California. Grasses and forbs introduced during and since the Spanish colonial period dominate, often with a large admixture of native California forbs. Among the introduced grasses that are most common are:

Avena barbata	slender wild oats
Avena fatua	common wild oats
Bromus diandrus [*B. rigidus*	rip-gut brome
Bromus hordeaceus [*B. mollis*]	soft chess brome
Bromus madritensis ssp. *rubens* [*B. rubens*]	red brome
Bromus tectorum	cheat grass
Hordeum murinum [*H. leporinum*]	foxtail barley
Lolium multiflorum	annual ryegrass
Vulpia myuros [*Festuca myuros; F. megalura*]	rat-tail fescue
Vulpia spp. [*Festuca*]	annual fescue

Native perennial grasses such as *Nassella pulchra*, *N. cernua*, and *Poa secunda* [*P. scabrella*], occur in some areas, but in most areas they are no longer dominant.

Fig. 11-5. Valley grassland at Laguna Lake Park in San Luis Obispo County. Native spring wildflowers grow here in grassland dominated by introduced annual grasses. Photo by David Keil.

Among the most common introduced forbs are species of *Brassica* (mustard) and *Erodium* (filaree). Many other introduced taxa are also found in valley grasslands. Often over 50 percent of the flora in a stand of valley grassland consists of introduced species. The remaining portion of the flora consists of native annual and perennial herbs, sometimes with a few native bunch grasses as well. In favorable growing seasons, these native species may be dominant or aspect-dominant in some areas. Valley grasslands may contain few or many native wildflowers (Fig. 11-5). Some of the most common natives are:

Amsinckia spp.	fiddleneck
Castilleja spp.	owl's clover
Cryptantha spp.	cryptantha
Delphinium spp.	larkspur
Eschscholzia californica	California poppy
Gilia spp.	gilia
Lasthenia californica [*L. chrysostoma*, *Baeria ch.*]	goldfields
Layia spp.	tidy tips
Linanthus spp.	linanthus
Lotus spp.	deervetch
Lupinus spp.	lupine
Phacelia spp.	phacelia
Plagiobothrys spp.	popcorn flower
Trifolium spp.	clover
Triphysaria spp.	owl's clover

Fig. 11-6. Northern coastal grassland community on marine terraces at Arroyo de la Cruz in San Luis Obispo County. This community is dominated by native perennial grasses

These are all spring-flowering species, and most are annuals. At or before the end of the spring growing season they set seed and die. In some areas vernal pools (Chapter 21) are scattered in these grasslands like islands of an archipelago. These seasonal pools of water dry out with their own characteristic floral displays as the spring comes to an end.

The summer aspect of valley grasslands is a golden-brown expanse of dead grasses and forbs. Not all plants die back to the ground level, however. Some species carry on only vegetative growth during spring and remain green into summer. These remain green even when growing in hard, nearly dry soil and flower in late summer and early autumn. The most conspicuous members of this second season of flowering are composites such as *Hemizonia* spp., *Madia* spp., and *Lagophylla ramosissima* (tarweeds). Other common summer-flowering species are *Eremocarpus setigerus* (turkey-mullein), *Trichostemma lanceolatum* (vinegar weed), and various species of *Eriogonum* (buckwheat). How they extract water from seemingly dry soils is unknown.

3. Northern Coastal Grassland Communities

Northern coastal grasslands or coastal prairies occur primarily as discontinuous patches from the San Francisco Bay region northward on coastal terraces and on glades and bald hills of the North Coast Ranges and Klamath Mountains (Fig. 11-1). Isolated occurrences occur farther to the south, as on the botanically rich coastal terraces near Arroyo de la Cruz in northwestern San Luis Obispo County (Fig. 11-6). These communities are located below 1000 m (3000 ft) elevation and generally within 100 km (60 mi) of the coast. The distribution of northern coastal grasslands is very similar to that of northern coastal scrub (Chapter 9). Stands of these two communities form a vegetational mosaic.

Northern coastal grasslands receive more rainfall and are not as hot and dry as the valley grasslands. Coastal and valley grasslands share some common species, and in the coastal regions from the San Francisco Bay area south to northern San Luis Obispo County there is a gradual replacement of the dominants of northern coastal grasslands by those of valley grassland communities.

Northern coastal grasslands were originally dominated by native perennial grasses. Today many introduced species occur together with the natives, and in some places the native grasses have largely been replaced by alien species. However, annual grasses are not as important in northern coastal grasslands as they are in valley grasslands. Modern northern coastal grasslands are largely perennial-dominated communities. These communities have a range similar to that of the northern coastal scrub and often form a vegetational mosaic with northern coastal scrub and other communities.

The history of human disturbance of northern coastal grasslands is long and varied. The native Indians periodically burned the grasslands. With the advent of European influence came grazing and other agricultural practices. Introduced weedy species became integrated into the grasslands, and one of these, Klamath weed (*Hypericum perforatum*) for a time became a dominant. Introduction of a beetle species that feeds upon Klamath weed resulted in decline of this species, and in the last 35 years there has been a re-establishment of grass dominance. Some present-day grassland areas of northern coastal areas are actually human artifacts resulting from clearing and conversion of forest or brushland areas to pasture. These are often mistaken for native grasslands.

The dominant grasses of the northern coastal grassland are perennial bunch grasses that vary from area to area under the influence of climate, soil type and distance from the ocean. Some of these grasses are native species, and others are fully naturalized aliens. Common grass species include:

Agrostis spp.	bent grass
Bromus carinatus	brome
Calamagrostis nutkaensis	reedgrass
Danthonia californica	California oatgrass
Deschampsia caespitosa	hairgrass
Festuca idahoensis	Idaho fescue
Festuca rubra	red fescue
Holcus lanatus	velvet grass
Hordeum brachyantherum	perennial barley
Koeleria macrantha [K. *cristata*]	junegrass
Nassella pulchra	purple needle grass
Poa spp.	bluegrass
Trisetum canescens	trisetum

They are often associated with species of *Carex* (sedges), *Juncus* (rushes) and a variety of perennial forbs such as *Iris douglasiana* (Douglas' iris), *Sisyrinchium bellum* (blue-eyed-grass), *Calochortus luteus* (Mariposa lily), *Lupinus formosus* (showy lupine), *Sanicula arctopoides* (footsteps of spring) and *Grindelia hirsutula* (gum plant). In addition to these perennials, assorted annual grasses that also occur in valley grassland communities are often present. A variety of forbs also occur in northern coastal grasslands.

4. Desert Grassland Communities

Desert grasslands consist of a series of grass dominated communities interspersed among desert woodlands and shrublands in transmontane California. These grasslands integrate with these other arid land communities, and the trees or shrubs of associated desert communties are usually scattered in the grassland areas. The transitional nature of desert grasslands makes distinguishing them from adjacent desert shrubland and woodland communities difficult and somewhat arbitrary. In fact, desert grasslands are sometimes referred to as steppes characterized by their dominant shrubs (e. g., Great Basin sagebrush steppe) indicating the transitional nature of these communities.

Desert grasslands are found interspersed with virtually all of the desert communities discussed in Chapters 18 and 19. They are usually dominated by perennial native bunch grasses 0.1 to 1 meters tall. These communities were more widespread in the past than they are today. Most areas of desert grasslands, as in the case of the valley and southern coastal grasslands, have been disturbed, and few pristine stands remain. Overgrazing during the 19th and early 20th centuries resulted in expansion of shrub species into the grasslands and/or replacement of native grasslands with introducted species. Cheatgrass (*Bromus tectorum*) has been a particularly invasive alien species. Species compositon of the desert grassland communities varies from desert region to desert region. The most common species include:

Achnatherum hymenoides	Indian ricegrass
Achnatherum thurberiana	Thurber needlegrass
Bromus madritenis ssp. *rubens*	Red brome
Bromus tectorum	Cheatgrass
Elymus elymoides	aquirreltail
Festuca idahoensis	Idaho fescue
Heterostipa comata	needle and thread grass
Koeleria macrantha	Junegrass
Leymus cinereus	Ashy wild-rye
Pascopyrum smithii	Western wheatgrass
Pleuraphis jamesii	Galleta
Pleuraphis rigida	Big galleta
Poa secunda	Blue grass
Pseudoroegneria spicata	Bluebunch wheatgrass

Grassland Communities—References

Baker, H. G. 1978. Invasion and replacement in Californian and Neotropical grasslands. Pp. 368–384 *in* J. R. Wilson (ed.), Plant relations in pastures. CSIRO, East Melbourne, Australia.

———. 1986. Patterns of plant invasion in North America. Pp. 44–57 *in* H. A. Mooney and J. A. Drake (eds.), Ecology of biological invasions of North America and Hawaii. Springer Verlag, New York.

Barry, W. J. 1972. California prairie ecosystems. Vol. 1: The Central Valley prairie. State of Calif. Resources Agency, Dept. Parks and Rec., Sacramento.

———. 1972. The Central Valley prairie. Calif. Dept. Parks and Recreation, Sacramento. 82 pp.

———. 1981. Jepson prairie—will it be preserved? Fremontia 9(1):7–11.

———. 1981. Native grasslands then and now. Fremontia 9(1):18 (map).

———. 1981. Selected bibliography on native grasses. Fremontia 9(1):19–20.

Bartholomew, B. 1970. Bare zone between California shrub and grassland communities, the role of animals. Science 170:1210–1212.

Bartolome, J. W. 1979. Germination and seedling establishment in California annual grassland. J. Ecol. 67:273–281.

———. 1981. *Stipa pulchra*, a survivor from the pristine prairie. Fremontia 9(1):3–6.

——— and B. Gemmill. 1981. The ecological status of *Stipa pulchra* (Poaceae) in California. Madroño 28:172–181.

Batzli, G. O., and F. A. Pitelka. Influence of meadow mouse populations on California grassland. Ecology 51:1027–1039.

Beetle, A. A. 1947. Distribution of the native grasses of California. Hilgardia 17:309–357.

Bentley, J. R., and M. W. Talbot. 1948. Annual-plant vegetation of the California foothills as related to range management. Ecology 29:72–79.

Biswell, H. H. 1956. Ecology of California grassland. J. Range Manage. 9:19–24.

Brown, D. E. 1982. Californian valley grassland. Pp. 132–135 *in* D. E. Brown (ed.). Biotic communities of the American Southwest—United States and Mexico. Desert Plants 4:1–341.

Broyles, P. 1987. A flora of Vina Plains Preserve, Tehama County, California. Madroño 34:209–227.

Burcham, L. T. 1957. California rangeland; an historico-ecological study of the range resources of California. Calif. Dept. Natural Resources, Div. Forestry, Sacramento. 261 pp.

Coupland, R. T. 1979. Grassland ecosystems of the world: analysis of grasslands and their uses. International Biol. Prog. 18. Cambridge Univ. Press, London.

Crampton, B. 1959. The grass genera *Orcuttia* and *Neostapfia*: a study in habitat and morphological specialization. Madroño 15:97–110.

Elliott, H. W., and J. D. Wehausen. 1974. Vegetational succession on coastal rangeland of Point Reyes peninsula. Madroño 22:231–238.

Evans, R. A., and J. A. Young. 1972. Competition within the grass community. Pp. 230–246 *in* V. B. Younger and C. M. McKell (eds.). The biology and utilization of grasses. Academic Press, N.Y.

Frenkel, R. E. 1977. Ruderal vegetation along some California roadsides. Univ. California Press, Berkeley.

Gould, F. W. and R. Moran. 1981. The grasses of Baja California, Mexico. San Diego Soc. Nat. Hist. Mem. 12.

Griggs, T. 1981. Life histories of vernal pool annual grasses. Fremontia 9(1):14–17.

Gulmon, S. L. 1979. Competition and coexistence: three annual grass species. Amer. Midl. Naturalist 101:403–416.

Halligan, J. 1973. Bare areas associated with shrub stands in grassland: the case of *Artemisia californica*. Bioscience 23:429–432.

———. 1974. Relationship between animal activity and bare areas associated with California sagebrush in annual grassland. J. Range Manage. 27:358–362.

Heady, H. F. 1958. Vegetational changes in the California annual type. Ecology 39:402–416.

———. 1972. Burning and the grasslands of California. Proc. Ann. Tall Timbers Fire Ecol. Conf. 12:97–107.

———. 1977. Valley grassland. Pp. 491–514 *in* M. G. Barbour and J. Major (eds.). Terrestrial vegetation of California. John Wiley and Sons, N.Y.

———, T. C. Foin, M. M. Hektner, D. W. Taylor, M. G. Barbour, and W. J. Barry. 1977. Coastal prairie and northern coastal scrub. Pp. 733–760 *in* M. G. Barbour and J. Major (eds.). Terrestrial vegetation of California. John Wiley and Sons, N.Y.

Holland, R. F., and S. K. Jain. 1977. Vernal pools. Pp. 515–533 *in* M. G. Barbour and J. Major (eds.). Terrestrial vegetation of California. John Wiley and Sons, N.Y.

Hoover, R. F. 1937. Endemism in the flora of the Great Valley of California. Ph.D. dissertation. Univ. Calif., Berkeley.

Huenneke, L. F., S. P. Hamburg, R. Koide, H. A. Mooney, and P. M. Vitousek. 1990. Effects of soil resources on plant invasion and community structure in Californian serpentine grassland. Ecology 71:478–491.

Huffaker, J. T. 1951. The return of native perennial bunchgrass following the removal of Klamath weed (*Hypericum perforatum* L.) by imported beetles. Ecology 32:443–458.

Hull, J. C., and C. H. Muller. 1977. The potential for dominance by *Stipa pulchra* in a California grassland. Amer. Midl. Naturalist 97:147–175.

Jackson, L. E. 1985. Ecological origins of California's mediterranean grasses. J. Biogeography 12:349-361.

Jones, M. B., and R. G. Woodmansee. 1979. Biogeochemical cycling in annual grassland ecosystems. Bot. Rev. 45:111–144.

Keeley, J. E. 1987. Distribution and stability of grasslands in the Los Angeles basin. Bull. So. Calif. Acad. Sci. 86:13–26.

———. 1988. Bibliographies on chaparral and the fire ecology of other Mediterranean systems, 2nd ed. Calif. Water Resource Cntr., Univ. Calif. Davis, Rep. No. 69. 328 pp.

———. 1989. The California valley grassland. Pp. 3–23 *in* A. A. Schoenherr (ed.), Endangered plant communities of southern California. Proceedings of the 15th Annual Symposium. Southern California Botanists Spec. Publ. 3.

Komarek, E. V. 1965. Fire ecology: grasslands and man. Proc. 4th Annual Tall Timbers Fire Ecol. Conf. 169–220.

Luckenbach, R. 1973. Pogogyne, polliwogs, and puddles—the ecology of California's vernal pools. Fremontia 1:9–13.

Major, J., and W. T. Pyott. 1966. Buried viable seeds in two California bunchgrass sites and their bearing on definition of a flora. Vegetatio 13:253–282.

McGown, R. L., and W. A. Williams. 1968. Competition for nutrients and light between the annual grassland species *Bromus mollis* and *Erodium botrys*. Ecology 49:981–990.

McNaughton, S. J. 1968. Structure and function in California grasslands. Ecology 49:962–972.

Minnich, R. A. 1983. Fire mosaics in southern California and northern Baja California. Science 219:1287–1294.

Murphy, D. D., and P. R. Ehrlich. 1989. Conservation biology of California's remnant native grasslands. Pp. 201–211 *in* L. F. Huenneke and H. Mooney (eds.), Grassland structure and function. Kluwer Academic Publ., Dordrecht.

Parker, V. T., and C. H. Muller. 1982. Vegetation and environmental change beneath isolated live oak trees (*Quercus agrifolia*) in a California annual grassland. Amer. Midl. Naturalist 107:69–81.

Parsons, D. J., and T. J. Stohlgren. 1989. Effects of varying fire regimes on annual grasslands in the southern Sierra Nevada of California. Madroño 36:154–168.

Pitt, M. D., and H. F. Heady. 1978. Responses of annual vegetation to temperature and rainfall patterns in northern California. Ecology 59:336–350.

Rossiter, R. C. 1966. Ecology of Mediterranean annual-type pasture. Adv. Agronomy 18:1-50.

Sampson, A. W., A. Chase, and D. W. Hedrick. 1951. California grasslands and range forage grasses. Calif. Agric. Exp. Stn. Bull. 724. 130 pp.

———, and B. S. Jesperson. 1963. California range brushlands and browse plants. Univ. Calif. Agric. Exp. Stn. Manual 33. 162 pp.

Schlising, R. A., and E. L. Sanders. 1982. Quantitative analysis of vegetation at the Richvale vernal pools, California. Amer. J. Bot. 69:734–742.

Sims, P. L. 1988. Grasslands. Pp. 265–286 *in* M. G. Barbour and W. D. Billings (eds.), North American Terrestrial Vegetation. Cambridge Univ. Press, Cambridge.

———, J. S. Singh, and W. K. Lauenroth. 1978. The structure and function of ten western North American grasslands. I. Abiotic and vegetational characteristics. J. Ecol. 66:251–285.

——— and ———. 1978. The structure and function of ten western North American grasslands. II. Intraseasonal dynamics in primary producer compartments. J. Ecol. 66:547–572.

——— and ———. 1978. The structure and function of ten western North American grasslands. III. Net primary production, turnover and efficiencies of energy capture and water use. J. Ecol. 66:573–597.

——— and ———. 1978. The structure and function of ten western North American grasslands. IV. Compartment transfers and energy flow within the ecosystem. J. Ecol. 66:983–1009.

Spedding, C. R. W. 1971, Grassland ecology. Clarendon Press, Oxford.

Talbot, M. W., H. H. Biswell, and A. L. Hormay. 1939. Fluctuations in the annual vegetation of California. Ecology 20:394–402.

Vogl, R. J. 1976. An introduction to the plant communities of the Santa Ana and San Jacinto Mountains. Pp. 77–98 *in* J. Latting (ed.). Plant communities of southern California. Calif. Native Plant Soc. Spec. Publ. 2.

Wells, P. V. 1962. Vegetation in relation to geological substratum and fire in the San Luis Obispo quadrangle, California. Ecol. Monogr. 32:79–103.

Wester, L. 1981. Composition of native grasslands in the San Joaquin Valley, California. Madroño 28:231–241.

White, K. L. 1967. Native bunchgrass (*Stipa pulchra*) on Hastings Reservation, California. Ecology 48:949–955.

Chapter 12

Closed-cone Coniferous Forest Communities

Several species of the coniferous genera *Pinus* and *Cupressus* bear cones that persist unopened for several to many years. These cones are **serotinous** (tardily opening); they differ from those of most other conifers in that the cone scales remain tightly appressed, and the enclosed seeds can remain alive in the cones for extended periods of time, as much as 40 years in *Cupressus abramsiana*. Serotinous cones require heat before the cone scales separate and release the seeds. The heat that causes the cones to open typically comes from a forest fire that kills the parent trees. The seeds are dispersed onto the newly burned ground where they germinate readily. Closed-cone coniferous forests are often associated with other fire-prone communities such as chaparral and coastal scrub.

Closed-cone coniferous forests occur as widely scattered stands throughout the mountains of California. They are more frequently encountered in the coastal mountain ranges of California than in the interior ranges. The land areas occupied by these communities are discontinuous and are separated from each other by wide expanses occupied by other kinds of plant communities (Fig. 12-1). The total land area occupied by closed-cone coniferous forests is rather small, totaling less than one percent of the state. Stands occur from near sea-level to about 2100 m (7000 ft) elevation.

The forests that occur near the coast generally have a rather mild climate because of coastal influences. Often fog and fog-drip play an important role in the distribution of these communities. Some closed-cone coniferous forests that are not located on the immediate coast (e.g., some stands of Sargent cypress) are located on ridge-tops where fog is frequent and provides a significant part of the annual precipitation. Nonetheless, these communities are usually strongly affected by seasonal drought and are fire prone.

Forests dominated by closed-cone conifers typically occur on rather poor soils. The parent materials are quite variable and in-

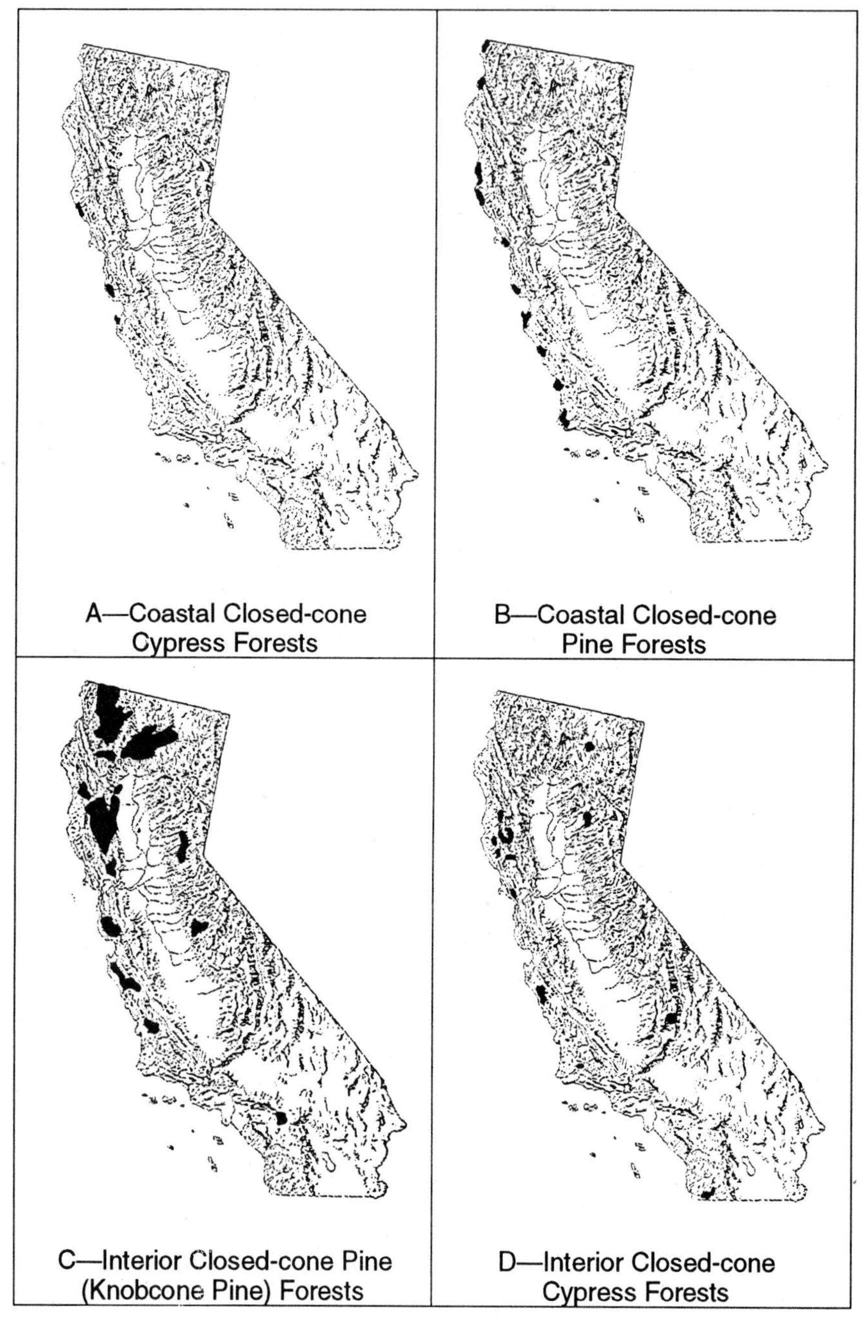

Fig. 12-1. Distribution of coastal closed cone pine, interior closed-cone pine, coastal closed-cone cypress, and interior closed-cone cypress communities in California.

clude sand, volcanic deposits, serpentine, shale, and diatomite. In some coastal areas on ancient marine terraces, the soils are strongly podsolized and may be underlain by an impervious hardpan. In these sterile soils, water is unable to penetrate the soil profile, and only the upper part of the soil is available to roots. Several of the closed-cone conifers are restricted to particular edaphic conditions; Sargent cypress, for instance, is almost wholly restricted to serpentine soils.

Communities dominated by closed-cone conifers range from tall forests with closed canopies to open woodlands or pygmy forests with dwarfed trees and tangles of shrubs. The trees vary in size, but they are often rather short and may be reproductively mature when quite small (sometimes one meter tall or less). Because of the involvement of fire in the reproduction of these trees, most stands are even aged with all members having originated following a fire. In many closed-cone conifer communities, only one species is dominant, often forming pure stands.

An understory of shrubs or sometimes small trees is often present. The species composition of the understory varies from one closed-cone coniferous forest to another and is strongly affected by the proximity of a forest stand to other communities. For example, in closed-cone coniferous forests that are contiguous with chaparral stands, chaparral species grow in close association with the conifers and form a part of the understory in the forest. In the central coastal area of California, chaparral shrubs such as *Arctostaphylos* spp., *Ceanothus* spp. and *Quercus* spp., form the understory in some stands dominated by *Pinus muricata* (Bishop pine), *Pinus attenuata* (knobcone pine) or *Cupressus sargentii*. Closed-cone coniferous forests also intergrade with such communities as coastal coniferous forests, mixed evergreen forests, coastal live oak woodlands, coastal scrub, and coastal dune scrub. In the northern part of the state, sharp ecotones may separate coastal prairies from closed-cone conifer forests.

The closed-cone habit evolved in response to periodic fires. The protection afforded to the seeds by the tightly appressed woody cone scales is often enough to keep many of the seeds alive as a fire sweeps through a forest (Fig. 12-2). The cone scales often provide such good insulation that even though the outer part of a cone is charred, the seeds inside are protected from the heat of the fire. The drawback of closed-cones is that they inhibit reproduction when fire is absent from the community. Some of the closed-cone conifers are comparatively short-lived trees. In the absence of fire, a stand of trees may become senescent as individual trees lose vigor and produce fewer and fewer new

Fig. 12-2. The thick, woody cone scales of knobcone pines (*P. attenuata*) protect the seeds inside the cone from the heat of a fast-moving forest fire. **A.** Unburned branch. **B.** Burned branch. Photos by David Keil.

cones. Without a fire such closed-cone conifers are likely to succumb to parasites and to be replaced by other species.

Some of the closed-cone conifers have cones that seldom ,if ever, open unless a fire sweeps through the forest. In others, however, the cones tardily open and release seeds, sometimes merely in response to the heat of a particularly warm day. This is common in areas where fires are infrequent. *Pinus radiata* (Monterey pine) is often called a semi-closed-cone pine because its cones do eventually open. Seedlings and saplings scattered in mature Monterey pine forests are an indication that seeds are released from the cones without fire. The degree to which cones remain unopened sometimes varies within a species. *Pinus muricata* (bishop pine) in southern California has tightly closed-cones. However, in northern California, their cones tardily open and release seeds.

Although fire is an important agent in the present-day maintenance of many closed-cone coniferous forest communities, frequent fires can also eliminate them. The trees must reach reproductive age and produce enough cones that an adequate seed supply is available before they are able to regenerate a stand following fire. The replacement of closed-cone conifer forests by chaparral after recurrent fires has been documented in historical times. Fossil records indicate that closed-cone conifer forests were formerly more extensive than they are today. During the moister portions of the Pleistocene, these forests were much more widely distributed in coastal parts of California than they are today. Apparently during and after the Xerothermic Period of the Holocene (8000–4000 years ago), many stands were eliminated and were replaced by more drought-tolerant and fire-tolerant communities. Many Indian tribes of California periodically set fires to enhance hunting opportunities, and these may have led to the local extirpation of closed-cone conifer stands.

Survey of California's Closed-cone Conifers

Serotinous cones have evolved independently in cypresses and several different groups of pines. In addition, the giant sequoias of the Sierra Nevada (*Sequoiadendron giganteum*) also has serotinous cones (Chapter 16). California's closed-cone cypresses are all rather closely related and somewhat difficult to tell apart:

Cupressus abramsiana	Santa Cruz cypress
Cupressus arizonica	
ssp. *arizonica* [*C. stephensonii*]	Cuyamaca cypress
ssp. *nevadensis* [*C. nevadensis*]	Piute cypress
Cupressus bakeri	Baker cypress
Cupressus forbesii	Tecate cypress
Cupressus goveniana	
ssp. *goveniana*	Gowen cypress
ssp. *pigmaea* [*C. pigmaea*]	pygmy cypress
Cupressus macnabiana	McNab cypress
Cupressus macrocarpa	Monterey cypress
Cupressus sargentii (Figs. 12-6, 12-3)	Sargent cypress

The closed-cone pines, on the other hand, represent three different lineages of pine species that have independently acquired serotiny:

Pinus attenuata (Figs. 12-2, 12-5)	knobcone pine
Pinus muricata (Fig. 12-3)	bishop pine
[*P. remorata*]	
Pinus radiata (Fig. 12-4)	Monterey pine
Pinus contorta	
ssp. *bolanderi*	pygmy pine
ssp. *contorta*	shore pine
Pinus coulteri (Fig. 12-8)	Coulter pine
Pinus torreyana	Torrey pine

In the discussion that follows, we examine the various closed-cone conifers of California (except giant sequoias).

Closed-cone Cypress Species

All of the cypresses are susceptible to fire and have no fire-resistance. Their resinous foliage burns readily, and the trees are killed by the fire. Only the seeds inside the cones survive to perpetuate the plants.

Cupressus abramsiana (Santa Cruz cypress) occurs above the fog belt in the Santa Cruz Mountains of Santa Cruz County and on Butano Ridge in San Mateo County. It is associated with chaparral, knobcone pine, and ponderosa pine forests. *Cupressus abramsiana* is closely related to *C. goveniana* and is sometimes considered to be a race of that species. As with *C. goveniana*, trees of *C. abramsiana* may be dwarfed on poor soils. Because of development and agricultural pressures, this species has been officially listed as endangered.

Cupressus arizonica is represented in California by two subspecies. Arizona cypress (*C. arizonica* ssp. *arizonica*) occurs from western Texas to southern California and northern Mexico. In California it is disjunct from the nearest Arizona populations and is restricted to the Cuyamaca Mountains of the Peninsular Range east of San Diego. It also occurs in northern Baja California. It is commonly referred to in California as the Cuyamaca cypress. These trees occur in close association with chaparral and Coulter pines. Piute cypress (*C. arizonica* ssp. *nevadensis*) occurs in the southern Sierra Nevada of Kern and Tulare counties. It grows in association with chaparral, foothill woodland, and piñon-juniper woodland species.

Cupressus bakeri (Baker cypress) has a northern distribution, occurring in the Klamath Mountains, the Cascades, and the northern Sierra Nevada. Its range also extends into southern Oregon. Baker cypress grows at higher elevations than the other California closed-cone cypresses. Its scattered groves are a minor part of the montane coniferous forests (Chapter 16), sometimes in association with *Pinus attenuata* (knobcone pine). At its highest elevation stand, about 2100 m (7000 ft) in the northern Sierra Nevada, the cypresses grow together with red firs.

Cupressus forbesii (Tecate cypress) occurs in four isolated groves in the Peninsular Ranges in Orange and San Diego counties and extends southward into the mountains of northern Baja California. It is associated with chaparral, and some populations have declined from the effects of repeated brush fires. Additional pressures have come from the rapid urban development of southern California.

Fig. 12-3. Forest of Sargent cypress on serpentine substrate in the Cuesta Botanical Area in the Santa Lucia Mountains of San Luis Obispo County. Associated species include *Arctostaphylos obispoensis* (right) and other chaparral shrubs. Photo by Robert F. Hoover.

Fig. 12-4. Foliage and persistent cones of *Cupressus sargentii*. Although the trees are killed by fire, many of the seeds survive within the hard, woody cones. Photo by David Keil.

Cupressus goveniana comprises two narrowly endemic subspecies. Gowen cypresses (*C. goveniana* ssp. *goveniana*) are restricted to the Monterey Peninsula, but these trees generally occur away from the immediate coast and at somewhat higher elevations than *C. macrocarpa* (up to about 300 m (1000 ft). They often grow in association with Monterey and bishop pines. Gowen cypresses vary in stature from stunted, shrublike trees on sterile soils to well-developed forest trees. Pygmy cypresses (*C. goveniana* ssp. *pigmaea*) are mostly restricted to the sterile soils of the Mendocino White Plains where they are one of the principal components of the Mendocino pygmy forests. On this sterile substrate, pygmy cypresses commonly grow only 1–3 m tall. On more fertile substrates, however, these trees are substantially larger, reaching heights of more than 40 m.

Cupressus macnabiana (McNab cypress) is distributed in the mountains that surround the Sacramento Valley, ranging from the North Coast Range to the southern Cascades and the northern Sierra Nevada. It often occurs on serpentine but is not restricted to this substrate. McNab cypress commonly occurs together with chaparral and foothill woodland species, and sometimes it grows together with *Pinus attenuata* (knobcone pine).

Cupressus macrocarpa (Monterey cypress) is among the most picturesque trees of California, occupying rocky habitats along the rugged coast of the Monterey Peninsula. Old cypress trees (200–300 years old) growing along the shore have been shaped into gnarled, unusual forms by the wind and salt spray. They are associated with several other coastal communities including Monterey pine forests, southern coastal scrub, northern coastal scrub, and seabluff coastal scrub. A considerable amount of residential development has occurred within the natural populations of Monterey cypresses. These trees are commonly planted in other areas of California, often as ornamentals or as windbreaks in agricultural areas, and they are a common feature of the California landscape. Several seemingly indigenous reproducing populations of Monterey cypress that occur outside the native range of the species represent early cultivation of these trees.

Cupressus sargentii (Sargent cypress) grows almost exclusively on serpentine outcrops in the Coast Ranges, occurring from northern Mendocino County to Santa Barbara County. It grows with a variety of associates, but most commonly with stands of serpentine chaparral (Chapter 10). In some locations it occurs with elements of mixed evergreen forests, redwood forests, foothill woodlands, riparian communities, and the lower-elevation fringe of montane coniferous forests. Sargent cypresses vary in size from tall shrubs on dry sites to well-developed trees in moist habitats.

A
B

Fig. 12-6. Wind-pruned bishop pines (*P. muricata*) on coastal bluff in Sonoma County. Photo by David Keil.

Closed-cone Pine Species

Pinus attenuata, *P. muricata*, and *P. radiata* are closely related and morphologically similar species of the subsection *Oocarpae* of the genus *Pinus*. with 2–3 needles per fascicle, usually between 5 and 12 cm in length. All three species commonly produce whorls of stout sessile cones on the branches (Fig. 12-2). These trees have thin bark and a moderate fire is sufficient to kill them. The ranges of these species are almost completely allopatric except for the Monterey Peninsula area where Monterey pine independently occurs together with both knobcone and bishop pines.

Pinus attenuata (knobcone pine; Fig. 12-2, 12-5) is the most widely distributed of the closed-cone pines in California. It is widely distributed in the mountains of northern California, extending southward in the Coast Ranges to San Luis Obispo County and in the Sierra Nevada to the vicinity of Yosemite National Park. Disjunct stands occur in the San Bernardino Mountains of the Transverse Range and the Santa Ana Mountains of the Peninsular Ranges. It extends northward into Oregon and one stand occurs in Baja California. Knobcone pine occurs from near sea level to over 1600 m (5200 ft). It often associates with chaparral and with trees of several different forest and woodland communities. In the mountains of northern California and Oregon, knobcone pine often occurs in dry sites in lower elevation portions of montane mixed coniferous forests, growing in associa-

Fig. 12-7. Forest of Monterey pines (*Pinus radiata*) in Cambria, San Luis Obispo County. Photo by Dave Melendy.

tion with such species as *Pinus ponderosa* (ponderosa pine) and *Calocedrus decurrens* (incense cedar). It commonly associates with *Pinus sabiniana* (foothill pine) and *Pinus coulteri* (Coulter pine) and with trees of mixed evergreen forests such as *Pseudotsuga menziesii* (Douglas-fir), *Lithocarpus densiflora* (tanbark oak), and *Quercus chrysolepis* (canyon live oak). It approaches the coast in the Monterey region where it grows with and occasionally hybridizes with *Pinus radiata* (Monterey pine).

Pinus muricata (bishop pine; Fig. 12-6) has a discontinuous coastal distribution from Santa Barbara County north to Humboldt County and occurs on Santa Cruz and Santa Rosa Islands. It also grows at one site in northern Baja California and on Cedros Island of the Baja California coast. There is considerable north-south variation in this species which has three geographical races that vary in the degree to which the cones are serotinous, oil chemistry, leaf color, and crossability. Stands occur in association with coastal scrub, chaparral, coastal live oak woodland, mixed evergreen forest, coastal redwood forest, and north coast coniferous forests. In northern California, bishop pines often occur immediately adjacent to the ocean, closer to the shore than any other tree species (Fig. 12-7). Southern stands often occur somewhat inland, surrounded by chaparral, coastal scrub, or mixed evergreen forest species. Fog-drip is often an important source of precipitation for bishop pines.

Pinus radiata (Monterey pine) occurs in three disjunct stands along the California coast at Cambria (San Luis Obispo County; Fig. 12-6), on the Monterey Peninsula (Monterey and Santa Cruz counties), and near Año Nuevo (San Mateo County) and on islands off the coast of Baja California. It occurs in association with chaparral, coastal scrub, coastal live oak woodland, grassland, and mixed evergreen forests. In some stands, as at Cambria, tall Monterey pines form a closed canopy forest with a short-tree understory of *Quercus agrifolia* (coast live oak) and *Heteromeles arbutifolia* (toyon) plus varied shrub and herb layers. Monterey pine is a semi-closed-cone species. Cones sometimes open in response to hot weather, and young pines can be found scattered in the forest. Although Monterey pines have one of the most restricted natural distributions of any California tree species, they have been widely planted in California as ornamentals and windbreaks. In several other regions of the world this species is grown as a timber tree. At present both cultivated and natural stands are threatened by pine pitch canker, a fatal disease. Urban development has occurred in native stands of Monterey pines in the Monterey and Cambria areas.

Pinus contorta represents a second pine lineage, *Pinus* subsection *Contorteae*. These trees have 2 short needles (mostly 2.5-6 cm long) and small cones. They are thin-barked trees that are easily killed by fire. *Pinus contorta* is a widespread and polymorphic species of western North America with four subspecies. Three of these occur in California and two of these have serotinous cones: ssp. *bolanderi* (pygmy pine) and ssp. *contorta* (shore pine). The third, ssp. *murrayana* (California lodgepole pine), has cones that open at maturity. Pygmy pines (*ssp. bolanderi*) seldom exceed 2 m in height. They are restricted to the pygmy forest of the Mendocino White Plains where they associate with bishop pines, pygmy cypresses and various ericaceous shrubs. Shore pines (ssp. *contorta*) also are sometimes dwarfed but may grow to 15 m in favorable situations. They range from Mendocino County north to southern Alaska along the immediate coast. In California the shore pines typically occur on coastal dunes and seabluffs, often within the salt spray zone. They also extend inland where they grade into adjacent coniferous forests. Not all races of shore pine have serotinous cones.

Pinus coulteri and *P. torreyana* are part of a third lineage characterized by 3 or 5 long needles (15–25 cm long) and massive cones. Both are semi-closed-cone pines that may retain seeds for several years in unopened or partially opened cones.

Pinus coulteri (Coulter pine, big-cone pine) ranges from the Peninsular Ranges of northern Baja California through southern California across the Transverse Ranges and up the South Coast Ranges to the San Francisco Bay area. It does not occur at all in the Sierra Nevada or in the northern half of the state. *Pinus coulteri* has both closed-cone and open-cone races. It occurs in a wide variety of habitats which are discussed in greater detail in

Fig. 12-8. *Pinus coulteri* (Coulter pine) in chaparral. Coulter pines that are associated with frequently burned communities such as chaparral often have serotinous cones. Photo by Robert F. Hoover.

Chapter 16. Closed-cone forms of Coulter pine are typically associated with fire-prone communities such as chaparral, closed-cone pine forests and closed-cone cypress forests. Coulter pines have thicker, more fire-resistant bark than either Sargent cypresses or knobcone pines and may survive fires that kill these associated species.

Pinus torreyana (Torrey pine) has the most geographically limited distribution of any pine in California. It is restricted to a single mainland population along the coast north of San Diego and a second population on Santa Rosa Island. Groves of Torrey pines are associated with chaparral, coastal scrub and coastal dune scrub. They occur on sandy soils or on diatomaceous shale. The mainland population has been severely impacted by the intense urban development in the San Diego area.

Classification of Closed-cone Conifer Forests

Many of the present-day closed-cone conifer forests are isolated relict stands. Their sporadic distribution makes classification of these communities difficult. They are much influenced by the composition of adjacent communities and by the local edaphic conditions. Several of the dominant species are very localized endemics. In the following system, these communities are broadly grouped on the basis of geographic and climatic features into coastal and interior closed-cone conifer forests.

1. Coastal Closed-cone Conifer Forests

Groves of closed-cone coniferous trees occur in scattered locations along the immediate coast of California from the Oregon border to San Diego (Figs. 12–1A, B). In northern California these forests always occur on the seaward side of the redwood forests and have a similar climatic regime. Both winters and summers are quite moderate and frosts are uncommon because of the nearby ocean. Rainfall varies from about 20 to 60 inches per year and additional precipitation occurs from fog drip. Soils are generally less fertile than in adjacent redwood forests. The closed-cone coniferous forests are frequently associated with grasslands or coastal scrub communities but usually receive somewhat greater precipitation than the shrub-dominated areas.

Coastal closed-cone conifer forests can be grouped by composition into three phases.

A. Coastal Cypress Forests

Coastal cypress forests occur only in San Mateo, Santa Cruz and Monterey Counties (Fig. 12–1A) . These forests occur as highly localized groves usually associated with podsolized soils on old beach terraces. Three narrowly endemic cypress species form these forests. *Cupressus macrocarpa* (Monterey cypress) occurs in two populations on and near the Monterey Peninsula. *Cupressus goveniana* ssp. *goveniana*(Gowen Cypress) is restricted to two populations on the Monterey Peninsula. *Cupressus abramsiana* (Santa Cruz cypress) is confined to four populations in the Santa Cruz Mountains.

B. Coastal Closed-cone Pine Forests

Several species of closed-cone pines occur in scattered groves along the California coastline and on the Channel Islands (Fig. 12–1B). *Pinus contorta* ssp. *contorta* (shore pine) occurs on stabilized sand dunes and coastal bluffs from Mendocino County north to Alaska. *Pinus muricata* (Bishop pine; Fig. 12-3) and *Pinus radiata* (Monterey pine; Fig. 12-4) both occur along the immediate coast. *Pinus muricata* occurs in scattered mainland populations from Santa Barbara County to Humboldt County and on Santa Cruz and Santa Rosa Islands. *Pinus torreyana* (Torrey pine) occurs in only two populations, one on the mainland along coastal bluffs of the Soledad valley north of San Diego and a second smaller population on Santa Rosa Island.

C. Pygmy Forests

These unusual plant assemblage composed of dwarfed closed-cone conifers occur on the Mendocino White Plains (Mendocino County) just inland from the immediate coast and on Huckleberry Hill on the Monterey peninsula in Monterey County. These small trees, which may be very old, are dwarfed as a result of very acidic, sterile, podsolized soils (known in the Mendocino County area as Blacklock soils) which are underlain by a shallow impervious hardpan. It is interesting to note that if the hardpan is broken through or if these trees are grown in favorable soils, they attain the proportions of ordinary trees. Common species in the Mendocino pygmy forests are *Pinus contorta* ssp. *bolanderi* (pygmy pine), *Pinus muricata* (Bishop pine), and *Cupressus goveniana* ssp. *pigmaea* [*C. pigmaea*] (pygmy cypress). The acidic soils support a sparse to dense growth of mostly ericaceous shrubs:

Arctostaphylos nummularia	creeping manzanita
Ledum glandulosum	Labrador-tea
Vaccinium ovatum	California huckleberry
Gaultheria shallon	salal
Rhododendron macrophyllum	rhododendron
Myrica californica	wax-myrtle

The ground is often carpeted with mosses, but herbaceous flowering plants are rather poorly represented. In the Monterey pygmy forests dwarfed forms of *Pinus radiata* (Monterey pine), *P. muricata* (bishop pine), and *Cupressus goveniana* ssp. *goveniana* (Gowen cypress) occur together. Here too the understory is a tangle of ericaceous shrubs.

2. Interior Closed-cone Conifer Forests

These communities are located on inland hills and mountain slopes away from the immediate coast (Fig. 12–1C, D). In these areas the dominant trees are not affected as greatly by the coastal environment as are those of coastal closed-cone conifer forests. The trees are usually not severely wind-pruned, and there is no salt-spray on the foliage. Temperature extremes are greater and as a result summer drought may be more extreme.

Two phases of interior closed-cone coniferous forests are recognized below, those dominated by *Pinus attenuata* (knobcone pine) and those dominated by various species of *Cupressus*. A third phase could be added, dominated by closed-cone forms of

Pinus coulteri (Coulter pine). We have chosen to treat these communities with the montane coniferous forest communities (Chapter 16).

A. Knobcone Pine Forests

Pinus attenuata (Knobcone pine; Fig. 12-2, 12-5) is the most widespread of the closed-cone pines in California. It approaches to within a few miles of the coast in Santa Cruz and San Luis Obispo counties but more often occurs well away from coastal influences (Fig. 12–1C). Some of these forests are essentially pure stands of *Pinus attenuata*, whereas others contain small to significant proportions of other pine species, Douglas-fir (*Pseudotsuga menziesii*) or other tree species. In portions of the North Coast Ranges, the Klamath-Siskiyou Mountains, the southern Cascades, and the Sierra Nevada, knobcone pines occur in lower elevation montane coniferous forests (Chapter 16) together with species such as *Pinus ponderosa* (Ponderosa pine), *P. lambertiana* (sugar pine), *Calocedrus decurrens* (Incense cedar), *Pseudotsuga menziesii*, and *Abies concolor* (White fir).

Knobcone pines often occur on relatively sterile substrates. In the northern portion of its range *P. attenuata* is very commonly found on serpentine soils. However in San Luis Obispo County knobcone pines occur on shale and nearby serpentine areas are occupied by *Cupressus sargentii*. In most cases the soils are shallow and rocky.

Open stands of knobcone pine frequently contain a sparse to rather dense understory of shrubs such as *Arctostaphylos* sp., *Adenostoma fasciculatum*, *Quercus* spp., etc. Dense stands may entirely lack an understory. Old stands have large accumulations of litter and fallen branches that burn readily.

B. Interior Cypress Forests

Several species of *Cupressus* occur in the mountains of California away from the immediate coast (Fig. 12-1D). These generally occur in disjunct stands varying in size from a few individuals to extensive populations (Fig. 12–4). Most occur on shallow infertile soils. The most widespread of these are *Cupressus macnabiana* (McNab cypress) and *C. sargentii* (Sargent cypress), both of which are largely restricted to serpentine soils. The remaining species are all very localized. *Cupressus forbesii* (Tecate cypress) and *Cupressus arizonica* ssp. *arizonica* [*C. stephensonii* (Cuyamaca cypress) are restricted to a few isolated groves in the Peninsular Ranges of Orange and San Diego County. *Cupressus arizonica* ssp. *nevadensis* [*C. nevadensis*] (Piute cypress) grows only in the southern Sierra Nevada in Tulare and

Kern Counties. *Cupressus bakeri* (Baker cypress) ranges from the Siskiyou Mountains of southern Oregon to the northern Sierra Nevada.

Because they occur over such a diversity of areas in California, few general observations can be made about the ecological relations of the interior cypress forests. Their common feature is of course the dominance by *Cupressus*. Fire is an important factor in maintaining some of these communities and is threatening to eliminate others. The Cuyamaca cypress is the rarest tree in California and is being replaced to a progressively greater extent by chaparral species following recurrent fires.

Human Impacts on Closed-cone Coniferous Forests

Communities that occupy small overall areas are potentially subject to major alteration by human activities. Many of the closed-cone conifer forests are particularly vulnerable. Those that occur in or near California's urban centers are subject to outright destruction, habitat degradation, population fragmentation, and other negative impacts. Even stands far from urban areas are vulnerable to careless or ignorant acts of destruction.

Human alteration of the fire regime has occurred in California since prehistoric times. Indians set fires that burned through chaparral, forests and other communities. This continued through colonial times up until the beginning of the 20th century when the prevailing attitudes toward fire reversed. As mentioned above the frequent fires of the past have eliminated some closed-cone conifer stands. In modern times the practice of fire suppression has become common, especially in and around urban areas.

Trees that are dependent on periodic fires may suffer from fire suppression as well as frequent fires. In the absence of fire, trees such as bishop pines and knobcone pines eventually begin to decline. Old trees produce fewer and fewer cones, and old cones gradually deteriorate. Parasites and diseases weaken the trees, and one by one they die or are blown down in storms. At the same time, the fuel load progressively increases to the point that when a fire finally does burn through the stand, it is often a particularly hot fire that consumes many of the remaining cones and the seeds they contain.

Stands of cypresses or pines on serpentine outcrops and some other substrates are vulnerable to the effects of mining. Commercially extractable ores of chromium, lead, mercury, and other heavy metals are often found in areas of serpentine, and extraction of the ores commonly results in removal of the vegetative cover.

Several of the California closed-cone conifers have long been on the decline and could be pushed to extinction. Cuyamaca cypress, Tecate cypress, and Torrey pine are all highly vulnerable to fire or other human-caused destruction. Careful conservation measures may be necessary to preserve these species.

Closed-cone Coniferous Forests—References

Axelrod, D. I. 1967. Evolution of the California closed-cone pine forest. Pp. 93–149 *in* R. N. Philbrick (ed.), Proceedings of the Symposium on the biology of the California Islands. Santa Barbara Bot. Gard., Santa Barbara.

———. 1976. History of the coniferous forests, California and Nevada. Univ. Calif. Publ. Bot. 70:1–62.

———. 1980. History of the maritime closed-cone pines, Alta and Baja California. Univ. Calif. Publ. Geol. Sci. 120:1–143.

Azavedo, J., and D. L. Morgan. 1974. Fog precipitation in coastal California forests. Ecology 55:1135–1141.

Borchert, M. 1985. Serotiny and cone-habit variation in populations of *Pinus coulteri* (Pinaceae) in the southern coast ranges of California. Madroño 32:29–48.

Brown, D. E. 1982. Relict conifer forests and woodlands. Pp. 70–71 *in* D. E. Brown (ed). Biotic communities of the American Southwest — United States and Mexico. Desert Plants 4:1–341.

Carlquist, S. 1965. Rare cypress clings to coast habitat. Nat. Hist. 74(8):38–43.

Cole, K. 1980. Geological control of vegetation in the Purisima Hills, California. Madroño 27:79–89.

Critchfield, W. B. 1957. Geographic variation in *Pinus contorta*. Maria Moors Cabot Found. Publ. 3:1–118. Harvard Univ., Cambridge.

———. 1967. Crossability and relationships of the closed-cone pines. Silvae Genet. 16:89–97.

Cylinder, P. D. 1995. The Monterey ecological staircase and subtypes of Monterey pine forest. Fremontia 23(1):7–13.

Deghi, G. S., T. Huffman, and J. W. Colver. 1995. California's native Monterey pine populations: potential for sustainability. Fremontia 23(1):14–23.

Fielding, J. M. 1953. Variation in Monterey pine. For. Timb. Bur. Bull. 31:1–43.

Forde, M. B. 1964. Variation in natural populations of *Pinus radiata* in California. Parts 1–4. New Zealand J. Bot. 2:213–257, 459–501.

———, 1966. *Pinus radiata* in California. New Zealand J. For. 11:20–42.

Griffin, J. R., and W. B. Critchfield. 1972. The distribution of forest trees in California. Pacific S.W. Forest and Range Exp.. Stn., Berkeley, Calif. 114 pp. (U.S.D.A. Serv. Res. Paper PSW 82).

———, and C. O. Stone. 1967. McNab cypress in northern California: a geographical review. Madroño 19:19–27.

Haller, J. R. 1967. A comparison of the mainland and island populations of Torrey pine. Pp. 79–88 *in* R. N. Philbrick (ed.). Proceedings of the symposium on the biology of the California Islands. Santa Barbara Bot. Gard., Santa Barbara.

———. 1986. Taxonomy and relationships of the mainland and island populations of *Pinus torreyana* (Pinaceae). Syst. Bot. 11:39–50.

Hardham, C. B. 1962. The Santa Lucia Sargent cypress groves and their associated northern hydrophilous and endemic species. Madroño 16:173–179.

Howell, J. T. 1941. The closed-cone pines of insular California. Leafl. West. Bot. 3:1–8.

Jenny, H., R. J. Arkley, and A. M. Schultz. 1969. The pygmy forest-podsol ecosystem and its dune associates of the Mendocino coast. Madroño 20:60–74.

Johnston, V. R. 1994. California forests and woodlands. A natural history. Univ. California Press, Berkeley & Los Angeles.

Keeley. 1988. Bibliographies on chaparral and the fire ecology of other Mediterranean systems, 2nd ed. Calif. Water Resource Cntr., Univ. Calif. Davis, Rep. No. 69. 328 pp.

Ledig, F. T. 1984. Gene conservation, endemics, and California's Torrey pine. Fremontia 12(3):9–13.

Libby, W. J. 1995. Native Monterey pine and domesticated radiata pine. Fremontia 23(1):24–28.

Linhart, Y. B. 1978. Maintenance of variation in cone morphology in California closed-cone pines. The roles of fire, squirrels and seed output. Southw. Naturalist 23:29–40.

———, B. Burr, and M. T. Conkle. 1967. The closed-cone pines of the northern Channel Islands. Pp. 151–177 *in* R. N. Philbrick (ed.). Proceedings of the symposium on the biology of the California Islands. Santa Barbara Bot. Gard., Santa Barbara.

Mason, H. L. 1930. The Santa Cruz Island pine. Madroño 2:8–10.

———. 1932. A phylogenetic series of the California closed-cone pines suggested by the fossil record. Madroño 2:49–56.

———. 1934. Pleistocene flora of the Tomales formation. Carnegie Inst. Wash. Publ. 415:81–179.

———. 1949. Evidence for the genetic submergence of *Pinus remorata*. Pp. 356–362 *in* G. L. Stebbins, G. G. Simpson and E. Mayr (eds.). Genetics, paleontology and evolution. Princeton Univ. Press, Princeton, N. J.

McMaster, G.S., and P. Zedler. 1981. Delayed seed dispersal in *Pinus torreyana* (Torrey pine). Oecologia 51:62–66.

McMillan, C. 1956. The edaphic restriction of *Cupressus* and *Pinus* in the Coast Ranges of central California. Ecol. Monogr. 26:177–212.

Millar, C. I. 1988. The California closed-cone pines (subsection *Oocarpae* Little and Critchfield): a taxonomic history and review. Taxon 35:657–670.

———, and W. B. Critchfield. 1988. Crossability and relationships of *Pinus muricata* (Pinaceae). Madroño 35:39–53.

Philbrick, R. N., and J. R. Haller. 1977. The southern California islands. Pp. 893–906 *in* M. G. Barbour and J. Major (eds.). Terrestrial vegetation of California. John Wiley and Sons, N.Y.

Sholars, R. E. 1982. The pygmy forest and associated plant communities of coastal Mendocino County, California. Published by the author. Mendocino, Calif.

———. 1984. The pygmy forest of Mendocino. Fremontia 12(3):3–8.

Stebbins, G. L., and J. Major. 1965. Endemism and speciation in the California flora. Ecol. Monogr. 35:1–35.

Stottlemyer, D. E., and E. W. Lathrop. 1981. Soil chemistry relationships of the tecate cypress in the Santa Ana Mountains, California. Aliso 10:59–69.

Vogl, R. J. 1973. Ecology of the knobcone pine in the Santa Ana Mountains, California. Ecol. Monogr. 43:125–143.

———, W. P. Armstrong, K. L. White, and K. L. Cole. 1977. The closed-cone pines and cypresses. Pp. 295–358 *in* M. G. Barbour and J. Major (eds.). Terrestrial vegetation of California. John Wiley and Sons, N.Y.

Wells, P. V. 1962. Vegetation in relation to geological substratum and fire in the San Luis Obispo quadrangle, California. Ecol. Monogr. 32: 79–103.

Westman, W. E. 1975. Edaphic climax pattern of the pygmy forest region of California. Ecol. Monogr. 45:109–159.

———, and R. H. Whittaker. 1975. The pygmy forest region of northern California: studies on biomass and primary productivity. J. Ecol. 63:493–520.

Wolf, C. B. 1948. Taxonomic and distributional studies of the New World cypresses. Aliso 1:1–250.

Chapter 13

Coastal Coniferous Forest Communities

The coastal coniferous forests of northern and central California (Figs. 13-1, 13-4) are the southern extension of the great coastal forests of the Pacific Northwest. These forests extend from the Alaskan coast south to central California. In Oregon, Washington, and British Columbia, coastal coniferous forests form extensive stands. Prior to the logging activities of the 19th and 20th centuries, these forests blanketed the coastal lowlands throughout the Pacific Northwest. The California portion of these forests is restricted to a narrow strip along the immediate coast and is rather discontinuous. These forests occupy about three percent of the state's land area.

The dominant members of the coastal coniferous forests are tall coniferous trees, sometimes exceeding 100 m in height. Several of the dominant trees are widespread in the Pacific Northwest and reach their southern limits along the coast of northern California (Fig. 13-3). These include *Abies grandis* (grand fir), *Picea sitchensis* (Sitka spruce), *Tsuga heterophylla* (western hemlock), and *Thuja plicata* (western red cedar). One of the common dominants, *Pseudotsuga menziesii* (Douglas-fir) is widespread both in the Pacific Northwestern forests and in California. The most well-known California component of these forests is *Sequoia sempervirens* (coastal redwood), a species that is almost entirely restricted to California.

In California the coastal coniferous forests are bordered to the east primarily by mixed evergreen forests with which they share some species (e.g., *Pseudotsuga menziesii*). In some areas the coastal coniferous forests occur immediately adjacent to the shore, and in others they are separated from the shore by bands of coastal prairie, northern coastal scrub, or coastal closed cone conifer forests. Because of the extensive logging activities of the past century and a half, many coastal coniferous forest stands are now in various stages of secondary succession. Successional species such as *Alnus rubra* [*A. oregona*] (red alder) and *Pseudo-*

Fig. 13-1. Distribution of north coast coniferous forest communities in California.

tsuga menziesii are often disproportionately represented in these stands relative to their original contribution to the community.

Coastal coniferous forests are treated here in two major groups. North coast coniferous forests form a narrow discontinuous band along the immediate coast from Mendocino County northward. Coastal redwood forests extend from Curry County, in southwestern Oregon to the Monterey-San Luis Obispo County line, and often extend inland into the Coast Ranges.

1. North Coast Coniferous Forests

The coniferous forests of the Pacific Northwest occur in the most mesic environments of western North America. Rainfall ranges from 100 to 250 cm (40 to 100 in) per year. Although the summers are generally dry, the summer drought in northern California is shorter in duration than elsewhere in the state and is ameliorated by frequently dense coastal fogs and cloud cover. The coastal coniferous forests of the Pacific Northwest are often considered to be a temperate rain forest. Temperatures are rather mild due to the proximity to the ocean and rarely reach the extreme highs and lows of areas further inland.

The area of California occupied by north coast coniferous forests (Fig 13-1) is small by comparison with the extensive lowland forests farther north. In California, these forests occupy coastal terraces and lower slopes of the North Coast Ranges and Klamath Mountains. North coast coniferous forests form a mosaic with other types of coastal vegetation including coastal prairies, northern coastal scrub, coastal redwood forests and closed cone coniferous forests. On mountain slopes, north coast coniferous forests grade into montane mixed coniferous forests and mixed evergreen forests.

Dominant species in these stands are *Abies grandis* (Figs. 13-2A, 13-3A), *Picea sitchensis* (Fig. 13-3B) and *Pseudotsuga menziesii* (Fig. 13-2B). On a local basis *Tsuga heterophylla* (Figs.. 13-2A, B, 13-3D), *Thuja plicata* (Figs. 13-2B, 13-3C), *Cupressus lawsoniana* [*Chamaecyparis l.*] (Port Orford cedar), and *Pinus muricata* are important coniferous associates or are locally dominant. These forests often contain a hardwood component as well, particularly in successional situations. Major hardwood components include *Acer circinatum* (vine maple), *Acer macrophyllum* (big leaf maple), *Alnus rubra*, *Populus balsamifera* ssp. *trichocarpa* [*P. trichocarpa*] (black cottonwood) and *Rhamnus purshiana* (cascara sagrada). Mature stands are very shady with a dense growth of mosses. Ferns are frequent and include such species as *Blechnum spicant* (deer fern), *Polystichum munitum* (sword fern), *Dryopteris arguta* (wood fern), and *Pteridium aqui-*

Fig. 13-2. North coast coniferous forests. **A.** *Tsuga heterophylla* (western hemlock) and *Abies grandis* (grand fir). Photo by Robert Rodin. **B.** *Thuja plicata* (western red cedar), *Tsuga heterophylla*, and *Pseudotsuga menziesii* (Douglas-fir). Photo by David Keil.

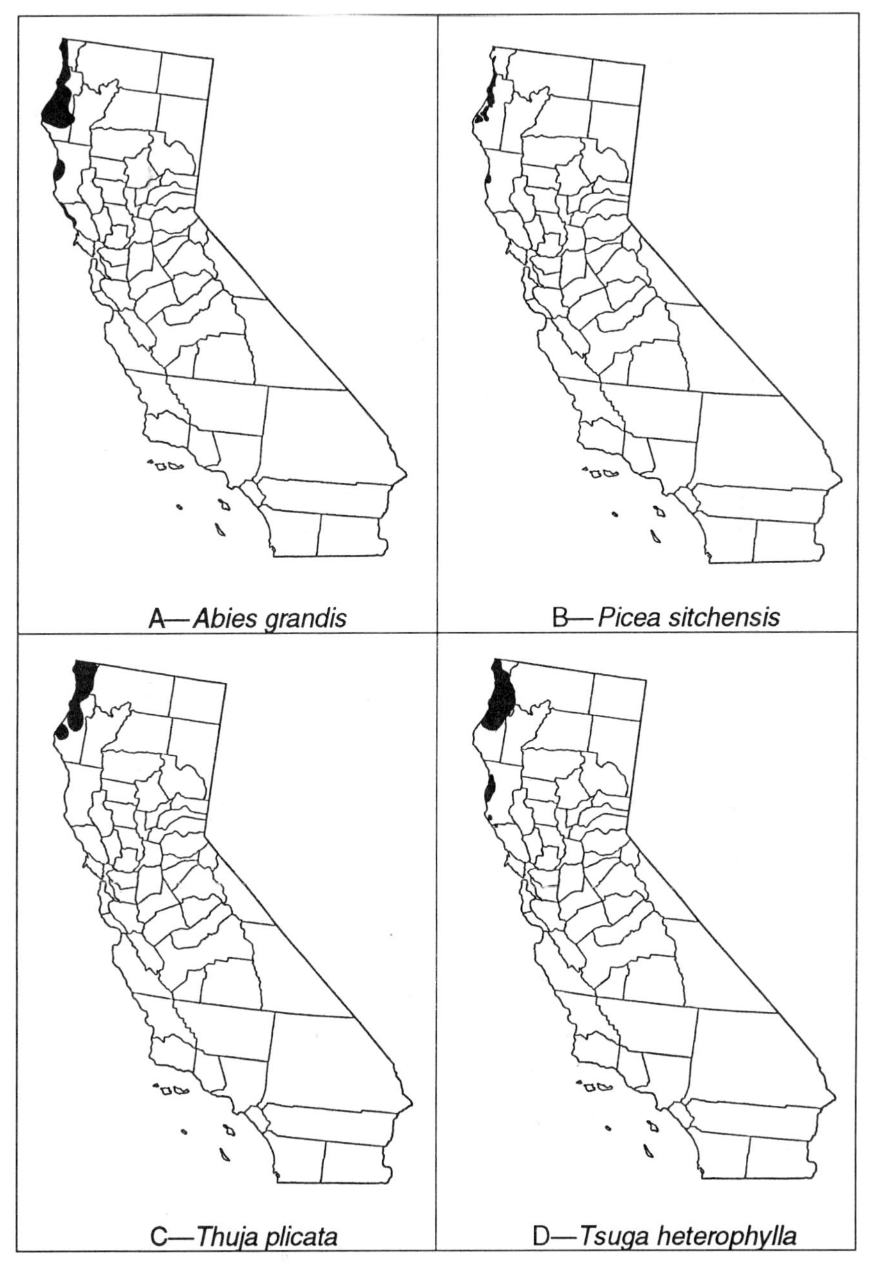

Fig. 13-3. Distribution of trees of the north coast coniferous forest communities in California. *Abies grandis* (Grand fir), *Picea sitchensis* (Sitka spruce), *Thuja plicata* (western red cedar), and *Tsuga heterophylla* (western hemlock). Maps adapted from Griffin and Critchfield (1972).

linum (bracken fern). A common low shrubby associate is *Gaultheria shallon* (salal).

2. Coastal Redwood Forests

Sequoia sempervirens (coast redwood) forms the tallest forest stands in the world. These impressive giants occur from the Klamath Mountains to the South Coast Ranges (Figs. 13-4, 13-5). In the northern part of their range, they form, or at one time formed, large continuous stands, either by themselves or in association with *Pseudotsuga menziesii* (Douglas-fir). In the South Coast Ranges redwoods are progressively restricted to canyon slopes and riparian corridors.

The distribution of coast redwoods is determined largely by their narrow ecological tolerances. These trees are intolerant of prolonged freezing temperatures and are susceptible to damage from summer drought. Because of this combination, the trees occur only in areas with moderate temperatures and year-round moisture availability. In northern California where precipitation may be as high as 250 cm (100 in) per year, and where the summer drought is comparatively short, stands occur both on ridges and in valleys. In the south where rainfall averages about 50 cm (20 in) per year, the trees are dependent on the enhanced soil moisture of canyons. Fog and stratus clouds are common during the summer and decrease the duration of daily moisture stress. Fog drip has been suggested as an important source of summertime moisture in redwood forests. Some ecologists have argued that at least some redwood stands are dependent on this source of summer moisture, whereas others have considered fog drip to be of minor significance.

Some redwood forests are dominated wholly by *Sequoia sempervirens*. The massive pillar-like trunks of the trees support a dense canopy that blocks most of the sunlight. The shaded forest floor has little or no woody understory. Because the branchlets and leaves of redwoods decay rather slowly, an accumulation of undecayed and partially decayed organic matter covers the soil surface. The seeds of redwoods are very small, and seedlings usually are unable to become established in this thick organic layer. In the most densely shaded areas, there is often little or no understory. On other sites, shade-tolerant herbaceous plants are common (Fig 13-5B). These include *Oxalis oregona* (redwood sorrel), *Aralia californica* (spikenard), *Claytonia sibirica* (candy flower), *Asarum caudatum* (wild-ginger), and several fern species. There may be scattered shade-tolerant shrubs such as *Rubus parviflorus* (thimbleberry), *Vaccinium* spp. (huckleberry, blueberry), *Berberis nervosa* (Oregon-grape) and *Gaultheria shallon* as well. Sites that have been disturbed or that are located

Fig. 13-4. Distribution of coastal redwood forest communities in California.

in ecotonal areas may have a more diverse herbaceous understory.

From Mendocino County northward, redwood forests intergrade with north coast coniferous forests. Although in mature stands *Sequoia* is either wholly dominant or is co-dominant with *Pseudotsuga*, in successional situations other trees are often present. *Tsuga heterophylla* is often seral following fires or other disturbances. Other species that may be present include *Abies grandis*, *Cupressus lawsoniana*, and *Thuja plicata*. If the entire forest cover is removed, as occurs when a stand is clear-cut, *Pseudotsuga* is often the primary successional tree. *Picea sitchensis* and *Alnus rubra*, both of which are more tolerant of salt spray than is *Sequoia*, sometimes border the redwood forests in a zone immediately adjacent to the ocean. South of the range of *Picea sitchensis*, a similar forest fringe is often formed by *Pinus muricata* (bishop pine).

Redwood forests also intergrade with mixed evergreen forests. In some areas the ecotone is broad, and in others, particularly in the South Coast Ranges, the ecotone is comparatively narrow. Rather extensive redwood-Douglas-fir forests occur in northern California. The hardwood components of mixed evergreen forests often form an understory in ecotonal areas under a tall canopy of redwoods or redwoods and Douglas-fir (Fig. 13-5A). Such species as *Umbellularia californica* (bay-laurel), *Lithocarpus densiflorus* (tanbark oak), *Acer macrophyllum* and *Arbutus menziesii* (madrone) often grow in ecotonal areas.

Redwoods are rather fire tolerant. The trees have a thick bark layer that insulates the living stem tissue from all but the hottest and most severe fires. Redwood trunks are often scarred or even hollowed out by forest fires. Even trees that are burned through are able to stump-sprout. As a result of their resistance to fire, coast redwoods can survive fires that burn away the accumulated litter and undergrowth. Recently burned slopes with mineral soils exposed are a good seedbed for germination of the tiny redwood seeds. It is interesting to note, and perhaps ecologically significant, that redwood forests where fires have been suppressed often develop a dense understory of hardwoods such as *Lithocarpus densiflorus*, *Umbellularia californica*, *Arbutus menziesii*, and *Acer macrophyllum*. The growth of these hardwoods can result in a buildup of fuel that would support a very hot, potentially devastating fire that could kill the redwoods.

The ability of redwoods to stump-sprout is significant in areas where the forest has been logged or burned. A ring of new trunks often develops around a cut stump or burned trunk, resulting in a towering multitrunked tree (Fig. 13-5C). Redwoods are unusual among gymnosperms in having this ability.

Fig. 13-5. **A.** Redwood forest in Sonoma County. *Sequoia sempervirens* (coast redwood) towers over woody understory of *Lithocarpus densiflorus* (tan-bark oak, left) and *Umbellularia californica* (California bay laurel, right). **B.** Dense herbaceous understory of sword fern (*Polystichum munitum*) and redwood sorrel (*Oxalis oregona*). **C.** Multiple trunks of *Sequoia sempervirens* around an old stump tell a story of logging and regrowth. Coast redwoods are unusual in their ability to sprout from the base. Photos by David Keil.

Another unusual feature is the ability of redwoods to tolerate burial of the lower portions of their trunks. Redwoods that grow in periodically flooded riparian areas often are subjected to accumulation of sediments around their trunks. Most conifers would die if this occurred, but redwoods are able to develop adventitious roots on the buried portions of their trunks. During the lifetimes of some redwoods, as much as 9 meters of sediments have accumulated around the trunks in layers as much as a meter thick. Sprouts and adventitious roots sometimes also develop from the trunk of a fallen redwood. This results in a row of new erect trunks. Eventually the fallen trunk may rot away, leaving behind a line of young trees.

Redwoods were at one time much more widespread than they are today. During the Tertiary Period, species of *Sequoia* ranged across North America and occurred in Greenland and Europe as well. As the climate became colder or drier in many regions (Chapter 4) the range of the redwoods gradually contracted until these trees occurred only in coastal California. Only in this region was the climate sufficiently free of severely cold winters and summer drought for redwoods to survive. Until a few thousand years ago, conditions were suitable for redwoods considerably south of their present range in California. Fossils of Pleistocene age have been found as far south as Los Angeles. However, the ice ages were followed by the hot, dry "Xerothermic Period". Conditions in southern California became too warm and dry and the range of the redwoods further contracted.

Human Impact On Coastal Coniferous Forests

The forests of the Pacific Northwest and northern California are among the most valuable timber resources in the United States and Canada. Douglas-fir and coast redwood are especially valued in construction. Many of the old-growth forests have been harvested for the lumber industry. Only about ten percent of the original redwood forests remain. Some forests have been permanently cleared and are now occupied by urban zones or used for agriculture. Early logging practices were carried out without a view to the future. The concept of sustained yield came later, after the depletion of much of the original timber.

Large areas of forest are now in secondary succession. The successional patterns depend to a considerable extent upon the nature of the agent that removed the original cover. Some forests have been clear-cut, and all of the mature trees have been removed. Others have been partially logged with individual trees left behind as seed trees. The logging activities in some areas have caused much soil damage and erosion. Often large amounts of broken branches and other debris have been left behind on the damaged slopes. Human-caused fires and fire-control practices have also had impacts. The ability of redwoods to sprout from stumps has enabled these trees to re-establish cover in some of the cleared forests. Hardwoods have invaded in some areas, particularly those with non-sprouting conifers. An introduced weedy shrub, *Cytisus scoparius* (Scotch broom) has invaded large areas of second-growth forest.

Many of the remaining forests are now managed to promote the growth of economically important species. Artificially selected trees with rapid growth and high timber-yields are used as seed sources. To reduce competition from economically undesirable hardwoods (e.g., *Lithocarpus densiflorus*), forest managers sometimes remove these plants by cutting or by herbicide treatment. The blanket spraying of non-selective herbicides to remove broad-leafed trees affects understory herbs and shrubs as well as the target species. Some managed forests are essentially tree-farms rather than natural communities. Forest-management practices are sometimes very controversial. Decisions are sometimes made on the basis of economic rather than biological factors. Since the mid-1980s, the remaining old growth forests of the Pacific Northwest have been the subject of much political debate, in part because of their importance as habitat for the northern spotted owl and the marbled murrelet.

Some areas of coastal coniferous forests have been set aside as preserves. The logging of large areas of virgin redwood forests led to the establishment of the Redwoods National Park to protect a remnant of the original forests. Other smaller areas are preserved in state parks.

Coastal Coniferous Forests—References

Axelrod, D. I. 1976. History of the coniferous forests, California and Nevada. Univ. Calif. Publ. Bot. 70:1–62.

Azavedo, J., and D. L. Morgan. 1974. Fog precipitation in coastal California forests. Ecology 55:1135–1141.

Barrows, K. 1984. Old–growth douglas–fir forests. Fremontia 11(4):20–23.

Becking, R. 1982. Pocket flora of the redwood forest. Island Press, Covelo, California.

Borchert, M., D. Segotta, and M. D. Purser. 1988. Coast redwood ecological types of southern Monterey County, California. U.S.D.A. For. Serv. Pacific Southwest For. & Range Exp. Stn. Gen. Tech. Rep. PSW-107. 27 pp.

Byers, H. R. 1953. Coast redwoods and fog drip. Ecology 34:192–193.

Cooper, W. S. 1917. Redwood, rainfall, and fog. Plant World 20:179–189.

Edmonds, R. L. 1982. Analysis of coniferous forest ecosystems in the western United States. US/IBP Synthesis Series 14. Hutchinson Ross Publ. Co., Stroudsburg, PA.

———. 1982. Introduction. Pp. 1–27 *in* R. L. Edmonds (ed.). Analysis of coniferous ecosystems in the western United States. US/IBP Synthesis Series 14. Hutchinson Ross Publ. Co., Stroudsburg, PA.

Finney, M. A., and R. E. Martin. 1992. Short fire intervals recorded by redwoods at Annadel State Park, California. Madroño 39:251–262.

Florence, R. G. 1965. Decline of old-growth redwood forests in relation to some soil microbiological processes. Ecology 46:52–64.

Franklin, J. F., and C. T. Dyrness. 1973. Natural vegetation of Oregon and Washington. U.S.D.A. Forest Serv., Pac. N.W. Forest and Range Exp. Stn. Tech. Rep. PNW-8. 417 pp.

———. 1988. Pacific Northwest forests. Pp. 103–130 *in* M. G. Barbour and W. D. Billings (eds.), North American Terrestrial Vegetation. Cambridge Univ. Press, Cambridge.

Fritz, E. 1957. California coast redwood, an annotated bibliography of 2003 references. Found. Amer. Resources Manage., Recorder Sunset Press, San Francisco. 267 pp.

Fujimori, T., S. Kawanabe, H. Saito, C. C. Grier, and T. Shidei. 1976. Biomass and production in forests of three major vegetation zones of the northwestern United States. J. Japanese For. Soc. 58:360–373.

Gardner, R. A. 1958. Soil-vegetation association in the redwood—Douglas- fir zone of California. Pp. 86–101 *in* First North Amer. Forest Soils Conf., Agric. Exp. Sta., Michigan State Univ.

Gholz, H. L. 1982. Environmental limits on aboveground net primary production, leaf area, and biomass in vegetation zones of the Pacific Northwest. Ecology 63:469–481.

Grier, C. C. 1977. Biomass, productivity, and nitrogen-phosphorus cycles in hemlock-spruce stands of the central Oregon coast. Pp. 71–81 *in* R. J. Zososki and W. A. Atkinson (eds.). Proceedings, conference on intensive management of western hemlock. Bull. 21, Univ. Washington Inst. Forest Products, Seattle.

———. 1978. A *Tsuga heterophylla — Picea sitchensis* ecosystem of coastal Oregon: decomposition and nutrient balance of fallen logs. Canadian J. For. Res. 8:198–206.

Griffin, J. R., and W. B. Critchfield. 1972. The distribution of forest trees in California. Pacific S.W. Forest and Range Exp. Stn., Berkeley, Calif. 114 pp. (U.S.D.A. Serv. Res. Paper PSW 82).

Johnston, V. R. 1994. California forests and woodlands. A natural history. Univ. California Press, Berkeley & Los Angeles.

Lenihan, J. M. 1990. Forest associations of Little Lost Man Creek, Humboldt County, California: reference-level in the hierarchical structure of old-growth coastal redwood vegetation. Madroño 37:69–87.

MacGinitie, H. D. 1933. Redwoods and frost. Science 78:190.

Marotz, G. A., and J. F. Lahey. 1975. Some stratus/fog statistics in contrasting coastal plant communities of California. J. Biogeogr. 2:289–295.

Muelder, D. W., and J. H. Hansen. 1961. Biotic factors in natural regeneration of *Sequoia sempervirens*. Internat. Union For. Res. Orgs., 13th Congress, Vienna. Proc. pt. 2, vol. 1 (21–4/1) 5 pp.

National Park Service. 1964. The redwoods. U.S. Dept. Interior, Natl. Park Serv. Prof. Rep. 52 pp.

Philbrick, R. N., and J. R. Haller. 1977. The southern California islands. Pp. 893–906 *in* M. G. Barbour and J. Major (eds.). Terrestrial vegetation of California. John Wiley and Sons, N.Y.

Shirley, J. C. 1942. The redwoods of coast and Sierra. 3rd ed. Univ. Calif. Press, Berkeley.

Waring, R. H., and J. F. Franklin. 1979. Evergreen coniferous forests of the Pacific Northwest. Science 204:1380–1386.

———, and J. Major. 1964. Some vegetation of the California coastal redwood region in relation to gradients of moisture, nutrients, light, and temperature. Ecol. Monogr. 34:167–215.

Whittaker, R. H. 1960. Vegetation of the Siskiyou Mountains, Oregon and California. Ecol. Monogr. 30:279–338.

Zinke, P. J. 1961. Chronology of the Bull Creek sediments and the associated redwood forests. Pp. 22–25 *in* Ann. Rep., Redwood Ecology Project, Wildland Research Center, Univ. California, Berkeley.

Chapter 14

Mixed Evergreen Forest Communities

Mixed evergreen forest communities are among the most characteristic of the Mediterranean type communities of California. These communities extend from southern Oregon southward through much of the Klamath-Siskiyou Mountains area and along the coastal mountains to San Diego County. Mixed evergreen forests also are scattered along the western slope of the Sierra Nevada (Fig. 14-1). In the northern parts of the state, mixed evergreen forests cover large areas along the inner flanks of the North Coast Range and Klamath Mountains from southern Oregon to Santa Cruz County. They generally occur inland from the coastal coniferous forests in areas that are warmer in summer and that receive less fog and precipitation. Further inland they grade into northern oak woodlands and foothill woodlands (communities characteristic of hot dry interior foothill regions of the coastal mountains). From Monterey County southward, mixed evergreen forests are mostly restricted to comparatively mesic north facing slopes and canyons. In the Sierra Nevada these communities are most common in mesic canyons adjacent to chaparral, foothill woodlands, and montane mixed coniferous forests. Mixed evergreen forests cover about 4.7 percent of California.

As the name mixed evergreen forest implies, these communities are dominated of tree species that retain their leaves throughout the year. Most of the dominant species are, indeed, evergreen, but a few are winter deciduous. The evergreen species are mostly sclerophyllous dicots, but some needle leafed conifers are also common. Most of the dominant hardwoods of these communities sprout vigorously from cut or burned trunks or stumps.

It is not surprising that, because of the wide distribution of mixed evergreen forests in California, the community composition varies greatly. This variation occurs both in a south to north direction and a west to east direction. However, it interesting that all of the stands of mixed evergreen forest, even those in the

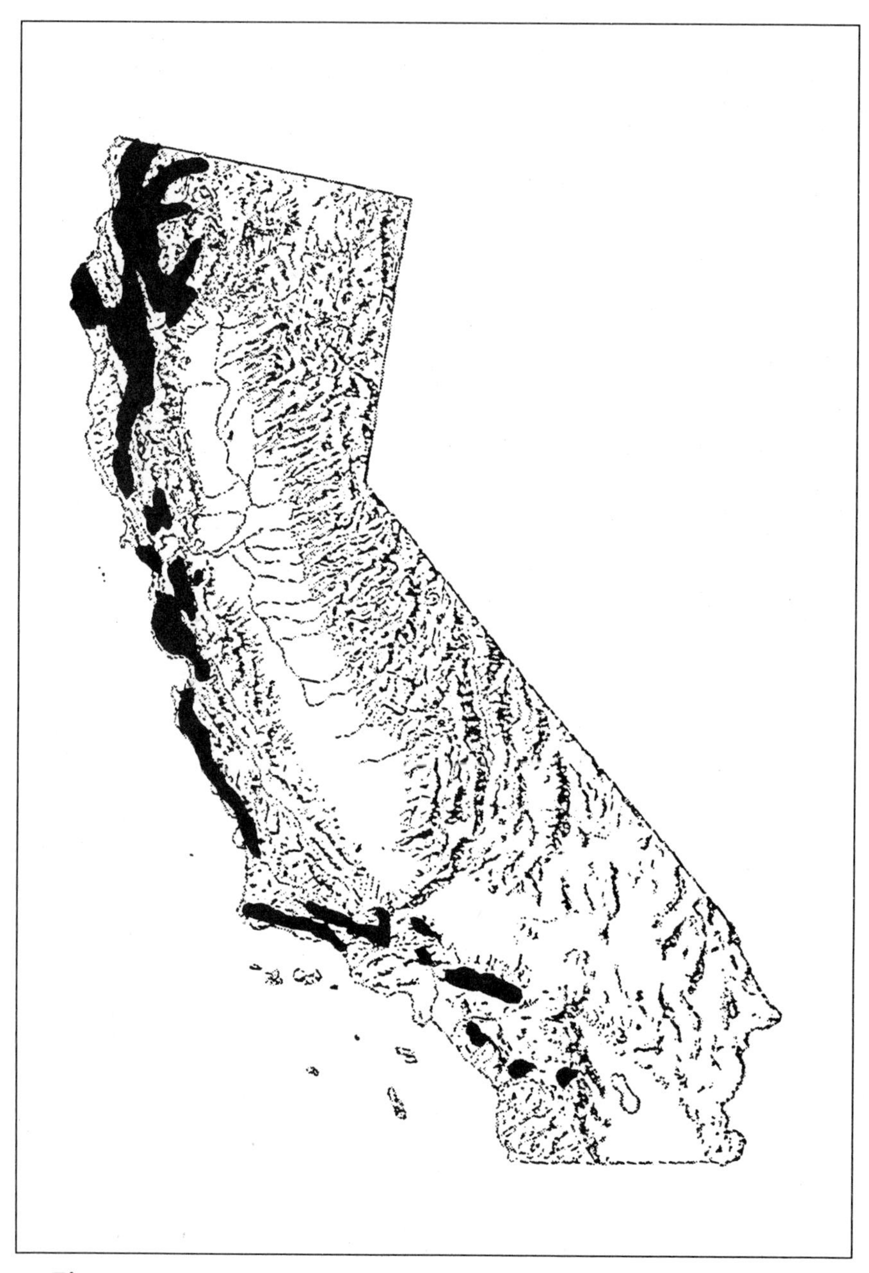

Fig. 14-1. Distribution of mixed evergreen forest communities in California.

the Sierra Nevada and southern California mountains, have some commonality in terms of physiognomy of the forest, species composition, and habitat conditions.

There is considerable site to site variation in species composition of mixed evergreen forests. The following are evergreen species that are dominants in at least some of California's mixed evergreen forests. However, not all occur together in any particular stand.

Arbutus menziesii Fig. 14-3B)	madrone
Chrysolepis chrysophylla (Fig. 14-4D) [*Castanopsis c.*]	giant chinquapin
Lithocarpus densiflorus (Fig. 14-3C)	tanbark oak:
Pinus coulteri (Fig. 14-8D)	Coulter pine
Pseudotsuga menziesii (Fig. 14-3A)	Douglas-fir
Quercus agrifolia (Fig. 14-4B)	coast live oak
Quercus chrysolepis (Fig. 14-4C)	canyon live oak
Quercus parvula var. *shrevei* (Fig. 14-8B)	Shreve oak
Umbellularia californica (Fig. 14-4A)	California bay laurel

Two common deciduous components of the mixed evergreen forest communities are *Acer macrophyllum* (big leafed maple; Fig. 14-3D) and *Quercus kelloggii* (black oak; Fig. 14-8A). Locally other evergreen or deciduous species may also be important components of the community.

1. Northern Mixed Evergreen Forests

Mixed evergreen forests in the northern half of California (from the Santa Cruz Mountains north to southern Oregon) often have *Pseudotsuga menziesii* as a major component of the community. In some areas, this species is the sole dominant, and such stands are often referred to as Douglas-fir forests. Some authors have classified these Douglas-fir dominated communities as a component of the coastal coniferous forests (Chapter 13). Douglas-fir is very commonly found in hardwood dominated communities as well (Fig. 14-2B).

In the Klamath Mountains and in southern Oregon, mixed evergreen forests are fairly continuous, but they are far from uniform. Topographic variation and parent materials interact in a complex fashion to produce a mosaic of intergrading forest types. Deep, well developed soils support forests dominated by Douglas-fir, usually with a herb dominated understory (Fig. 14-5). The moistest sites (Fig. 14-6, 14-7) support an association of Douglas-fir with *Cupressus lawsoniana* [*Chamaecyparis l.*] (Port Orford cedar), *Chrysolepis sempervirens* (giant chinquapin), and *Taxus brevifolia* (yew).

Fig. 14-2. Northern mixed evergreen forests. **A.** Second-growth forest in Sonoma County recovering from a forest fire. The scattered tall coast redwoods (*Sequoia sempervirens*) survived the fire. **B.** Forest in Mendocino County dominated by *Lithocarpus densiflorus* (tan-bark oak) and *Pseudotsuga menziesii* (Douglas-fir) with scattered *Acer macrophyllum* (big-leaf maple), *Arbutus menziesii* (madrone) and *Sequoia sempervirens*. Photos by David Keil.

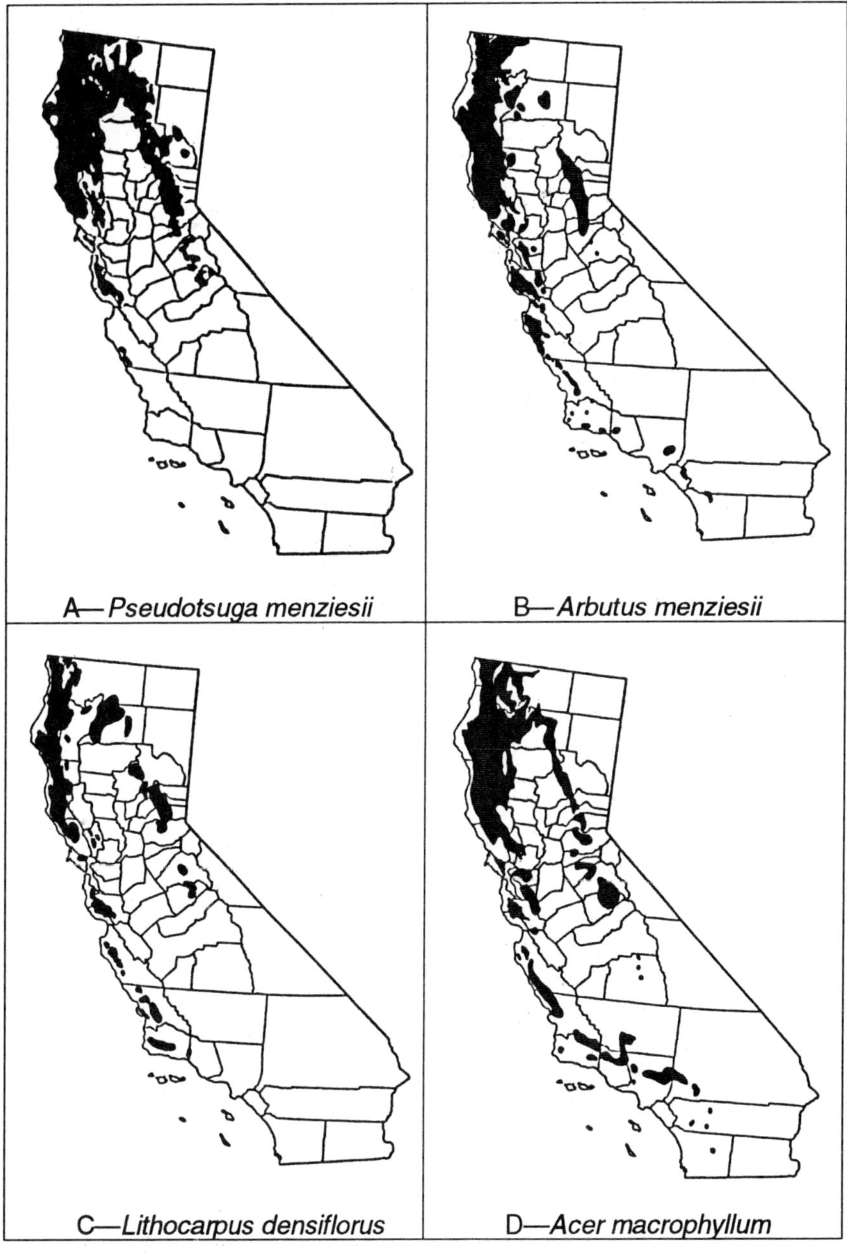

Fig. 14-3. Distribution of *Pseudotsuga menziesii* (Douglas-fir), *Arbutus menziesii* (madrone), *Lithocarpus densiflorus* (tan oak), and *Acer macrophyllum* (big-leaf maple) in California. Maps modified from Griffin and Critchfield, 1972).

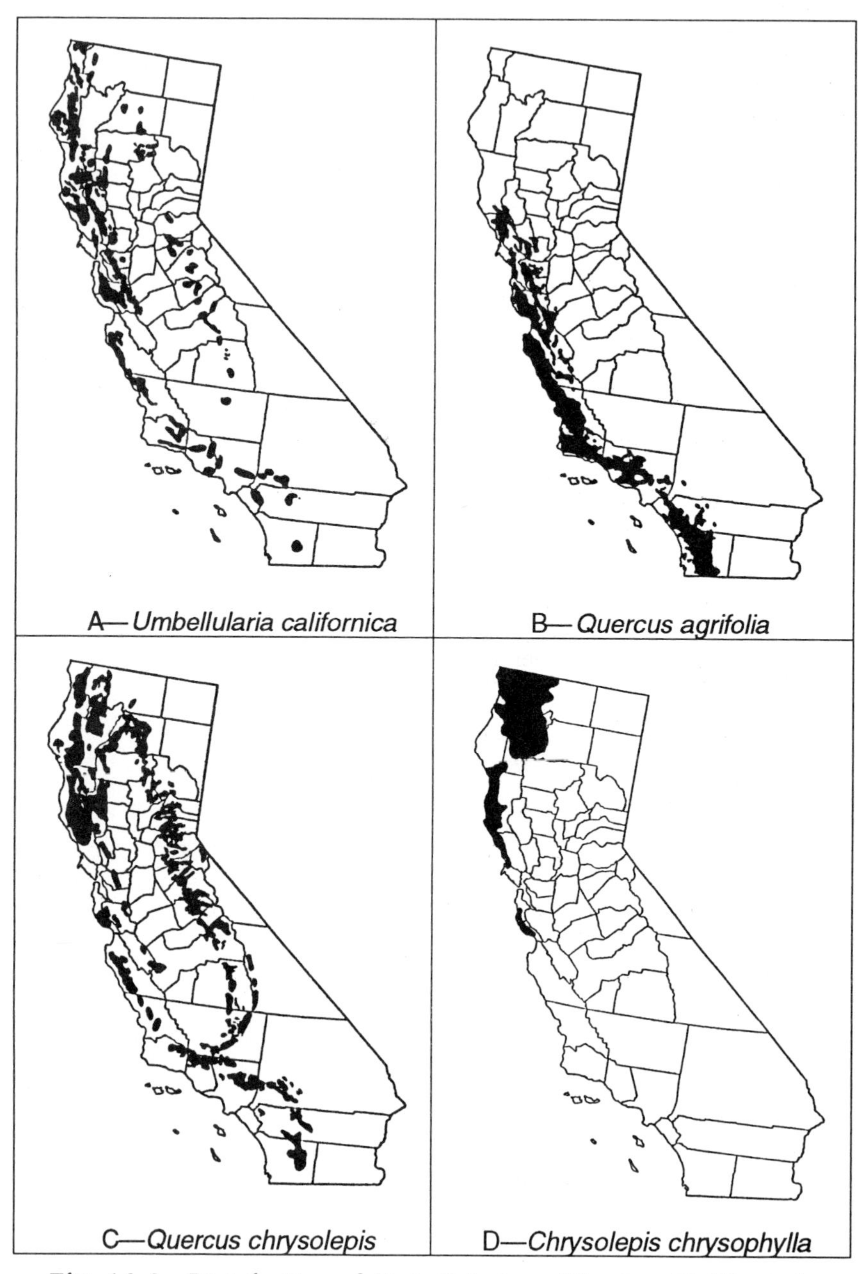

Fig. 14-4. Distribution of *Umbellularia californica* (California bay-laurel), *Quercus agrifolia* (coast live oak), *Quercus chrysolepis* (canyon live oak), and *Chrysolepis chrysophylla* (chinquapin) in California. Maps modified from Griffin and Critchfield, 1972).

Fig. 14-5. Mixed evergreen forest in Sonoma County. Dominant trees include *Pseudotsuga menziesii* (Douglas-fir), *Lithocarpus densiflorus* (tanbark oak), and *Umbellularia californica* (California bay-laurel). *Polystichum munitum* (sword fern) is abundant in the understory. Photo by V. L. Holland.

On shallower and somewhat drier soils, *Lithocarpus densiflorus* and *Arbutus menziesii* form mixed stands with *Pseudotsuga menziesii* (Fig. 14-2B). Sometimes additional tree species occur in these stands such as:

Chrysolepis chrysophylla	giant chinquapin
Pinus lambertiana	sugar pine
Pinus ponderosa	ponderosa pine
Quercus chrysolepis	canyon live oak
Quercus kelloggii	black oak

Understory shrubs in these communities include *Ceanothus* spp. (ceanothus), *Arctostaphylos* spp. (manzanita), and *Quercus vaccinifolia* (huckleberry oak). These shrubs may form patches of chaparral on dry sites with shallow soil.

Soils derived from ultrabasic parent material (serpentine, etc.) support conifer dominated communities with few if any sclerophyllous hardwoods. Dominance varies among several coniferous species:

Fig. 14-6. Mixed evergreen forest in the Klamath Mountains of Siskiyou County. Dominant trees are *Pseudotsuga menziesii* (Douglas-fir) and *Chrysolepis chrysophylla* (giant chinquapin). The dense understory is dominated by *Taxus brevifolia* (yew) and saplings of *Chrysolepis*. Photo by Todd Keeler-Wolf (from Keeler-Wolf 1988), reprinted by permission of California Botanical Society.

Calocedrus decurrens [*Libocedrus d.*]	incense cedar
Pseudotsuga menziesii	Douglas-fir
Pinus attenuata	knobcone pine
Pinus jeffreyi	Jeffrey pine
Pinus lambertiana	sugar pine
Pinus monticola	western white pine

These forests are often rather open, and shrubs are often a major portion of the community. The driest, steepest slopes are dominated mainly by canyon live oak.

In extreme northern California, just inland from the coast redwood forest, *Tsuga heterophylla* (western hemlock), *Abies grandis* (grand fir), and *Thuja plicata* (western red cedar) grow together with *Pseudotsuga menziesii, Lithocarpus densiflorus, Umbellularia californica* and other species more typical of the mixed evergreen forest. These communities have been considered both as a "*Tsuga* phase" of the mixed evergreen forest communities and as a southern extension of the great coniferous forests of the Pacific Northwest. These areas are somewhat ecotonal between the coniferous forests of the northwest (Chapter 13) and the sclerophyll-dominated mixed evergreen forest.

Fig. 14-7. Mixed evergreen forest in the Klamath Mountains of Siskiyou County. *Chrysolepis chrysophylla* (giant chinquapin) and *Pseudotsuga menziesii* (Douglas-fir), Photo by Todd Keeler-Wolf (from Keeler-Wolf 1988), reprinted by permission of California Botanical Society.

Mixed evergreen forest communities of the north coast ranges are located in a zone between the more mesic coastal coniferous forests and the more xeric northern oak woodlands (see Chapter 15). Here too, mixed evergreen forests occur on a variety of substrates and species composition often changes from one type of parent material to another. In the north coast ranges, Douglas-fir is often the dominant species, as it is in the Klamath Mountains. Often it is associated with a mixture of other species, particularly tan bark oak and madrone. Douglas-fir is also a component of the coastal coniferous forests, and the ecotone between the mixed evergreen forest and the coastal coniferous forest is often rather indistinct. On the eastern slopes of the north coast ranges, *Quercus garryana* (Oregon oak) is often a component of the mixed evergreen forest and in the drier valleys it assumes dominance.

2. Mixed Evergreen Forests of Central and Southern California

In central California *Pseudotsuga menziesii* is progressively restricted to mesic sites and in southern California it is entirely absent. In these areas hardwoods are usually the dominant species of the mixed evergreen forests. At lower elevations *Quercus agrifolia* (coast live oak) is often dominant or grows in association with other hardwoods:

Acer macrophyllum	big leafed maple
Arbutus menziesii	madrone
Lithocarpus densiflorus	tan bark oak
Quercus kelloggii	black oak
Umbellularia californica	California bay laurel

In stands where coast live oak grows in pure stands, we have treated these communities as coast live oak woodlands (Chapter 15). They could also be treated here as well as a reduced form of the mixed evergreen forests. There are also sites, both in the North and South Coast Ranges where California bay laurel occurs in pure stands. These are also a reduced phase of the mixed evergreen forests.

At higher elevations, such as above 700 m (2200 ft) in the Santa Lucia Mountains, and on canyon slopes bordering the redwood forests, the dominant species of the mixed evergreen forest include such species as:

Acer macrophyllum	big leafed maple
Arbutus menziesii	madrone
Lithocarpus densiflorus	tan bark oak
Pinus coulteri	Coulter pine
Quercus agrifolia	coast live oak
Quercus chrysolepis	canyon live oak
Quercus parvula var. *shrevei*	Shreve oak

In rocky upland sites in the Santa Lucia Mountains, *Abies bracteata* (Santa Lucia fir) forms scattered stands in association with mixed evergreen forests. Often an assortment of shrubs such as *Ceanothus* spp. and *Arctostaphylos* spp. form an open understory. Understory herbs are sometimes fairly dense in mesic canyon sites where ferns may predominate and very sparse

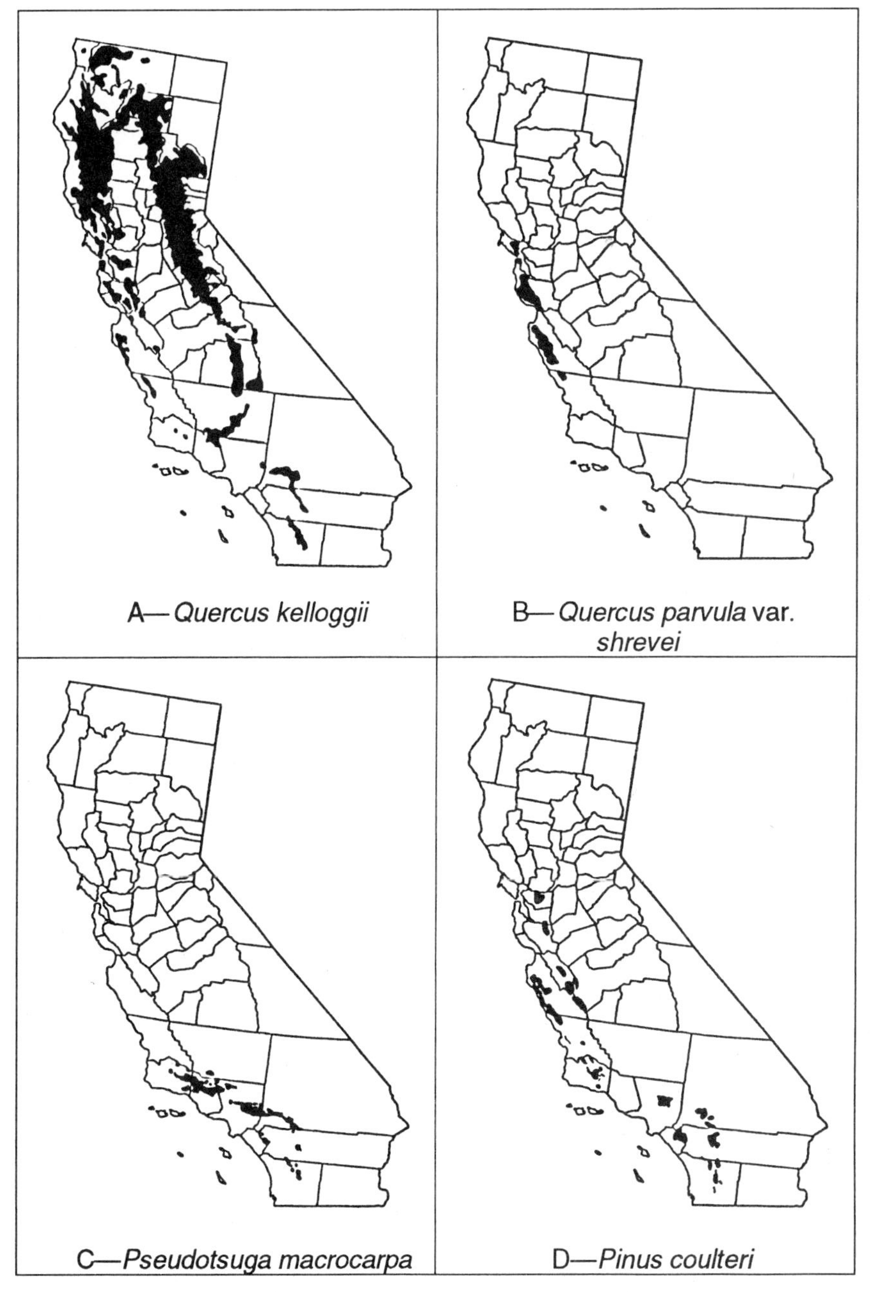

Fig. 14-8. Distribution of *Quercus kelloggii* (black oak), *Quercus parvula* var. *shrevei* (Shreve oak), *Pseudotsuga macrocarpa* (big cone Douglas-fir), and *Pinus coulteri* (Coulter pine) in California. Maps modified from Griffin and Critchfield, 1972).

Fig. 14-9. Mixed evergreen forest in the Liebre Mountains of northern Los Angeles County. Dominant trees are *Pseudotsuga macrocarpa* (big-cone Douglas-fir), *Quercus chrysolepis* (canyon live oak), and *Q. kelloggii* (black oak). The latter are the deciduous trees in the center. Photo by David Keil.

on open slopes. In southern California the distribution of similar forests is progressively restricted to mesic canyon sites.

Mixed evergreen forest communities are associated ecotonally with several other communities of central and southern California. In many areas mixed evergreen forests occur on mesic north facing slopes or in canyons. On more exposed slopes, such communities as chaparral or foothill woodland replace the mixed evergreen forests. Such ecotones are sometimes rather sharp. In the Santa Lucia Mountains of Monterey County, mixed evergreen forest dominants often form an understory in the coastal redwood forests, particularly in canyon areas. Mixed evergreen forest species such as *Acer macrophyllum*, *Quercus agrifolia*, and *Umbellularia californica*, sometimes also become a part of riparian communities.

Although few peaks in the South Coast Ranges are high enough to support stands of montane coniferous forest, such is not the case in the Transverse Ranges of southern California. There, localized stands of mixed evergreen forest intergrade with montane coniferous forests. These relatively small stands of mixed evergreen forest are sometimes dominated by only one or two species and are not always treated as ecologically equivalent to the larger more diverse mixed evergreen forest communities farther north. Dominants in these stands are usually such species as:

Pinus coulteri	Coulter pine
Pseudotsuga macrocarpa	big cone Douglas-fir
Quercus agrifolia	coast live oak
Quercus chrysolepis	canyon live oak
Quercus kelloggii	black oak
Quercus wislizenii	interior live oak
Umbellularia californica	California bay laurel

In the mountains of southern California, *Pseudotsuga macrocarpa* forms a component of several communities. In mixed evergreen forest stands (Figs. 14-8C, 14-9), this species often grows together with canyon live oak, and sometimes with big leafed maple. In some sites big cone Douglas-fir is also a component of montane coniferous forests, and in others it is closely associated with such communities as chaparral and piñon juniper woodland.

3. Sierran Mixed Hardwood Forests

The western slopes of the Sierra Nevada, at elevations below the montane mixed coniferous forests, bear several different communities. In some areas, such as in the vicinity of Sequoia National Park, a mixed hardwood conifer forest occupies a zone between the foothill woodlands and the mixed coniferous forests. This hardwood conifer forest resembles the mixed evergreen forests of the coastal mountains. Dominants include:

Acer macrophyllum	big leafed maple
Pinus ponderosa	ponderosa pine
Quercus chrysolepis	canyon live oak
Quercus kelloggii	black oak
Umbellularia californica	California bay laurel

On some slopes the forest grades into chaparral or foothill woodland communities, and in canyon areas into a mixed conifer-riparian dominated community. Moist rocky canyon sites in the Sierra Nevada are often dominated by canyon live oak, black oak, and in some areas by Douglas-fir. For example, the canyons around Yosemite Valley support a mixed evergreen forest composed of almost pure stands of canyon live oak with a few Douglas-firs. In the northern portion of the Sierra Nevada, such species as *Arbutus menziesii*, *Lithocarpus densiflorus* and *Torreya californica* (California-nutmeg) join the mixed hardwood communities.

Human Impacts on Mixed Evergreen Forests

Human influence in the mixed evergreen forest communities has taken several forms. Prior to the coming of the Spanish and other later settlers, the local Indian tribes occasionally set fires that burned into the forests. Subsequently, the attitude of residents toward fire has often led to the suppression of both human-caused and natural fires. The change from fire promotion to fire suppression has resulted in changes in the community structure in many areas. The density of the vegetation has increased, and species that require bare ground for seedling establishment have had a lowered rate of reproductive success. This pattern is particularly noticeable in the ecotonal areas where mixed evergreen species grow with *Sequoia sempervirens* (coast redwood). The coast redwood seeds seldom become established in the absence of a bare mineral soil layer, such as is available immediately after a fire (see Chapter 13). In the absence of fire, such mixed evergreen tree species as tanbark oak and California bay laurel are able to attain larger size and greater density than they would if fires were frequent. If fires are suppressed for a long enough time, these mixed evergreen forest species may successionally replace the redwoods in at least a part of the redwoods' range.

Logging is a second major area of human disturbance. Douglas-fir is economically an extremely important timber tree in California. This species has been logged in many areas of the state, and as a result, many areas where Douglas-fir occurs are in various stages of succession. Most of the hardwoods of the mixed evergreen forest are economically less desirable than Douglas-fir, and seral communities, where these plants have replaced the logged Douglas-fir, are often further modified to make Douglas-fir regeneration more rapid. Both clearing of the hardwoods and application of herbicides have been used to modify these communities. Broad spectrum application of herbicides affects all broad leafed species and is the most destructive of these forestry practices.

Mixed Evergreen Forests–References

Axelrod, D. I. 1976. Evolution of the Santa Lucia fir (*Abies bracteata*) ecosystem. Ann. Missouri Bot. Gard. 63:24–41.

Barbour, M. G. 1988. California upland forests. Pp. 131–164 *in* M. G. Barbour and W. D. Billings (eds.), North American Terrestrial Vegetation. Cambridge Univ. Press, Cambridge.

Bolton, R. B., and R. J. Vogl. 1969. Ecological requirements of *Pseudotsuga macrocarpa* in the Santa Ana Mountains, California. J. Forestry 67:112–116.

Brown, D. E. 1982. Californian evergreen forest and woodland. Pp. 66–69 *in* D. E. Brown (ed.). Biotic communities of the American Southwest — United States and Mexico. Desert Plants 4:1–341.

Campbell, B. 1980. Some mixed hardwood communities of the coastal ranges of southern California. Phytocoenologia 8:297–320.

Cooper, W. S. 1922. The broad sclerophyll vegetation of California. An ecological study of the chaparral and its related communities. Carnegie Inst. Wash. Publ. 319. 124 pp.

Gause, G. W. 1966. Silvical characteristics of bigcone Douglas-fir (*Pseudotsuga macrocarpa* (Vasey) Mayr). U.S.D.A. Forest Service Res. Paper PSW-31. 10 pp.

Griffin, J. R., and W. B. Critchfield. 1972. The distribution of forest trees in California. Pacific S.W. Forest and Range Exp. Stn., Berkeley, Calif. 114 pp. (U.S.D.A. Serv. Res. Paper PSW 82).

Johnston, V. R. 1994. California forests and woodlands. A natural history. Univ. California Press, Berkeley & Los Angeles.

Keeler-Wolf, T. 1988. The role of *Chrysolepis chrysophylla* (Fagaceae) in the *Pseudotsuga*-hardwood forest of the Klamath Mountains of California. Madroño 35:285–308.

Keeley. 1988. Bibliographies on chaparral and the fire ecology of other Mediterranean systems, 2nd ed. Calif. Water Resource Cntr., Univ. Calif. Davis, Rep. No. 69. 328 pp.

McDonald, P. M., and E. E. Littrell. 1976. The bigcone Douglas fir–canyon live oak community in southern California. Madroño 23:310–320.

Minnich, R. A. 1976. Vegetation of the San Bernardino Mountains. Pp. 99–124 in J. Latting (ed.). Plant communities of southern California. Calif. Native Plant Soc. Spec. Publ. 2.

———. 1980. Wildfire and the geographic relationships between canyon live oak, Coulter pine, and bigcone Douglas-fir forests. Pp. 55–61 *in* T. R. Plumb (tech. coord.), Proceedings of the symposium on the Ecology, management, and utilization of California oaks. Pacific S.W. Forest and Range Exp. Stn., Berkeley.

———. 1982. *Pseudotsuga macrocarpa* in Baja California? Madroño 29:22–31.

Myatt, R. G. 1980. Canyon live oak vegetation in the Sierra Nevada. Pp. 86–91 *in* T. R. Plumb (tech. coord.), Proceedings of the symposium on the Ecology, management, and utilization of California oaks. Pacific S.W. Forest and Range Exp. Stn., Berkeley.

Navah, Z. 1967. Mediterranean ecosystems and vegetation types in California and Israel. Ecology 48:445–459.

Sawyer, J. O., D. A. Thornburgh, and J. R. Griffin. 1977. Mixed evergreen forest. Pp. 359–381 in M. G. Barbour and J. Major (eds.). Terrestrial vegetation of California. John Wiley and Sons, N.Y

Wainwright, T. C., and M. G. Barbour. 1984. Characteristics of mixed evergreen forest in the Sonoma Mountains of California. Madroño 31:219–230.

Waring, R. H. 1969. Forest plants of the eastern Siskiyous: their environmental and vegetational distribution. Northwest Sci. 43:1–17.

———, and J. Major. 1964. Some vegetation of the California coastal redwood region in relation to gradients of moisture, nutrients, light, and temperature. Ecol. Monogr. 34:167–215.

Whittaker, R. H. 1960. Vegetation of the Siskiyou Mountains, Oregon and California. Ecol. Monogr. 30:279–338.

Chapter 15

Oak Woodland Communities

Oak woodlands form a characteristic vegetational cover in the foothills of cismontane California mountains. They cover several million hectares around and in the Central Valley (especially the mesic Sacramento Valley) and in the foothills of the southern California mountains (Fig. 15-1). Roughly ten percent of California is covered by oak woodlands. While there are oak woodlands in other areas of the United States, the oak woodland communities of California are unique in that they are almost entirely restricted to the state, and most of the dominant trees are endemic to California. Many think that oak woodlands should be declared California's state vegetation type because they are so characteristic and unique to the state.

Oak woodlands occur at elevations ranging from about 10 to 1500 m (30 to 5000 ft) where summers are warm and dry and winters rather mild. Average annual precipitation is about 50 cm (20 in) ranging from less than 25 cm (10 in) in dry sections of the South Coast Ranges to over 100 cm (40 in) in northern California. Oak woodlands are transitional between the grasslands of hot, dry valleys and the montane forests of moist, cool uplands. In interior mountain ranges, oak woodlands grade into montane mixed coniferous forests (Chapter 16), and in coastal mountains, they grade into mixed evergreen forests.

Oak woodlands are dominated by trees (mostly oaks) 5–21 m (15–70 ft) tall. These woodlands vary from open savannas to dense, closed-canopy communities. The most common woodland type consists of scattered trees and shrubs with an understory of grasses and forbs. The shrubs, often species that also occur in chaparral or coastal scrub communities, may grow both under and between the trees. However, in savanna woodlands shrubs are often entirely absent, and the ground cover is essentially the same as that of grasslands (Chapter 11). In mesic, favorable sites, the trees may be dense and form a closed canopy with only shade-tolerant species comprising the understory.

Factors that cause variation in composition and structure of these communities include latitude, elevation, slope aspect, and local conditions of soil, precipitation, moisture availability and air temperature. Generally oak woodlands are most open where moisture is limited and densest where adequate water supplies are available. The woodlands are often denser on north-facing than on south-facing slopes. Woodlands of upper elevation sites are often denser and have greater species diversity than those of low elevations.

Oak woodlands are dominated by several species of deciduous or evergreen oaks and often also have a coniferous component. Most frequently the woodlands are dominated by one to three species. The following are dominants in a significant part of California's oak woodlands:

Quercus agrifolia (Figs. 15-3A, 15-4A)	coast live oak
Quercus douglasii (Figs. 15-3B, 15-4B)	blue oak
Quercus engelmannii (Fig. 15-11A)	Engelmann oak
Quercus garryana (Figs. 15-7C, 15-10)	Oregon oak
Quercus kelloggii (Fig. 15-7D)	black oak
Quercus lobata ((Figs. 15-4D, 15-5)	valley oak
Quercus wislizenii (Fig. 15-7A)	interior live oak
Pinus sabiniana (Figs. 15-3C, 15-4C)	foothill pine

Several other species of trees or large shrubs occur together with the oaks as codominants or subdominants in some parts of the state. These include:

Aesculus californica (Fig. 15-7B)	buckeye
Juglans californica (Fig. 15-13B)	California black walnut
Cercis occidentalis	redbud
Juniperus californica	California juniper
Pinus coulteri	Coulter pine

California's oak woodland communities are rather variable from place to place in the state. The species listed above occur in various combinations, and nowhere in California do all of them occur together. We have chosen to group the various combinations into six types of oak woodland communities: (1) coastal live oak woodlands; (2) valley oak woodlands; (3) foothill woodlands; (4) northern oak woodlands; (5) southern oak woodlands; and (6) island oak woodlands. These intergrade somewhat with each other and with adjacent communities. There is considerable internal variation in each group. Our grouping of this complex of communities is based upon dominant species, geographic location and environmental conditions. The distributions of oak woodland communities in California are shown on Figs. 15-1 and 15-2.

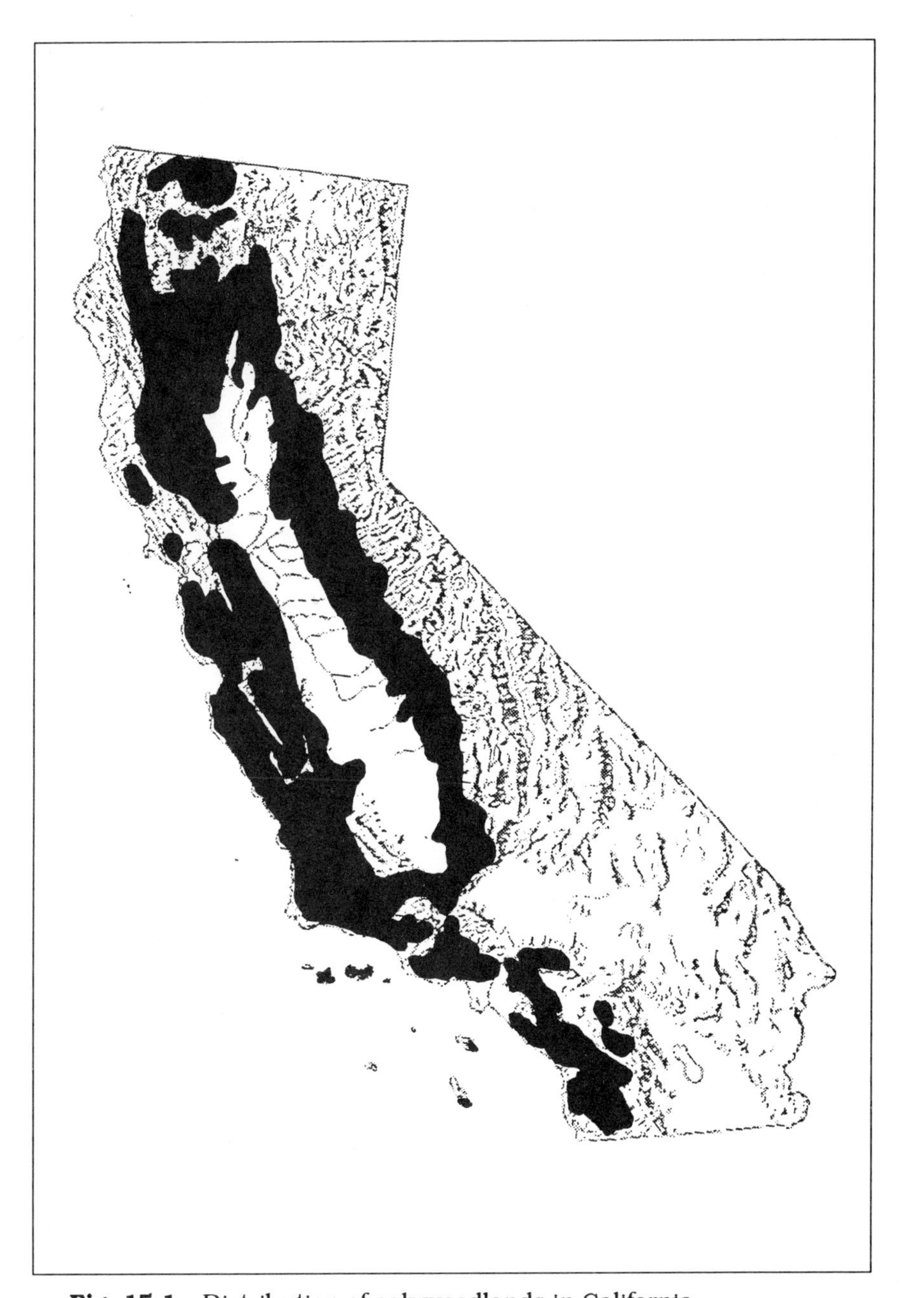

Fig. 15-1. Distribution of oak woodlands in California

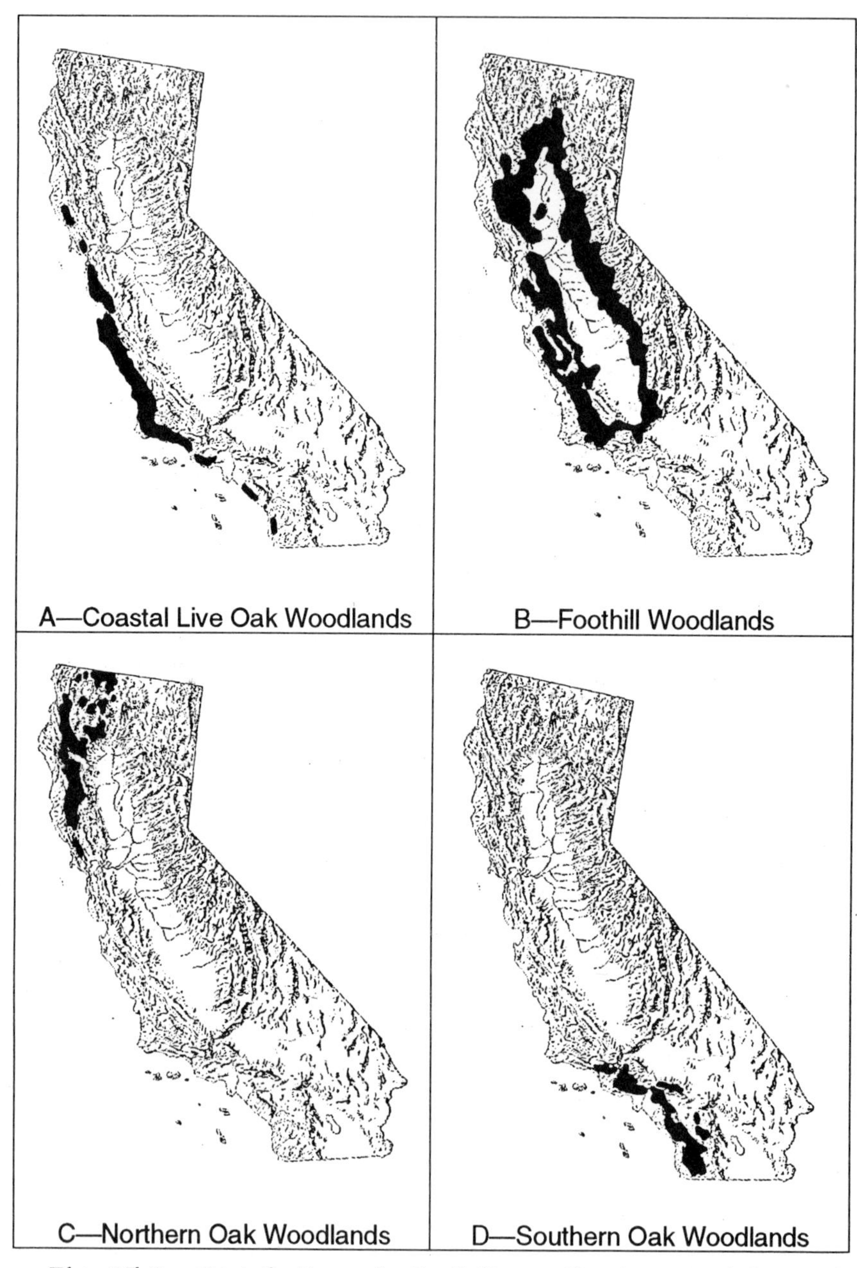

Fig. 15-2. Distribution of foothill woodland, coastal live oak woodland, southern oak woodland and northern oak woodland communities.

Dominant Species of California's Oak Woodlands

In our classification of oak woodlands, the combination of dominant trees can be used as indicator species to distinguish one community from another. These are discussed and compared below in terms of distribution and habitat requirements.

Quercus douglasii (blue oak; Figs. 15-3B, 15-4B), a deciduous white oak, is the most conspicuous and common tree in foothill (oak) woodlands surrounding the Central Valley. Its range extends from Shasta County in the north to the Liebre Mountains of Los Angeles County and the Santa Ynez Valley of Santa Barbara County (both in the Transverse Ranges). Although a few trees grow within three miles of the coast in Santa Barbara County, the main distribution of blue oak is in the interior foothills where climatic conditions are hotter and drier. *Quercus douglasii* is usually considered the most drought tolerant of the California oak trees.

Blue oak does not extend into southern California but is replaced ecologically by a semi-evergreen white oak, *Quercus engelmannii* (Engelmann oak; Fig. 15-11A) south of the Transverse Ranges. Engelmann oak has a limited distribution in southern California and adjacent Baja California and is an integral component of the southern oak woodlands.

In northern California, blue oak overlaps with, and is ecologically replaced by, another deciduous white oak *Quercus garryana* (Oregon oak; Figs. 15-7C, 15-10). Oregon oak is the only oak species in California that extends very far beyond the State's boundary. It occurs as far north as southwestern British Columbia. Oregon oak forms oak woodlands in Oregon that are similar in appearance to those in California that we call the northern oak woodlands.

In deep, alluvial soils, *Quercus lobata* (valley oak; Figs. 15-4D, 15-5), another deciduous white oak, becomes dominant. Its distribution pattern is very similar to that of blue oak, extending from Shasta County in the north to Los Angeles County in the south. This endemic oak is the dominant tree in the oak savannas that extend into the Central Valley as well as some of the smaller interior valleys. In many of these areas, valley oak grows as the only tree in the woodland. These communities are referred to as valley oak woodlands. Valley oak is also a component of the foothill woodland communities.

Common associates of these white oaks are two evergreen black oaks. *Quercus agrifolia* (coast live oak; Figs. 15-3A, 15-4A) is restricted to the coastal mountains from Sonoma County south into northern Baja California. It is an important component of

Fig. 15-3. **A.** Coast live oak (*Quercus agrifolia*) in foothill woodland near Santa Margarita in San Luis Obispo County. Photo by Paula Kleintjes. **B.** Blue oak (*Quercus douglasii*) in foothill woodland near Shell Creek in La Panza Mountains of San Luis Obispo County. Photo by Paula Kleintjes. **C.** Foothill pine (*Pinus sabiniana*) in foothill woodland in La Panza Mountains of San Luis Obispo County. Photo by David Keil.

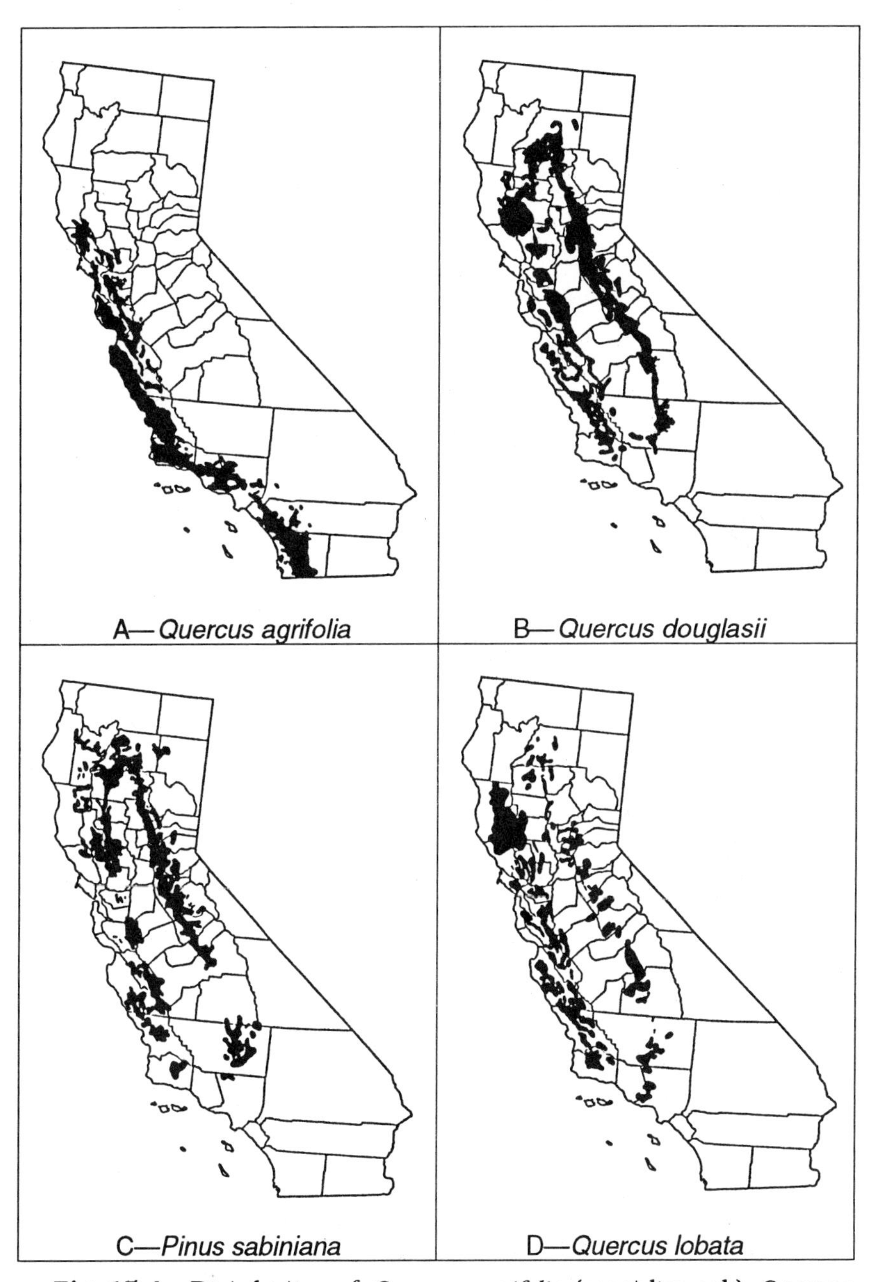

Fig. 15-4. Distribution of *Quercus agrifolia* (coast live oak), *Quercus douglasii* (blue oak), *Pinus sabiniana* (foothill pine), and *Quercus lobata* (valley oak). Maps redrawn from Griffin and Critchfield (1972).

A
B

the oak woodlands in the central coast and southern California. In many areas along the immediate coast, coast live oak is the only tree present. These communities are referred to as coastal live oak woodlands. In the interior coastal mountains, coast live oak is part of the foothill woodland communities and mixes with blue oak and foothill pine.

Quercus wislizenii (interior live oak; Fig. 15-7A) is largely a Californian species, but like coast live oak, it occurs in northern Baja California. As the name implies, interior live oak is more common in inland sites and is an important component of the foothill woodland in the Sierra Nevada-Cascade foothills and of the southern oak woodland in the foothills of the Transverse and Peninsular Ranges of southern California. It is widespread in the coast ranges as well. In the North Coast Ranges it occurs in a variety of communities, but in the South Coast Ranges where it is mostly represented by a shrubby race, var. *frutescens*, it most commonly occurs in chaparral or as a component of forest understories. In some of the higher elevations of foothill woodland, interior live oak may completely dominate localized areas to the exclusion of blue oak.

Pinus sabiniana (foothill pine; Figs. 15-3C, 15-4C) is also an integral part of California's foothill woodlands. This endemic pine has a distribution pattern that is almost identical to that of blue oak. It very frequently is an associate of blue oak except for a 88 km-wide (55 mile) gap between the Kings River and South Fork of the Tule River (most of Tulare County) in the southern Sierra Nevada. In this area blue oak occurs as the only tree in lower elevations of the foothill woodland and mixes with interior live oak at higher elevations.

Several additional oaks play a minor role in the oak woodlands of California. *Quercus kelloggii* (black oak; Fig. 15-7D) enters the upper elevational portions of foothill woodlands in scattered locations and sometimes is a dominant or codominant northern oak woodlands. *Quercus chrysolepis* (canyon live oak) commonly grows in rocky canyon areas and occasionally forms a minor part of foothill woodland and northern oak woodland communities. *Quercus johntuckeri* (desert scrub oak) and *Q. berberidifolia* (scrub oak) both typically grow as shrubs but on occasion reach the stature of trees. Both occasionally form part of the overstory in foothill woodland communities (as in the La Panza Mountains of San Luis Obispo County).

Fig. 15-5. Valley oak (*Quercus lobata*) typically occurs in the deep alluvial soils of valleys or gently sloping hills. **A.** Valley oak woodland in northern Santa Barbara County. Photo by David Keil. **B.** A large valley oak in winter-deciduous condition. Photo by Robert F. Hoover.

Survey of Oak Woodland Communities

1. Coastal Live Oak Woodlands

Probably the most mesic of the foothill woodland communities are the coastal live oak woodlands (Figs. 15-2A, 15-5). These communities are dominated by *Quercus agrifolia*, which is often the only tree species present. The communities are restricted to coastal areas from Sonoma County south into Baja California. Coastal live oak woodlands are variable. In mesic areas such as north- facing slopes and canyons, these communities are dense and sometimes intergrade with mixed evergreen forests (chapter 13). In these areas, such trees as *Umbellularia californica* (California bay-laurel), *Arbutus menziesii* (madrone) and *Acer macrophyllum* (big-leaf maple) can sometimes be found growing with coast live oaks.

Because *Quercus agrifolia* is an evergreen species with a dense canopy, the environment at the ground level in a dense stand is very shady (Fig. 15-5B). Understory vegetation may be sparse or absent. Typical understory plants in areas of dense coast live oak woodlands are shade tolerant shrubs such as *Rubus ursinus* (wild blackberry), *Symphoricarpos mollis* (snowberry), *Heteromeles arbutifolia* (toyon) and *Toxicodendron diversilobum* (poison oak); and herbaceous plants such as *Pteridium aquilinum* (bracken fern), *Polypodium californicum* (polypody fern), *Pholistoma auritum* (fiesta flower) and *Claytonia perfoliata* [*Montia p.*] (miner's lettuce).

In drier, more exposed areas where soils are usually shallower, coast live oaks are more scattered and form an open woodland. In these areas the shrubby and herbaceous understory varies significantly. Where coast live oak woodlands intergrade with grasslands, the understory consists almost entirely of grassland species with few shrubs. In other areas (usually on somewhat steeper slopes), there is a diversity of shrubs under and between the trees and a sparser herbaceous cover. In areas where coast live oak woodlands intergrade with chaparral, typical chaparral species such as *Arctostaphylos* spp. (manzanita), *Adenostoma fasciculatum* (chamise), *Ribes* spp. (gooseberries and currants) and *Ceanothus* spp. (ceanothus) form the understory. In areas where coast live oak woodlands intergrade with coastal scrub, typical understory species are *Mimulus aurantiacus* (bush monkeyflower), *Baccharis pilularis* (coyote bush), *Salvia mellifera* (black sage) and *Artemisia californica* (California sagebrush). Poison oak seems to be a constant associate in all cases. This shrub understory is quite variable from place to place. In some areas the shrubs are so dense the community appears similar to a chaparral community with taller coast live oaks growing in it. In other areas, the shrubs are scattered or only locally dense.

Fig. 15-6. Coastal live oak woodland. **A.** The understory vegetation in sunny areas between the groups of oaks in this Santa Barbara County stand is composed of coastal scrub species including California sagebrush and purple sage. Photo by David Keil. **B.** Shaded interior of woodland with sparse understory of shade-tolerant herbs and much oak leaf litter. Photo by Robert Rodin.

An interesting phase of the coastal live oak woodlands occurs on stabilized sand dunes along the coast of central California. These communities are dominated by a dwarfed form of coast live oak sometimes referred to as pygmy oak. Whether these plants are genetically or environmentally dwarfed has not been determined. Certainly the infertile unstable soils along with salt-laden winds, must play a role in its growth pattern. Coast live oaks are sometimes also extremely dwarfed on some coastal slopes on other parent materials where exposed to severe coastal winds but reach more normal proportions in nearby sheltered sites.

2. Valley Oak Woodlands

On alluvial terraces of large valleys and on low rolling hills from Lake Shasta to northern Los Angeles County, *Quercus lobata* once formed extensive woodlands. Valley oak woodlands were mostly restricted to deep alluvial soils at low elevations. In the Central Valley, these communities formed belts varying in width from a few hundred meters to a few kilometers. These bands of valley oak woodland often paralleled the riparian communities. Valley oak, the largest of California's oaks, often is the only tree present in deep alluvial soils of valleys. In these areas, it formed woodlands that range from forests to open savannas. The understory was largely composed of grasses and forbs.

Most of these original valley oak woodlands are now gone. Early pioneers where quick to realize that the presence of these huge valley oaks were indicative of prime agricultural soils. Consequently, the trees have been cut, and the land is now in cultivation. Immense relict trees can still be seen scattered in some of these valley areas, suggesting that these valley oak woodlands must have been impressive. A few stands remain in areas around dwellings or in parks, but most are not undergoing any regeneration. The original extent of these woodlands is difficult to determine because many stands were cut for firewood or cleared for agriculture before any accurate records were kept.

Valley oak woodlands grade into foothill woodlands. Where this occurs valley oaks tend to occupy alluvial valley soils, whereas other oak species often occur on slopes on shallower soils.

3. Foothill Woodlands

Foothill woodland communities (Figs. 15-2B, 15-8, 15-9) are the best known and most widespread of California's oak woodlands. These communities occur in the foothills of both the

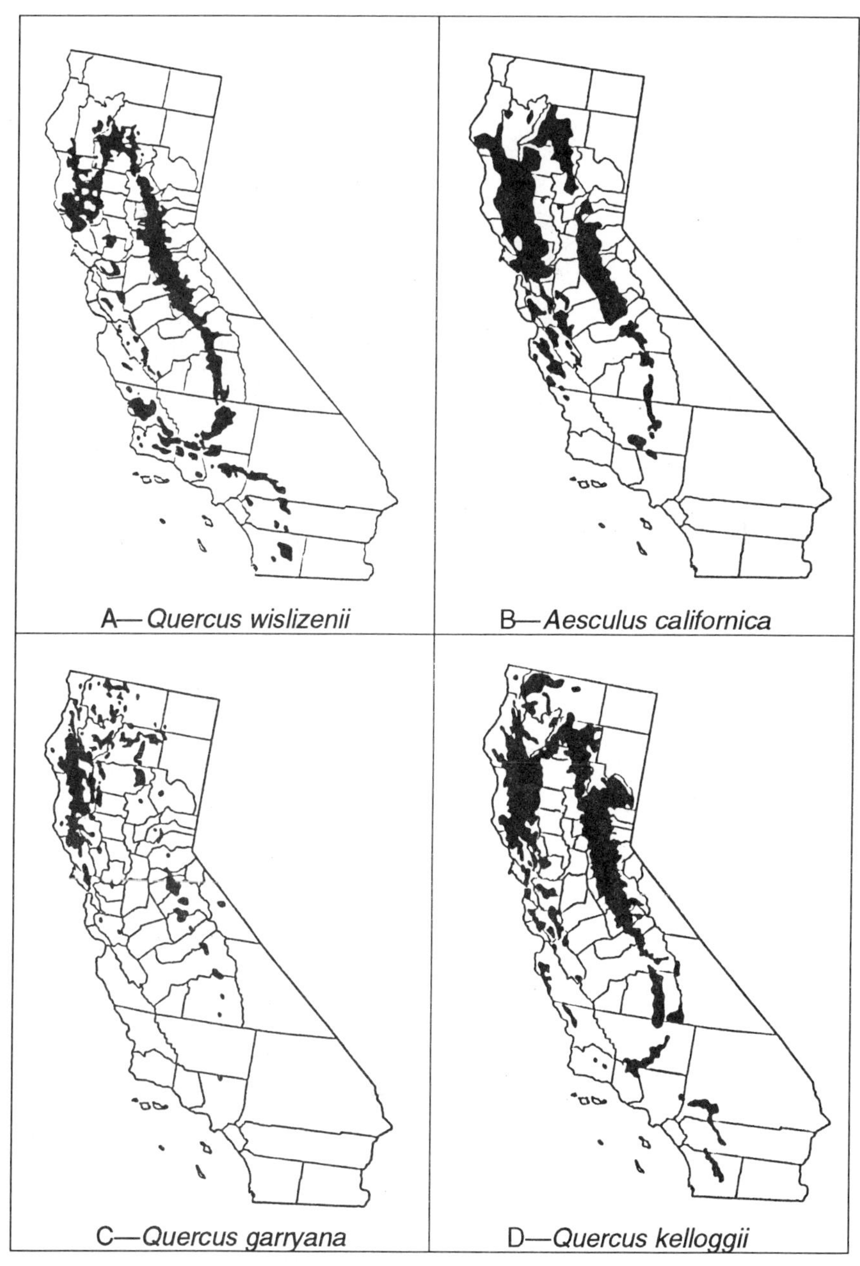

Fig. 15-7. Distribution of *Quercus wislizenii* (interior live oak), *Aesculus californica* (California buckeye), *Quercus garryana* (Oregon oak), and *Quercus kelloggii* (black oak). Maps redrawn from Griffin and Critchfield (1972).

Fig. 15-8. Foothill woodland in Sierra Nevada foothills in Kern County. The woodland is dominated by *Quercus douglasii)* (blue oak), *Q. wislizenii* (interior live oak), *Aesculus californica* (California buckeye), and *Pinus sabiniana* (foothill pine). Photo by David Keil.

from about 120 to 1200 m (400 to 5000 ft) in the Coast Ranges, Transverse Ranges, Sierra Nevada, Cascades, and Klamath-Siskiyou mountains.

Foothill woodlands consist of tree species (mostly oaks) 5 to 21 m (15 to 70 ft) tall in an open to somewhat dense woodland with scattered shrubs and a ground cover of valley grassland species and woodland herbs. Dominant trees in most stands are *Quercus douglasii* and *Pinus sabiniana*. *Quercus lobata* is also a common associate throughout the geographic range of the association but is usually restricted to deep, alluvial soils (Fig. 15-9A). *Quercus agrifolia* is an important component in the Coast Ranges, and *Quercus wislizenii* is a significant component in the interior foothills like those of the Sierra Nevada-Cascade axis and in the North Coast Ranges. *Aesculus californica* (California buckeye) is also a common tree (or large shrub) in many areas. Other common shrubs or small trees include:

Arctostaphylos spp.	manzanita
Ceanothus cuneatus	buckbrush
Ceanothus spp.	ceanothus
Cercis occidentalis	redbud
Eriodictyon californicum	yerba santa
Rhamnus californica	California coffeeberry
Ribes quercetorum	gooseberry

Fig. 15-9. Foothill woodland at Shell Creek in the La Panza Mountains of San Luis Obispo County. **A.** Valley oaks (*Quercus lobata*) occupy the deep alluvial soil of valley areas. Lupines and other wildflowers dominate the valley floor. **B.** Blue oaks (*Q. douglasii*) and foothill pines (*Pinus sabiniana*) occur on the nearby hillsides, along with scattered California junipers (*Juniperus californica*). Photos by Paula Kleintjes.

Common herbaceous species include many of the same species as those listed for the California grasslands (Chapter 11) plus an assortment of more shade-tolerant woodland species. Herb composition varies greatly depending on the density of the tree cover and the slope exposure. Nearby sites may differ markedly in the understory composition and development.

Foothill woodland communities vary from extremely open savanna communities to dense closed canopy forests. The trees are most widely scattered in the low elevation foothills bordering the Central Valley. In these areas blue oak is often the only tree species present in the foothill woodland. These areas of blue oak populations are often referred to as blue oak woodlands. With increased elevation tree species diversity and density increases significantly, initially in canyons and on north-facing slopes. In these areas, adjacent south-facing slopes may support open woodlands of blue oak or grasslands and chaparral. Near the upper-elevational limits of foothill woodlands these communities integrate with montane mixed coniferous forests in Sierra Nevada and Cascade Range or with mixed evergreen forests in the coastal mountains. The relationships of foothill woodlands to other communities may be very complex. In upper-elevation foothill woodlands trees such as *Quercus chrysolepis* (canyon live-oak), *Quercus kelloggii* and *Umbellularia californica* (California bay-laurel) may become part of the community.

In xeric foothill woodland areas where climatic conditions are extreme, blue oak often grows in pure stands. This phase of the foothill woodland communities occurs at low elevations of the interior South Coast Range such as the Temblor Range and also in part of the Transverse Range and the southern end of the Sierra Nevada (Tehachapi and Piute Mountains). Blue oak woodlands are often the first woodland communities above valley grasslands in the Sierra Nevada foothills.

In foothills bordering desert regions, such as the Temblor and Tehachapi Mountains, blue oak overlaps with species of the Great Basin Floristic Province. For example, southern stands of blue oak occur together with California juniper (*Juniperus californica*) in the interior south Coast Ranges (Fig. 15-9B) and in the Transverse Range (Mt. Abel and Mt. Pinos area). In addition to the junipers, *Quercus johntuckeri* (desert scrub oak) sometimes grows together with blue oak, and the two species frequently hybridize. Intermediates are referred to as *Quercus* x *alvordiana*, and sometimes form an important part of the woodland. In addition, blue oak sometimes is associated with such unlikely species as *Yucca brevifolia* (Joshua tree), *Pinus monophylla* (piñon pine), *Artemisia tridentata* (Great Basin sagebrush) in the southern Sierra Nevada or the Transverse Ranges, and *Juniperus occidentalis* western juniper) in northern California.

Fig. 15-10. Northern oak woodland in North Coast Range of Sonoma County. The dominant trees are *Quercus garryana* (Oregon oak). Photo by V. L. Holland.

In portions of the southern Sierra Nevada, another xeric phase of the foothill woodlands occurs in the vicinity of Walker Pass. In this region there is a broad ecotone between transmontane desert and desert woodland communities and cismontane foothill woodland and chaparral communities. The western slope of Walker Pass, though technically located in cismontane California, is situated in the rain shadow of the Tehachapi and Greenhorn Mountains [parts of the southern Sierra Nevada]. The area is quite dry and has cold winters. The open foothill woodlands in this area are dominated by foothill pine and interior live oak. Blue oak is only a minor component, and in some areas it is completely absent. Piñon pine, California juniper, and Joshua trees sometimes occur together with the foothill woodland species. The understory composition ranges from an open woodland chaparral to a mixture of Great Basin sagebrush, rabbitbush (*Chrysothamnus nauseosus*) and other desert shrubs.

4. Northern Oak Woodland Communities

Northern oak woodlands (Fig. 15-10) occur from about Napa County to Trinity County and Humboldt County in the foothills of the North Coast Ranges. They occur in the warm, dry regions to the interior of the coastal coniferous forests and mixed evergreen forests. These communities usually form a more or less open woodland and differ from foothill woodlands mostly because *Quercus garryana* is the common deciduous white oak instead of blue oak. The herbaceous and shrubby understories are very similar to those of foothill woodlands both in structure and composition, and foothill pine is common in both.

Some Oregon oak stands in northern California are very similar to the oak woodlands of southern Oregon. Northern oak woodlands seem to occur under more mesic conditions than the foothill woodlands to the south. As a result, species common to moister sites such as *Quercus kelloggii*, *Quercus chrysolepis* and *Arbutus menziesii* (madrone) are often found mixed with the Oregon oak. However, Oregon oak is also found with *Juniperus occidentalis* var. *occidentalis* (western juniper) in the drier, volcanic soils of the Cascade Mountains.

5. Southern Oak Woodlands

From about Ventura County southward, significant floristic changes occur in the California oak woodlands. There is not much change in the introduced species of forbs and grasses, but the native shrubs and herbs become more typical of southern California. Blue oak, foothill pine, California buckeye and valley oak, which are dominants in the foothill woodlands of central and northern California, do not extend into southern California. Only coast live oak and interior live oak occur in both communities. The dominant trees of the southern oak woodlands are *Quercus engelmannii* (Engelmann oak; Fig. 15-11A), *Q. agrifolia*, *Q. wislizenii*, and *Juglans californica* (California black walnut; Fig. 15-11B). Engelmann oak is a semideciduous white oak that is an ecological homologue of blue oak and replaces blue oak in southern California. Interior live oak usually occurs at higher elevations of the interior mountains of southern California and often is associated with rock outcrops. Coast live oak grows in moister sites, especially near the coast, but extends farther inland in southern California than it does in the remainder of its range. It often forms mixed stands with Engelmann oak in the foothills of the Peninsular ranges.

Understory composition in southern oak woodlands is similar to that of foothill woodlands in which grassland species predominate. In open stands there are scattered shrubs typical of southern California chaparral between the trees. These include *Rhus integrifolia* (lemonade berry), *Rhus ovata* (sugar sumac), *Rhus trilobata* (squaw bush) along with various species of *Ceanothus*, *Ribes*, and *Arctostaphylos*.

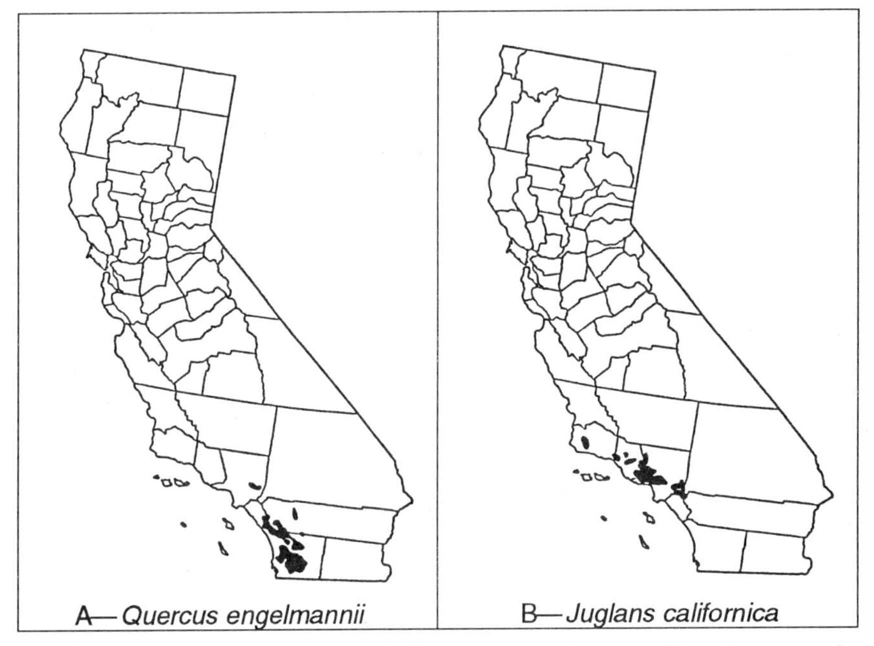

Fig. 15-11. Distribution of *Quercus engelmannii* (Engelmann oak) and *Juglans californica* (California black walnut). Maps redrawn from Griffin and Critchfield (1972).

Many areas of southern oak woodland have been destroyed by the expansion of urban areas in southern California. Formerly extensive stands in the Los Angeles area are now almost all gone.

6. Island Oak Woodlands

These communities (Fig. 15-12A) are only found on the islands offshore from southern California and Baja California. The dominant trees in upland sites are usually *Quercus tomentella* (island live oak) and *Lyonothamnus floribundus* (island ironwood; Fig. 15-12B), both of which are endemic to the islands. Island live oak is a close relative of canyon live oak (*Q. chrysolepis*). A few typical canyon live oaks occur on the islands, but most seem to be influenced by hybridization with the island live oak. *Quercus agrifolia* is also common on the islands, usually in canyons and on lower slopes, and forms open to dense woodland communities by itself or together with other species. The island oak woodland communities are closely associated with island chaparral, and the two communities share several species. Common associated small trees and shrubs are:

Fig. 15-12. A. Island oak woodlands on Santa Catalina Island. Dominants include *Quercus* ×*macdonaldii* (MacDonald oak) and *Q. berberidifolia* (scrub oak). **B.** Grove of *Lyonothamnus floribundus* (island ironwood) on Santa Catalina Island. Photos by David Keil.

Arctostaphylos insularis	island manzanita
Cercocarpus betuloides var. *blancheae*	island mountain-mahogany
Comarostaphylis diversifolia	summer-holly
Heteromeles arbutifolia var. *macrocarpa*	island toyon
Prunus ilicifolia ssp. *lyonii* [P. lyonii]	Catalina cherry
Quercus berberidifolia	scrub oak
Quercus xmacdonaldii	MacDonald oak
Rhamnus pirifolia [*R. crocea* ssp. *pirifolia*]	island redberry
Xylococcus bicolor	mission-manzanita

Island oak woodland communities occur on only the larger of the Channel Islands. The smaller of the islands have a more depauperate set of plant communities.

Human Impacts on Oak Woodland Communities

Human activities in California's oak woodlands have had several effects. These communities constitute a grazing resource second only to the California grasslands. Beginning with the colonization of California during the Mission Period (1769–1824) and especially during the last century, marked changes have occurred in the grasslands and oak woodlands of California, brought about by the introduction of domestic grazing animals and accompanying land management practices. One of the best known of these changes is the replacement of perennial bunch grasses by introduced annual grasses.

Less well known until recently is the lack of successful regeneration of many of the oaks (especially the valley oak and blue oak), evidenced by the absence of oak seedlings and saplings Fig. 15-12). Few areas in California can be found where any reproduction has occurred since about 1900. Even in some areas where small trees 5–10 cm in diameter are found, analysis of growth rings often shows the trees to be at least seventy years old.

A lack of viable acorn production is not the cause since even in poor acorn years there are probably plenty of acorns to assure regeneration. Nor is germination a problem. Blue oaks have a very high germination percentage; on occasion acorns will germinate while still on the tree. The problem arises after the acorn falls. At this point its chance of surviving and establishing a new tree approaches zero in most areas. On rangeland, acorns and seedlings are avidly eaten by both cattle and sheep; many of those that are not eaten are eliminated by trampling. In many areas of California the only surviving blue oak seedlings to be found are along road cuts and like places where no grazing occurs. But even in natural areas where no sheep and cattle have

Fig. 15-13. Young blue oak trees in ungrazed roadside in La Panza Mountains of San Luis Obispo County. In much of California few young blue oaks have become established in the past 100 years. Photo by Paula Kleintjes.

grazed for many years, an overabundance of deer causes the destruction of young blue oaks. Man has increased the food supply and habitat for deer by clearing forested areas and has exterminated their natural predators; some wildlife biologists believe that certain areas of the state have been overstocked with deer for the last thirty years and more.

Unfortunately, deer and livestock are not the only herbivores that have increased in numbers. When annual grasses replaced the perennial bunch grasses, there was a tremendous increase in overall annual seed production. The annual grasses produce many more seeds than do the perennials they replaced. As a consequence of the increase in available seeds, seed-eating animals such as ground squirrels and field mice, as well as birds such as finches and sparrows, have increased in numbers. Along with the increase of food supply for these animals, man has reduced the population of natural predators.

Other small animals, such as pocket gophers, have benefited from disturbances of the landscape and have responded with an increase in population. Pocket gophers are among the most significant predators on oak seedlings. Several species such as pocket gophers and ground squirrels have increased so markedly in the foothills in recent years that they are considered pests.

Man has responded by applying poisons to kill them. These poisons, however, create additional problems, killing non-target herbivorous animals and carnivores, some of which are already threatened by human activities.

The change of vegetation from perennial to annual grass may also increase the difficulties of blue oak survival. A seedling 100 to 200 years ago probably competed successfully with scattered bunch grasses and native annuals for nutrients and water. However, today's seedling must compete with much more aggressive introduced annual weedy species, and blue oak seedlings are not successful in this new competition. These aggressive, introduced species that now dominate the herbaceous cover in oak woodlands are a significant factor in reducing the survival rate of the oak trees.

No one factor has caused the decline in oak regeneration in the California foothills. Clearly the introduction of cattle and sheep grazing in California have resulted in major ecosystem changes in the foothill woodlands and grasslands. However, prior to the introduction of livestock in California, California grasslands and oak woodlands supported herds of grazing animals such as pronghorn antelope, tule elk and deer. Thus, some argue that domestic livestock grazing has just replaced the grazing by these native animals. However, patterns of grazing have changed. Native grazers passed through an area, heavily grazed it and left, allowing the vegetation to recover. In contrast, domestic livestock graze areas extensively throughout the year, and this pattern continues year after year. Recent work suggests that changing the grazing patterns of livestock to more closely follow those of historic native grazing animals may be a better land-use practice and may be less detrimental to oak regeneration.

In addition to the problems discussed above, many areas of oak woodlands have been cleared for agriculture, urban development or other human uses of the land. Oak trees, especially blue oak, have been eradicated in some areas for the purpose of increasing livestock rangeland forage, for use as cordwood, and to increase water flow from watersheds. Some investigators question the wisdom of such management practices. Recent studies have shown that blue oaks are beneficial trees that improve and enrich rangeland soils and increase both production and nutritional quality of rangeland forage. These trees also reduce erosion of steep slopes and provide essential habitat for numerous animal species. Many areas once covered by extensive oak woodlands now have few trees or none at all.

In addition to controversial land management programs, large residential tracts have been developed in the oak woodlands. Poor management during the construction phase of these developments has resulted in loss of large numbers of trees. Others have succumbed to mistreatment and improper care by homeowners. For example, many homeowners have irrigated the oaks during the summer when the soil is normally dry, and as a result the trees have died of root rot.

Although oak woodlands cover a large part of California at present, the outlook for these communities is not optimistic. As more and more trees die of natural causes or from human disturbance, they are not being replaced. A time may come when some of California's oaks become endangered.

Oak Woodland Communities—References

Allen-Diaz, B. H., and J. W. Bartolome. 1992. Survival of *Quercus douglasii* (Fagaceae) seedlings under the influence of fire and grazing. Madroño 39:47–53.

———, and B. A. Holtzman. Blue oak communities in California. Madroño 38:80–95.

Axelrod, D. I. 1983. Biogeography of oaks in the Arcto-Tertiary Province. Ann. Missouri Bot. Gard. 70:629–657.

Baker, G. A., P. W. Rundel, and D. J. Parsons. 1981. Ecological relationships of *Quercus douglasii* (Fagaceae) in the foothill zone of Sequoia National Park, California. Madroño 28:1–12.

Barbour, M. G. 1988. California upland forests. Pp. 131–164 *in* M. G. Barbour and W. D. Billings (eds.), North American Terrestrial Vegetation. Cambridge Univ. Press, Cambridge.

Borchert, M., F. W. Davis, J. Michaelsen, and L. D. Oyler. 1989. Interactions of factors affecting seedling recruitment of blue oak (*Quercus douglasii*) in California. Ecology 70:389–404.

———, ———, and B. Allen-Diaz. 1991. Environmental relationships of herbs in blue oak (*Quercus douglasii*) woodlands of central coastal California. Madroño 38:249–266.

———, N. D. Cunha, P. C. Krosse, and M. L. Lawrence. 1993. Blue oak plant communities of southern San Luis Obispo and northern Santa Barbara counties, California. U.S.D.A., Pac. S.W. Res. Sta. Gen. Tech. Rep. PSW-GTR-139. 49 pp.

Brown, D. E. 1982. Californian evergreen forest and woodland. Pp. 66–69 *in* D. E. Brown (ed.). Biotic communities of the American Southwest — United States and Mexico. Desert Plants 4:1–341.

Callaway, R. M. 1992. Effect of shrubs on recruitment of *Quercus douglasii* and *Quercus lobata* in California. Ecology 73:2119–2128.

———, and C. M. D'Antonio. 1991. Shrub facilitation of coast live oak establishment in Central California. Madroño 38:158–169.

———, and D. W. Davis. 1993. Vegetation dynamics, fire, and the physical environment in coastal central California. Ecology 74:1567–1578.

———, N. M. Nadkarni, and B. E. Mahall. 1991. Facilitation and interference of *Quercus douglasii* on understory productivity in Central California. Ecology 72:1484–1499.

Cooper, W. S. 1926. Vegetational development upon alluvial fans in the vicinity of Palo Alto, California. Ecology 7:1–30.

Gordon, D. R., and K. J. Rice. 1993. Competitive effects of grassland annuals on soil water and blue oak (*Quercus douglasii*) seedlings. Ecology 74:68–82.

Graves, G. W. 1932. Ecological relationships of *Pinus sabiniana*. Bot. Gaz. 94:106–133.

Griffin, J. R. 1971. Oak regeneration in the upper Carmel Valley, California. Ecology 52:862–868.

———. 1976. Regeneration in *Quercus lobata* savannas, Santa Lucia Mountains, California. Amer. Midl. Naturalist 95:422–435.

———. 1977. Oak woodland. Pp. 383–415 *in* M. G. Barbour and J. Major (eds.). Terrestrial vegetation of California. John Wiley and Sons, N.Y.

———, and W. B. Critchfield. 1972. The distribution of forest trees in California. Pacific S.W. Forest and Range Exp. Stn., Berkeley, Calif. 114 pp. (U.S.D.A. Serv. Res. Paper PSW 82).

Haggerty, P. K. 1994. Damage and recovery in southern Sierra Nevada foothill oak woodland after a severe ground fire. Madroño 41:185–198.

Holland, V. L. 1973. A study of soil and vegetation under *Quercus douglasii* H. & A. compared to open grassland. Ph.D. dissertation, Univ. Calif., Berkeley. 369 p.

———. 1976. In defense of blue oaks. Fremontia 4:3–8.

———. 1986. Coastal Oak Woodland *in* Guide to Wildlife Habitats of California. CDF-Forest and Rangeland Resources Assessment Program Publication. In press.

———. 1980. Effect of blue oak on rangeland forage production in central California. Pp. 314–318 *in* Plumb, T.R. (tech. coord.), Proceedings of symposium on ecology, management and utilization of California oaks. Pacific S.W. Forest and Range Exp. Stn., Berkeley.

———. and Jimmy Morton. 1980. Effect of blue oak on nutritional quality of rangeland forage in central California. Pp. 319–322 *in* Plumb, T.R. (tech. coord.), Proceedings of symposium on ecology, management and utilization of California oaks. Pacific S.W. Forest and Range Exp. Stn., Berkeley.

Hanes, T. L. 1976. Vegetation types of the San Gabriel Mountains. Pp. 65–76 *in* J. Latting (ed.). Plant communities of southern California. Calif. Native Pl. Soc. Spec. Publ. 2. Pacific Horticulture 43(1):13–17.

Johnston, V. R. 1970. Sierra Nevada. Houghton, Mifflin Co., Boston.

———. 1994. California forests and woodlands. A natural history. Univ. California Press, Berkeley & Los Angeles.

Keeley. 1988. Bibliographies on chaparral and the fire ecology of other Mediterranean systems, 2nd ed. Calif. Water Resource Cntr., Univ. Calif. Davis, Rep. No. 69. 328 pp.

———. 1990. Demographic structure of California black walnut (*Juglans californica*: Juglandaceae) woodlands in southern California. Madroño 37:237–248.

Kylver, F. D. 1931. Major plant communities in a transect of the Sierra Nevada Mountains of California. Ecology 12:1–17.

Little, E. L. 1976. Atlas of United States trees. Vol. 3. Minor western hardwoods. U.S.D.A. Forest Service Misc. Publ. 1314. 13 pp. + 210 maps.

Marañon, T., and J. W. Bartolome. 1994. Coast live oak (*Quercus agrifolia*) effects on grassland biomass and diversity, Madroño 41:39–52.

McClaran, M. P., and J. W. Bartolome. 1989. Effect of *Quercus douglasii* (Fagaceae) on herbaceous understory along a rainfall gradient. Madroño 36:141–153.

Mensing, S. A. 1992. The impact of European settlement on blue oak (*Quercus douglasii*) regeneration and recruitment in the Tehachapi Mountains, California. Madroño 39:36–46.

Minnich, R. A. 1976. Vegetation of the San Bernardino Mountains. Pp. 99–124 *in* J. Latting (ed.). Plant communities of southern California. Calif. Native Plant Soc. Spec. Publ. 2.

———. 1980. Wildfire and the geographic relationships between canyon live oak, Coulter pine, and bigcone Douglas-fir forests. Pp. 55–61 *in* T. R. Plumb (tech. coord.), Proceedings of the symposium on the Ecology, management, and utilization of California oaks. Pacific S.W. Forest and Range Exp. Stn., Berkeley.

Mullally, D. P. 1992. Distribution and environmental relations of California black walnut (*Juglans californica*) in the eastern Santa Susana Mountains, Los Angeles County. Crossosoma 18(2):1–18.

Parker, V. T., and C. H. Muller. 1982. Vegetation and environmental change beneath isolated live oak trees (*Quercus agrifolia*) in a California annual grassland. Amer. Midl. Naturalist 107:69–81.

Parsons, D. J. 1981. The historical role of fire in the foothill communities of Sequoia National Park. Madroño 28:111–120.

Philbrick, R. N., and J. R. Haller. 1977. The southern California islands. Pp. 893–906 *in* M. G. Barbour and J. Major (eds.). Terrestrial vegetation of California. John Wiley and Sons, N.Y.

Plumb, T. R. (tech. coord.). 1980. Proceedings of the symposium on the Ecology, management, and utilization of California oaks. Pacific S.W. Forest and Range Exp. Stn., Berkeley. 368 pp.

———. 1980. Response of oaks to fire. Pp. 202–215 *in* T. R. Plumb (tech. coord.), Proceedings of the symposium on the Ecology, management, and utilization of California oaks. Pacific S.W. Forest and Range Exp. Stn., Berkeley.

Quinn, R. D.. 1990. The status of walnut forests and woodlands (*Juglans californica*) in southern California. Pp. 42–54 in A. A. Schoenherr (ed.), Endangered plant communities of southern California. Proceedings of the 15th Annual Symposium. Southern California Botanists spec. publ. 3.

Rundel, P. W. 1980. Adaptations of Mediterranean-climate oaks to environmental stress. Pp. 43–54 *in* T. R. Plumb (tech. coord.), Proceedings of the symposium on the Ecology, management, and utilization of California oaks. Pacific S.W. Forest and Range Exp. Stn., Berkeley.

Saenz, L. and J. O. Sawyer. 1986. Grasslands as compared to adjacent *Quercus garryana* woodland understories exposed to different grazing regimes. Madroño 33:40–46.

Smith, F. E. 1982. The changing face of the San Joaquin Valley. Fremontia 10(1):24–27.

Snow, G. E. 1980. The fire resistance of Engelmann and coast live oak seedlings. Pp. 62–66 *in* T. R. Plumb (tech. coord.), Proceedings of the symposium on the Ecology, management, and utilization of California oaks. Pacific S.W. Forest and Range Exp. Stn., Berkeley.

Sugihara. N. G., L. J. Reed, and J. M. Lenihan. 1987. Vegetation of the bald hills oak woodlands, Redwood National Park, California. Madroño 34:193–208.

Thilenius, J. F. 1968. The *Quercus garryana* forests of the Willamette Valley, Oregon. Ecology 49:1124–1133.

Vasek, F. C. 1966. The distribution and taxonomy of three western junipers. Brittonia 18:350–372.

Waring, R. H. 1969. Forest plants of the eastern Siskiyous: their environmental and vegetational distribution. Northwest Sci. 43:1–17.

Wells, P. V. 1962. Vegetation in relation to geological substratum and fire in the San Luis Obispo quadrangle, California. Ecol. Monogr. 32:79–103.

White, K. L. 1966. Structure and composition of foothill woodland in central coastal California. Ecology 47:229–237.

Whitney, S. 1979. A Sierra Club naturalist's guide to the Sierra Nevada. Sierra Club Books, San Francisco.

Whittaker, R. H. 1960. Vegetation of the Siskiyou Mountains, Oregon and California. Ecol. Monogr. 30:279–338.

Wolfe, J. A. 1980. Neogene history of California oaks. Pp. 3–6 *in* T. R. Plumb (tech. coord.), Proceedings of the symposium on the Ecology, management, and utilization of California oaks. Pacific S.W. Forest and Range Exp. Stn., Berkeley.

Chapter 16

Montane Coniferous Forest Communities

A diversity of coniferous forest communities occurs in montane regions of California (Fig. 16-1). These forests occur above about 400 m (1200 ft) in northern California, 600 m (2000 ft) in central California, 1500 m (5000 ft) in southern California and 1800 m (6000 ft) in the desert mountains of transmontane California. Montane coniferous forests cover a large portion of California, occupying about 20 percent of the state's land area. Montane forests are complex and change in composition and structure from north to south, from west to east, and from low to high elevations. Along these gradients, various environmental features such as precipitation, temperature, frost, soils, wind, frequency of fire, and levels of solar radiation change. Forest vegetation changes along with these environmental variables.

Montane coniferous forests are best developed on deep, well-drained soils. These soils result from weathering of bedrock below and decomposition of litter from the plants above. Conifer needles make up the bulk of the litter. Topography and elevation greatly affect the rate of soil formation. Steep slopes generally have shallow rocky soils. Broad montane valleys are areas where deep soils have accumulated. Soil moisture levels in valley areas may be so high that trees are excluded and meadow communities develop. At extremely high elevations, (in subalpine and alpine zones) soils are often very shallow and rocky. Here plant growth is slow, litter deposition is sparse, and decomposition is inhibited by cold temperatures. Chemical weathering is slow, although mechanical weathering of soil by alternate freezing and thawing may be significant. Such soils have low water- and nutrient-holding capacities.

Soil fungi are the primary decomposers in the coniferous forest ecosystem. Chemicals released from decaying litter of conifers tend to be acidic which inhibits the growth of bacteria, the common decomposers in many ecosystems. Mycorrhizal fungi are also critical in California's coniferous forests. Many trees

Fig. 16-1. Distribution of montane coniferous forest communities in California.

could not survive without mycorrhizal fungi growing in association with their roots. The vast network of hyphae that radiate out from the mycorrhizal association permeates the soil in a coniferous forest and serves as an absorbing surface for nutrients and water transmitted to the trees. These hyphae are much more efficient at taking up water and nutrients from infertile mountain soils than roots of conifers by themselves.

Classification of montane coniferous forest communities is not easy. These communities are variable from site to site and from region to region. They also vary along elevational gradients. We have chosen to group the montane coniferous forests into four major community types. These community types more or less correspond to elevational zones. Although they usually can be recognized, their shared boundaries (ecotones) are seldom distinct, and they overlap and integrate with each other. The groups that we have recognized are listed below. **(1) Mixed coniferous forests** dominate lower montane zones ranging in elevation from 400 to 1800 m (1200 to 6000 ft) in northern California and from 1200 to 2150 m (4000 to 8000 ft) in southern California. **(2) Red fir forests** dominate upper montane zones on deep, well-drained soils from 1800 to 2150 m (6000 to 7000 ft) elevation in the north and from 2450 to 2700 m (8000 to 9000 ft) in the southern Sierra Nevada, but are absent from the southern California mountains.. **(3) Lodgepole pine forests** are typical of glacial basins in the lower subalpine zone, forming open stands at elevations from 1800 to 2450 m (6000 to 8000 ft) in the north and from 2450 to 3350 m (8000 to 10,000 ft) in the south. **(4) Subalpine forests** consist of dwarfed or shrubby trees occupying rocky slopes above about 2450 m (8000 ft) in the north and 2900 m (9400 ft) in the south. Above the subalpine zone is the treeless alpine zone (see Chapter 17).

Low- and middle-elevation montane coniferous forests (montane mixed coniferous forests and red fir forests) are generally dominated by tall coniferous trees ranging from 75 to 200 ft (Fig. 16-2). Above 2450 m (8000 ft) trees generally become progressively shorter.

Understory vegetation consists of a diversity of shrubs, herbs, and, in some areas, smaller trees. The understory may be locally dense and multilayered or very sparse. Species composition of the understory varies with latitude, locality and elevation. Understory composition in cismontane montane coniferous forests is distinctly different from that of transmontane forests. In transmontane coniferous forests, high desert plants often form a significant portion of the understory.

As a general rule, coniferous forest communities on transmontane slopes occupy higher elevation sites than do comparable

Fig. 16-2. Montane mixed coniferous forest. A giant sugar pine (*Pinus lambertiana*) towers over a forest of white fir (*Abies concolor*), incense cedar (*Calocedrus decurrens*), and ponderosa pine (*Pinus ponderosa*). Photo by V. L. Holland.

communities on cismontane slopes. Transmontane forest communities are "telescoped" together in upper elevation zones. Slopes at elevations that in cismontane areas would be vegetated by montane mixed coniferous forests are occupied in transmontane areas by desert woodland or desert scrub communities. One or more zones may be completely absent on a particular transmontane slope.

Montane coniferous forests are bordered by a variety of other communities. In the northern part of the state these forests grade into coastal coniferous and mixed evergreen forests near the coast and foothill woodlands to the east. The Cascades-Sierra Nevada forests occur just above the foothill woodlands and the Sierran mixed hardwood forests. In southern California montane coniferous forests may be bordered directly by chaparral communities. On transmontane slopes piñon pine and juniper woodlands or high desert communities border or intergrade with the coniferous forests. Ecotones between montane forests and these other communities may be broad or very abrupt. In general montane coniferous forests occur at higher elevations than do other forest communities of California but local topographic and edaphic factors sometimes make ecotonal transitions and zonation very complicated. Moist shaded canyons, for instance, may be occupied by coniferous forest species with chaparral or woodland communities on adjacent higher elevation sites.

1. Montane Mixed Coniferous Forests

Montane mixed coniferous forest communities cover about 10% of California. They are best developed in cismontane California, where they occur from 1000 to 2300 m (3000 to 7000 ft) in the Klamath Mountains, the Cascades and the Sierra Nevada and at higher elevations (2000 to 2700 m; 6000 to 8000 ft) in the Transverse and Peninsular Ranges of southern California. These forest communities are fairly well developed in the North Coast Ranges but are only of scattered occurrence from Sonoma County to Santa Barbara County in the South Coast Range.

Montane mixed coniferous forests also occur on the east side of the Sierra Nevada-Cascade axis in transmontane California but are not as well developed as on the western slope and often grade rather abruptly into high desert communities. Transmontane mixed coniferous forests are restricted to high elevations (2000 to 2700 m; 6000 to 8000 ft) except along streams where they may occur at lower elevations. Species diversity is lower than on the western mountain slopes, and forest trees often have high desert species as an understory. Isolated montane coniferous forest communities also occur in some desert mountain ranges, such as the Clark, Kingston, and New York Mountains, but they are absent from other desert ranges such as the White Mountains (Fig. 16-22).

Montane mixed coniferous forests are extremely variable. Community composition and habitat vary from north to south, from east to west, and from low to high elevations. Variations in topography, geology and site history play important parts in determining community structure. Precipitation varies in quantity, quality ,and seasonality. Temperature conditions range from warm to rather cold. Duration of the growing season varies

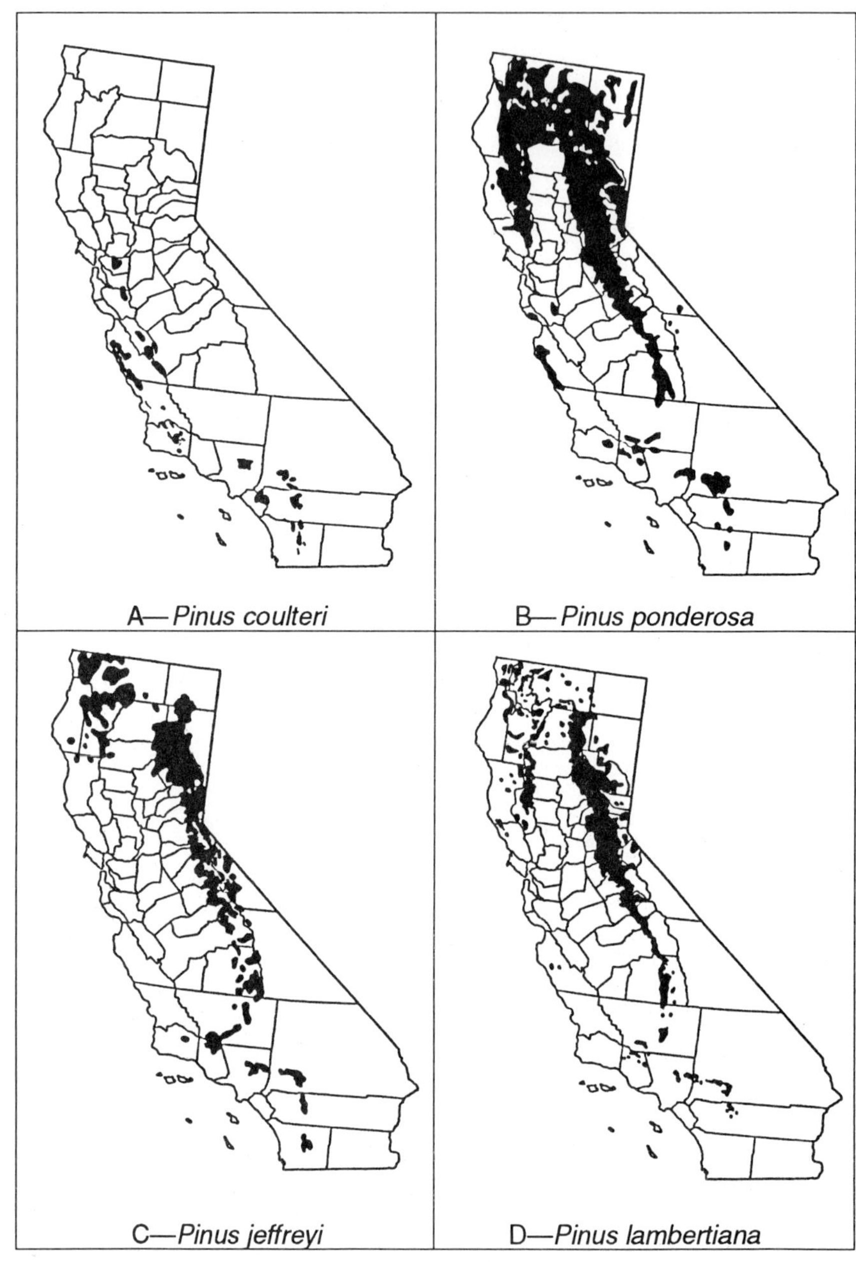

Fig. 16-3. Distribution of *Pinus coulteri* (Coulter pine), *Pinus ponderosa* (ponderosa pine), *Pinus jeffreyi* (Jeffrey pine), and *Pinus lambertiana* (sugar pine). Modified from Griffin and Critchfield (1972).

with latitude, elevation, topography, and depth of winter snowpack. In some situations the forests are composed of a mixture of conifers and hardwoods, sometimes with several co-dominants. Elsewhere one species may occur in almost pure stands with few other tree species in association. The forests also vary from being rather open with a park-like appearance to forming a dense closed canopy. The understory varies from lush and diverse to quite sparse.

Because montane mixed coniferous forests are so variable we have chosen to recognize several phases. These are characterized both by dominant species and by habitat characteristics. The forests grade into one another forming an almost unlimited number of combinations, and no two stands are ever exactly alike in all their features. However, the grouping of the diverse forests can provide some insight into the complexities of the forests and of recognizable patterns in all the diversity. Ecological relationships among the various groups can also be examined.

The communities that we are treating here as montane mixed coniferous forests are sometimes grouped together as "yellow pine forests". The name, yellow pine, is mostly used in California for *Pinus ponderosa* (also known as ponderosa pine) although *Pinus coulteri* (Coulter pine, big-cone pine) and *Pinus jeffreyi* (Jeffrey pine) are also yellow pines. The name, "yellow pine forest," is often used for a variety of forest communities in which yellow pines grow together with other conifers, and sometimes even for communities in which the pines are absent. We prefer to recognize the latter communities as phases of the montane mixed coniferous forests. Although communities dominated by Coulter pine, ponderosa pine, and Jeffrey pine are physiognomically similar, we are treating them separately because of the significant ecological differences among them. Therefore, we do not use the name "yellow pine forest".

A. Coulter Pine Forests

Pinus coulteri (Coulter pine, big-cone pine) ranges from the Peninsular Ranges of northern Baja California through southern California across the Transverse Ranges and up the South Coast Ranges to the San Francisco Bay area (Fig. 16-3A). It does not occur at all in the Sierra Nevada or in the northern half of the state.

Coulter pine forest communities are difficult to characterize. As a dominant tree species, *Pinus coulteri*, grows in a diversity of habitats, some of which are clearly outside the scope of the montane mixed coniferous forests, some of which are a part of these forests, and many of which are in one way or another transitional. Its affiliation with the forests of the South Coast

Ranges is variable and complex. It is a common component of the scattered stands of montane mixed coniferous forest that occur along the Central Coast where it grows with *Pinus ponderosa*, and *Pinus lambertiana* (sugar pine). It often occurs at rather low elevations along the Central Coast (230 to 1000 m; 700 to 3000 ft) and is frequently a part of mixed evergreen forests (Chapter 14). This species has wide range of tolerance to edaphic conditions (e.g., in San Luis Obispo County it occurs on serpentine soils in association with *Cupressus sargentii* [Sargent cypress] and on soils derived from Monterey shale with *Pinus attenuata* [knobcone pine]). The trees are sometimes so widely spaced that they form a woodland rather than a forest community. It sometimes extends into oak woodlands in association with or in place of *Pinus sabiniana* (foothill pine).

In southern California, where it is most abundant, *Pinus coulteri* is usually found at elevations above 1000 m (3000 ft). In the Peninsular Ranges and parts of the Transverse Ranges, Coulter pine forests are often the lowest zone of montane mixed coniferous forests and grade into chaparral or southern oak woodland at lower elevations and well-developed montane mixed coniferous forest at higher elevations.

Pinus coulteri usually forms an open forest with an understory of chaparral shrubs and/or pineland herbaceous species. Common associated shrubs are manzanitas (*Arctostaphylos glandulosa*, *A. pringlei* ssp. *drupacea*, *A. pungens*), *Ceanothus integerrimus* (deerbrush) and other *Ceanothus* species, and *Cercocarpus betuloides* (mountain-mahogany). At higher elevations it may be associated with ponderosa pine, Jeffrey pine, *Quercus kelloggii* (black oak), *Q. chrysolepis* (canyon live oak) and other species typical of montane mixed coniferous forests. In hot, dry sites, it may form a depauperate open forest in association with only black oak or canyon live oak. Stands of Coulter pine often grow near stands of *Pseudotsuga macrocarpa* (big-cone Douglas-fir), but the two seldom occur in direct association. Coulter pine tends to grow on dry hillsides and ridges, whereas big-cone Douglas-fir more often occurs in mesic canyons.

B. Ponderosa Pine Forests

In terms of abundance and range, *Pinus ponderosa* is the dominant tree of western North America. It occurs from the Pacific Coast to the Rocky Mountains and from southwestern Canada to northern Mexico. Within California, ponderosa pine is widespread in lower and middle elevation forests (Fig. 16-3B) where it dominates warm xeric sites in the lower montane zone (Fig. 16–4) just above foothill woodland or chaparral communities. These warm dry areas range from about 600 to 1800 m (2000 to

Fig. 16-4. Park-like stand of ponderosa pines in the Sierra Nevada. Photo by V. L. Holland.

6000 ft) in northern California and 1200 to 2150 m (4000 to 7000 ft) in southern California. In canyon areas it may extend as individual scattered trees or narrow bands into foothill woodland or chaparral communities.

Common associates of *Pinus ponderosa* in lower elevation montane coniferous forests are *Quercus kelloggii* (black oak; Fig. 16-5A) and *Calocedrus decurrens* [*Libocedrus d.*] (incense cedar; Fig. 16-5D). Understory vegetation is quite variable both in its composition and its percent ground cover. Chaparral shrubs are often scattered in ponderosa pine forests, and these shrubs may form well-developed chaparral communities on slopes next to the pine forests. In various areas, both in coastal and interior mountains, mixed hardwood forests intergrade with ponderosa pine forests. Herb cover is usually low but diverse in its composition.

Ponderosa pine often grows in almost pure stands. It is a valuable timber tree, and forests are often managed to promote its growth. It requires full sun for establishment and growth. Older stands are often invaded by shade-tolerant species such as *Abies concolor* (white fir), a species with less commercial value. Removal of the firs and thinning of the young pines are both used to increase the timber yield.

Fire suppression in ponderosa pine forests also leads to the decline of *P. ponderosa* from dominance. Failure to understand the interrelationship of fire and reproduction of ponderosa pine has led to major successional changes in some of California's forests. Increasing maturity and shade also leads to a change in the understory structure, particularly to a decrease in the shrub cover.

On cismontane slopes *Pinus ponderosa* is replaced above 1500 m (5000 ft) in the north and 1800 m (6000 ft) in the south by *Pinus jeffreyi*. Jeffrey pine also replaces ponderosa pine on the transmontane side of the Sierra from Carson Pass south and in the Transverse and Peninsular Ranges of southern California. In mesic sites at middle and high elevations, ponderosa pine forests grade into more diverse montane mixed coniferous forests. *Pinus lambertiana* (sugar pine), *Pseudotsuga menziesii* (Douglas-fir) and *Abies concolor* (white fir) join ponderosa pine as important components of the forest.

C. Jeffrey Pine Forests

In many areas of California *Pinus jeffreyi* (Fig. 16-3C) dominates montane coniferous forests, sometimes occurring in almost pure stands. Unlike its close relative ponderosa pine, Jeffrey pine is essentially a California species: it ranges only a short distance beyond the state's borders into the mountains of northern Baja California, the Klamath Mountains of southern Oregon, and the Sierra Nevada of extreme western Nevada. It is closely related to and resembles ponderosa pine but is more tolerant of drought, low temperatures and deep snow. As a result Jeffrey pine often replaces ponderosa pine on transmontane mountain slopes in both northern and southern California and at higher elevations on western (cismontane) slopes from 1520 to 1830 m (4600–5500 ft) in the north and 2130 to 2740 m (6400–8300 ft) in the south.

On cismontane slopes Jeffrey pine is often a component of upper montane mixed coniferous forests with incense cedar, black oak, white fir, sugar pine and often ponderosa pine. Western slopes dominated by Jeffrey pine are largely confined to warm, dry regions of the southern Sierra (Kern Plateau), and to the Transverse and Peninsular Ranges of southern California. In these areas, Jeffrey pine often forms relatively pure stands between 2590 and 2740 m (6800 and 8300 ft). On the western slope of the northern Sierra, Jeffrey pines usually occur on dry, rocky exposed sites where they are sometimes stunted and twisted by adverse soil and moisture conditions and exposure to wind. Krummholz (contorted) forms of Jeffrey pine occur in some subalpine forests. Under favorable conditions and on deep forest soils Jeffrey pines are large trees ranging from 18 to 55 m (60 to 180 ft) in height and 1.2 to 2.1 m (4 to 7 ft) in trunk diameter.

Jeffrey pine declines in importance in the northern part of California. It is generally not well represented on the Modoc Plateau, in the Klamath Mountains, north of the Pit River in the Cascade Mountains, or in the North Coast Ranges. In the northern part of its range, Jeffrey pine occurs mainly on serpentine soils. Its extension into the Klamath Mountains of Oregon is almost exclusively on serpentine soils. *Pinus jeffreyi* is much more tolerant of this substrate than is its close relative, ponderosa pine.

On the east side of the Sierra, *Pinus jeffreyi* is the dominant yellow pine, often to the exclusion of *Pinus ponderosa*. In the north, it is often mixed with *Pinus ponderosa*, *Abies concolor* and *Juniperus occidentalis* (western juniper), but from the Lake Tahoe Basin (where it reaches its maximum development) southward, it usually dominates forests below 2450 m (8000 ft) and often occurs in pure stands. In these areas, Jeffrey pine forests often form a distinct zone between red fir forests at higher elevations and piñon pine-juniper woodlands or Great Basin sagebrush scrub at lower elevations (Fig. 16-6C).

In southern California Jeffrey pine increases in importance in montane mixed coniferous forests, and it replaces ponderosa pine, particularly in the interior regions. It is usually the only yellow pine on the transmontane slopes of the southern California mountains. In spite of their ecological differences, Jeffrey pine and ponderosa pine are very similar in appearance, overlap in many areas, and occasionally hybridize.

Understory vegetation of Jeffrey pine forests is extremely variable depending of geographical location and habitat features. Common understory shrubs in cismontane Jeffrey pine forests include:

Arctostaphylos patula	greenleaf manzanita
Ceanothus cordulatus	mountain whitethornsnowbush
Ceanothus prostratus	prostrate ceanothus
Ceanothus velutinus	velvet-leaf ceanothus
Chamaebatia foliolosa	mountain-misery
Chrysolepis sempervirens [*Castanopsis s.*]	mountain chinquapin
Eriogonum wrightii	buckwheat
Lupinus elatus	lupine
Prunus virginiana var. *demissa*	western choke cherry
Symphoricarpos parishii	snowberry

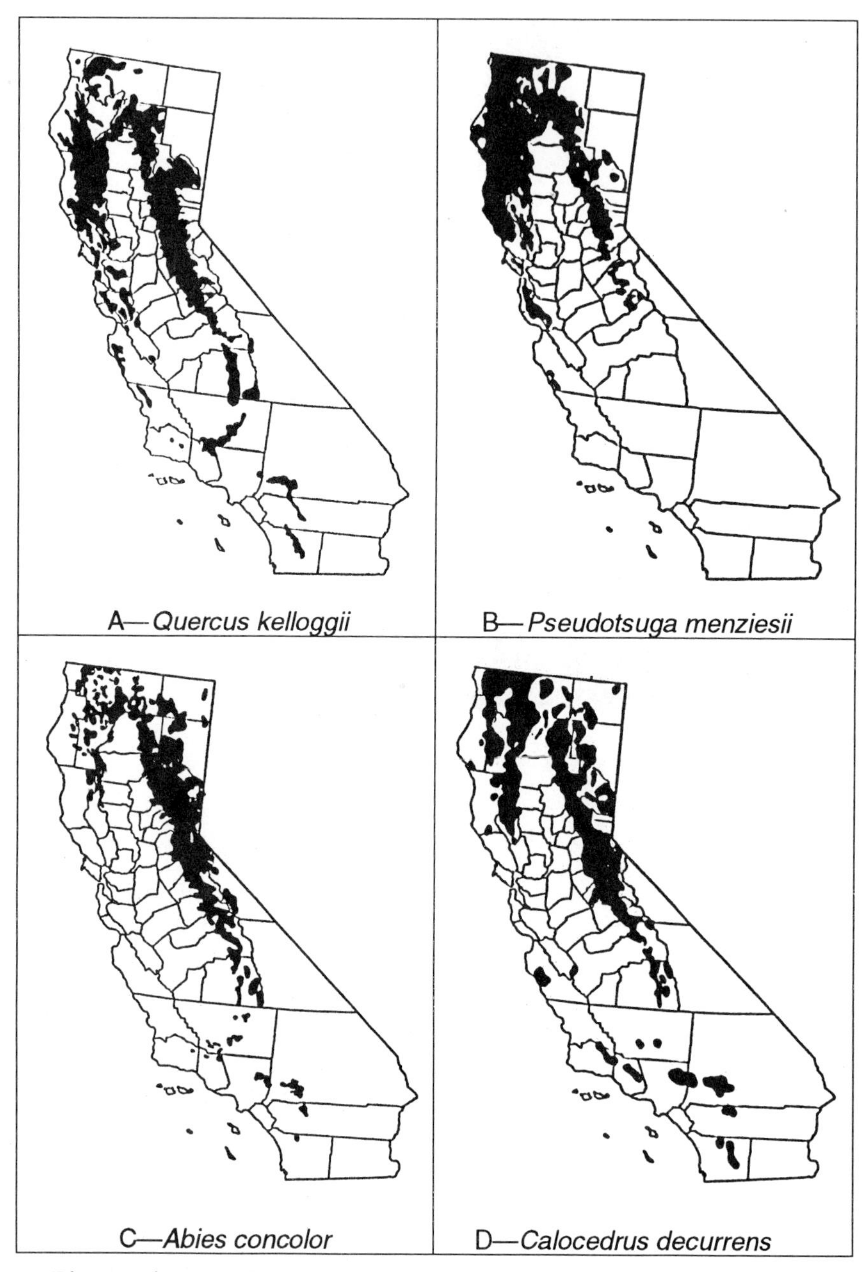

Fig. 16-5. Distribution of *Quercus kelloggii* (black oak), *Pseudotsuga menziesii* (Douglas-fir), *Abies concolor* (white fir), and *Calocedrus decurrens* (incense cedar) in California. Modified from Griffin and Critchfield (1972).

Fig. 16-6. Montane mixed coniferous forests. **A.** Mixed coniferous forest in Sierra Nevada of Tulare County. Dominant trees include ponderosa pine and white fir. The understory is a dense growth of mountain misery. **B.** Mixed coniferous forest in the Sierra Nevada of Tulare County. Trees include ponderosa pine, white fir, and sugar pine. The understory is a thorny growth of mountain whitethorn. **C.** Forest of Jeffrey pine in Mono County. The understory is dominated by Great Basin sagebrush. A, B, photos by David Keil; C, photo by V. L. Holland.

Some of these species also extend into transmontane California where they are joined by understory species from the deserts:

Artemisia tridentata	Great Basin sagebrush
Cercocarpus ledifolius	curl-leaf mountain-mahogany
Chrysothamnus nauseosus	rabbitbush
Chrysothamnus parryi	Parry's rabbitbush
Leptodactylon pungens	granite-gilia
Prunus emarginata	bitter cherry
Purshia tridentata	bitterbrush
Tetradymia canescens	horsebrush

D. Mixed Conifer Forests

Montane mixed coniferous forests often are composed of a mixture of 4 to 6 coniferous tree species, often with one or more hardwoods (Figs. 16-2, 16-6A, B). *Pinus ponderosa* (yellow pine, ponderosa pine, Fig. 16-3B) and *Abies concolor* (white fir; Fig. 16-5C) dominate most western slope stands, the former at lower elevations on warm, dry sites, the latter on higher, cooler, moister ones. *Calocedrus decurrens* (incense cedar, Fig. 16-5D), *Pinus lambertiana* (sugar pine, Fig. 16-3D), and *Quercus kelloggii* (black oak, Fig. 16-5A) are important associates in most stands. All five of these species extend into the mountains of southern Oregon, but well-developed stands containing all of them are only found in California and are best represented in the Sierra Nevada. *Pinus jeffreyi* (Jeffrey pine; Fig. 16-3C) is also present in many western slope stands and dominates some. It is sometimes dominant between 2135 and 2750 m (6400 and 8250 ft) on both cismontane and transmontane slopes and is much more abundant in southern California where it is usually the only yellow pine on the transmontane slopes.

Pseudotsuga menziesii (Douglas-fir; Fig. 16-5B) is an important component of mixed coniferous forests in mesic sites in northern California in the Cascades, the Klamath Mountains, the northern part of the Sierra Nevada and the North Coast Ranges. Douglas-fir is also a dominant component of coastal redwood forest communities (Chapter 13) and of mixed evergreen forest communities (Chapter 14). Although Douglas-fir is widespread in western mountain ranges from southern Canada to Mexico, in California it does not occur south of Fresno County in the Sierra Nevada and is absent from the mountains of southern California.

Four other coniferous trees are of localized significance in montane mixed coniferous forests. *Pinus attenuata* (knobcone pine; Fig. 16-7B) is a closed-cone pine that occurs in arid coniferous forests, often on poor soils, from the Transverse Range to northern California. In some areas it forms stands by itself (discussed in Chapter 12), and in others it occurs in the lower elevation portions of the mixed coniferous forests. *Pseudotsuga macrocarpa* (big-cone Douglas-fir; Fig. 16-7C) occurs in lower elevation portions of mixed coniferous forests in southern California. *Pinus coulteri* (Fig. 16-3A) is also sometimes a component of these communities as well, especially in southern California. *Sequoiadendron giganteum* (giant sequoia; Fig. 16-7A) is restricted to the Sierra Nevada where it dominates some stands.

Black oak is the most common hardwood in montane mixed coniferous forests, ranging from 300 to 2450 m (1000 to 8000 ft) on western mountain slopes. It is most common on warm dry sites where it commonly occurs with ponderosa pine. *Quercus chrysolepis* (canyon live oak) is the other common oak in these communities and ranges up to 2450 m (8000 ft). As its name implies, it is most common on canyon slopes and rocky stream banks where is sometimes forms almost pure stands. Douglas-fir is a common associate in these mesic canyon habitats. *Acer macrophyllum* (big leaf maple), *Cornus nuttallii* (mountain dogwood), *Arbutus menziesii* (madrone) and *Umbellularia californica* (California bay-laurel) often occur in these communities especially in northern California. All four species extend into and become significant components of the mixed evergreen forests of California especially in the north. Most of the other hardwood tree species found within montane mixed coniferous forests are associated with riparian habitats. These include:

Acer macrophyllum	big-leaf maple
Alnus incana ssp. *tenuifolia* [*A. tenuifolia*]	mountain alder
Alnus rhombifolia	white alder
Populus balsamifera ssp. *trichocarpa* [*P. trichocarpa*]	black cottonwood
Populus tremuloides	quaking aspen
Betula occidentalis	water birch
Salix spp.	willows

Understory shrubs and herbs are extremely variable and diverse. Some are restricted to montane coniferous forests, and some are widespread in a variety of different communities.

Among the more common understory shrubs of cismontane slopes include:

Amelanchier alnifolia	serviceberry
Arctostaphylos spp.	manzanita
Ceanothus spp.	ceanothus
Chamaebatia foliolosa (Fig. 16-6A)	mountain-misery
Chrysolepis sempervirens [*Castanopsis s.*]	mountain chinquapin
Corylus cornuta	hazelnut

Prunus emarginata	bitter cherry
Rhamnus ruber	Sierra coffeeberry
Ribes spp.	gooseberries and currants
Salix spp.	willows
Symphoricarpos spp.	snowberries

Common understory shrubs of the eastern (transmontane) slopes include:

Artemisia tridentata	Great Basin sagebrush
Ceanothus prostratus	prostrate ceanothus
Cercocarpus ledifolius	curl-leaf mountain mahogany
Chrysothamnus spp.	rabbitbush
Eriogonum spp.	buckwheats
Leptodactylon pungens	granite-gilia
Purshia tridentata	bitterbrush

In addition to the understory of shrubs, there is also a diversity of herbaceous species. Herb composition is extremely variable depending on location in California as well as type of microhabitat conditions (e.g., shady forests, open forests, damp forests, dry forests). Herbaceous understory composition on western slopes differs markedly from that on eastern slopes. The western slopes have cismontane species common to the California Floristic Province, whereas the eastern slopes have mostly species with affinities for the Great Basin Floristic Province.

E. Giant Sequoia Forests

Sequoiadendron giganteum (big tree, giant sequoia) is an endemic California species restricted to seventy-five groves extending along the west slope of the Sierra Nevada from Tulare to Placer Counties (Fig. 16-7A). These groves occur on mesic, unglaciated ridges between 1370 and 2560 m (4500 to 8400 ft) and are closely associated with white fir forests. Giant sequoia forests are treated separately from the white fir forests because giant sequoia is usually a dominant tree where it grows and because special public attention has been given to this monarch tree species over the years. The majority of giant sequoia groves occur in Tulare County and southern Fresno County. The largest groves in Sequoia and Kings Canyon National Parks cover approximately 1000 hectares (2500 acres) each and have about 20,000 mature giant sequoias apiece. The eight groves north of the Kings River (Fresno County) are much smaller with the northernmost one in Placer County being the smallest (covering about three acres with six living trees).

Sequoiadendron giganteum is a relict species that was more widespread in the Tertiary Period than it is today. Because the

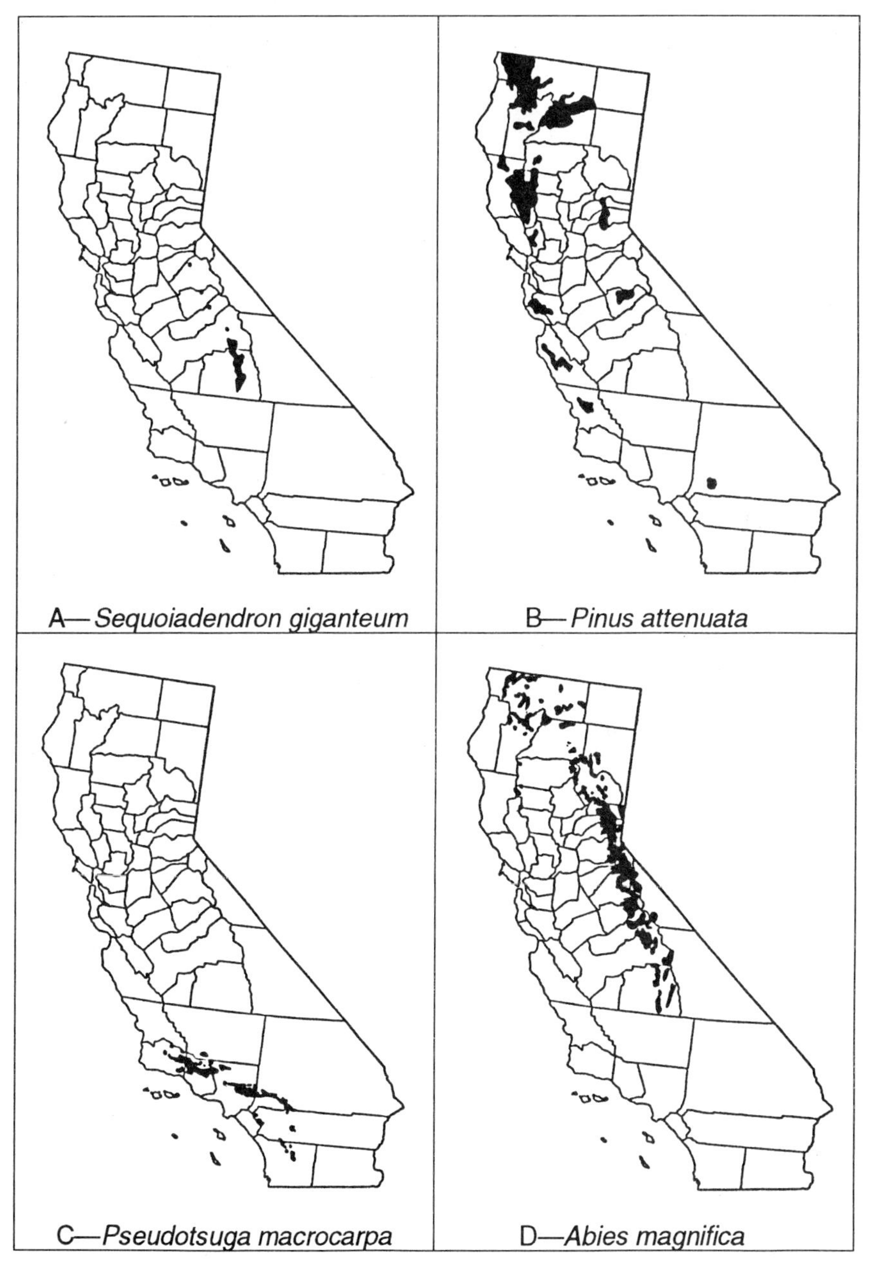

Fig. 16-7. Distribution of *Sequoiadendron giganteum* (giant sequoia), *Pinus attenuata* (knobcone pine), *Pseudotsuga macrocarpa* (big cone Douglas-fir), and *Abies magnifica* (red fir), in California. Modified from Griffin and Critchfield (1972).

Fig. 16-8. A pair of giant sequoias (*Sequoiadendron giganteum*) at the edge of a meadow. The tree on the right shows a large fire scar. The smaller trees are white firs (*Abies concolor*). Photo by V. L. Holland.

winters have gradually become cooler and the summers hotter and drier, communities of more drought tolerant plants have replaced the mesic forests of the past. Some species have survived in spite of the climatic changes but are restricted to areas that are mesic and have reliable summer moister. The restriction of the giant sequoia to sites along the western slope of the Sierra is apparently due to these climatic changes. The further restriction of these trees to scattered groves has been attributed to glacial activity during the Pleistocene epoch (2.5 million to 12,000 years ago) and to a warming trend in the climate after the last ice age. It has been suggested that giant sequoias

Fig. 16-9. Base of giant sequoia partially hollowed out by fire. Such trees can stand for many years after a fire. The wood is very decay-resistant. Photo by David Keil.

were eliminated from glaciated valleys and that once the glaciers melted, they were unable to reinvade the glaciated areas. As a result, the groves are now restricted to unglaciated ridges. Present data indicate that the existing groves are not enlarging or contracting. However, few groves have enough young trees to maintain the present density of mature giant sequoias in the future.

Ecologically, the single most important factor controlling the maintenance of the present giant sequoia groves seems to be availability of soil moisture during the dry summers. Since giant sequoia seedlings are not as drought tolerant as those of some other conifers, they must be able to send a root system down deep enough to tap reliable moisture during the first summer. Because root systems of giant sequoias are typically shallow, extending down only about 1.5 m (5 ft), seedlings as well as mature trees are dependent upon adequate soil moisture at these depths in summer. Lateral root systems in these trees are extensive in the upper soil horizons for 30 to 45 m (100 to 150 ft) out from the trunk of mature trees. The entire root system may extend through and affect as much as four acres.

The major factor causing death of giant sequoias is the susceptibility to falling over. The root system is so shallow that these giants become vulnerable to undercutting by floodwater. A modern problem related to this is human foot traffic on the ground and root system near the tree trunks. This causes soil compaction and low oxygen availability. Once the root system is weakened, a heavy accumulation of snow on the crown during the winters can cause the trees to be top heavy and topple over. In the parks, the trees are often fenced to keep the public from trampling and compressing the ground and root system immediately around the tree trunks.

Fire is also another important factor affecting the growth and reproduction of giant sequoias. These trees have several features that enable them to survive forest fires. Mature giant sequoias are protected from fire by fibrous bark 60 to 120 cm (2 to 4 ft) thick. The thick bark does not burn readily and insulates the living tissues of the inner bark from the heat of a fire. Even if a fire is hot enough and lasts long enough to penetrate the bark, the trees may survive if some bark and living vascular tissue remain (Figs. 16-8, 16-9). Giant sequoias hollowed out by fire are evident in the forests and are sometimes used as winter dens for black bears. In addition,the trunks of mature giant sequoias are self pruning and may not bear branches or foliage for 30 m (100 ft) or more above the ground. A fire sweeping through the forest may pass beneath the elevated canopy of the trees without causing any more damage than a charred layer of outer bark.

The cones of *Sequoiadendron giganteum* are much like those of the closed cone pines and cypresses (Chapter 12). They are tardily dehiscent and serotinous, opening when exposed to the heat of a forest fire. Following a forest fire millions of seeds rain down onto the forest floor. The tiny giant sequoia seeds must fall on partially burned or bare mineral soil for germination to be successful. Otherwise the seeds sift down into the soil litter and decompose. Successful germination requires that the seeds lie within a centimeter or so of a bare mineral soil surface. Mineral soil is commonly exposed by periodic fires that destroy the litter and open up the forest to more sunlight. Additionally, partially burned soil can hold up to 270 percent more available water and is more nutrient-rich than unburned soil and forms a much better seed bed.

Because fire is essential for the successful reproduction of the giant sequoias, fire suppression in the last century or so has reduced establishment of new trees (Fig. 16-10). Without fire one can expect a very gradual decline of the giant sequoia groves and a trend toward white fir forest. This of course will take centuries because of the longevity of the giant sequoias. They typically live for 2000 to 3000 years with the oldest known tree reaching 3300 years.

Fig. 16-10. Interior of a giant sequoia grove in Tulare County. White firs form a subcanopy beneath the towering giants. In the absence of fire, shade-tolerant white firs become progressively more dense. Seedlings of giant sequoias are unable to become established in the thick forest floor litter. Periodic forest fires kill most of the fir trees and provide the mineral soil required for establishment of young giant sequoias. Photo by David Keil.

Fires sometimes start in the tops of giant sequoias. Because these trees tower over other trees of the forest, they are natural lightning rods. Large individuals are repeatedly struck by lightning during their long lifetimes and form gnarled crowns, evidence of lightning damage. Occasionally lightning ignites a fire in the canopy of a tree. Such a fire may smolder for days.

Giant sequoias are the most massive trees in the world. These huge trees are much more massive than the next largest tree in the montane forests of the Sierra, the sugar pine. While sugar pine may reach 60 m (200 ft) tall and 2.1 m (7 ft) wide at the base, mature giant sequoias average about 75 m (250 ft) tall and 4.5 m (15 ft) thick at the base. Exceptional trees exceed 90 m (300 ft) tall but are generally slightly shorter than the coast redwoods (Chapter 13).

Giant sequoias are able to live so long and reach such huge dimensions because of their resistance to fire, insects, fungal activity and disease. The presence of large amounts of tannins in the wood and bark apparently inhibits parasites in living trees and slows decay of fallen trees. Fallen trunks may take several centuries to fully decay. Because the only real difference between the giant sequoia forest phase and the white fir forest phase is the presence and domination of the giant sequoia, other components of these communities are very similar both in structure and in species composition. The common subordinate trees and shrubs are listed in the discussion below of the white fir forest.

F. White Fir Forests

Abies concolor (white fir; Fig. 16-7) is an important tree of higher elevation montane mixed coniferous forests in mountains from southern Oregon to southern California and eastward into the Rocky Mountains from southern Idaho and Wyoming to Arizona and New Mexico. The California race of white fir (*A. concolor* var. *lowiana*) is a common component of montane mixed coniferous forests in the mountains of both northern and southern California at elevations of 800 to 1000 m (2400 to 3000 ft) in the Klamath Mountains, 1250 to 2200 m (4000 to 7000 ft) in the Sierra Nevada and 1675 to 2600 m (5000 to 8000 ft) in southern California.

Toward the upper limits of its elevational range and the upper limits of montane mixed coniferous forests, white fir often forms a distinct zone just below the red fir forests. In this zone white fir typically accounts for 80 percent or more of the tree overstory. This is the most mesic of montane mixed coniferous forest communities with 100 to 150 cm (40 to 60 in) of precipitation per year, about half of which falls as snow. Only red fir forests average more precipitation. As a result, white fir forests are often

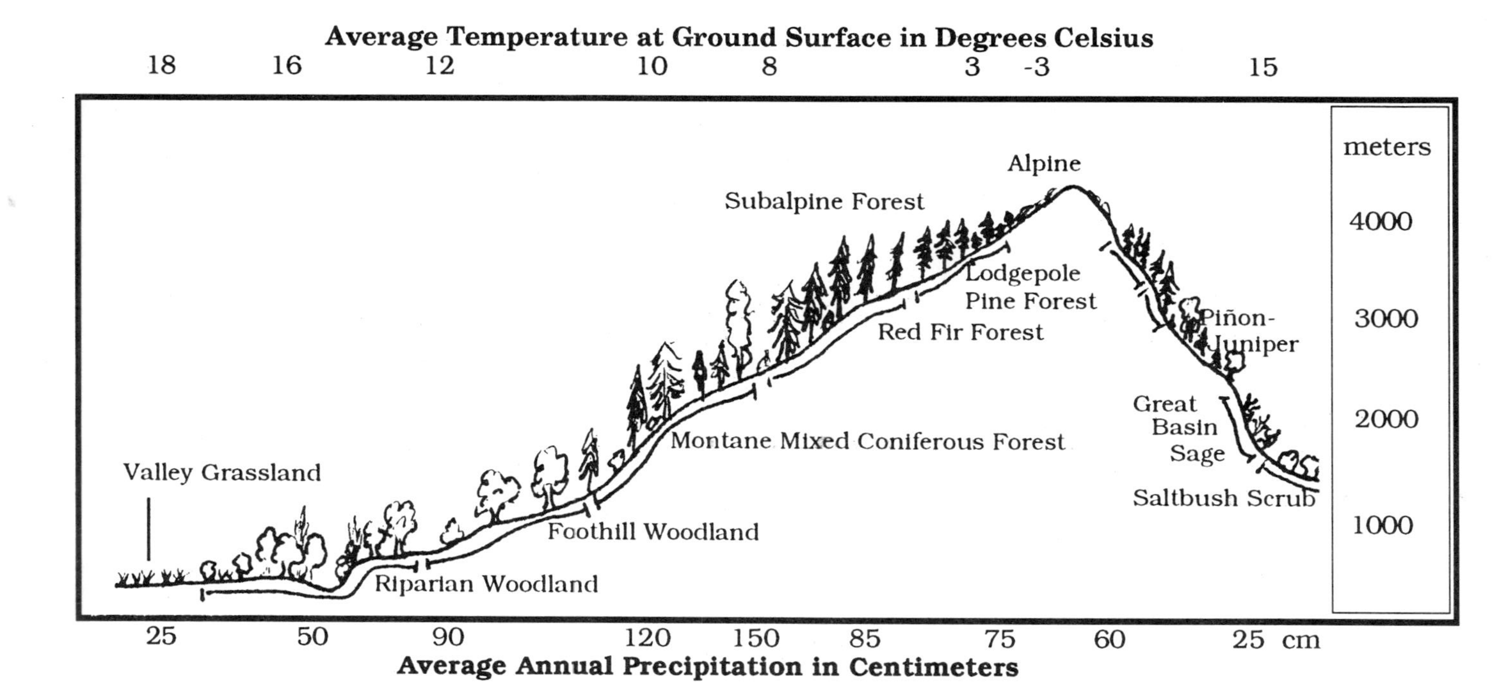

Fig. 16-11. Generalized transect across the Sierra Nevada through Yosemite National Park. Illustrates distribution of plant communities, average temperature, and average precipitation with respect to topography and elevation. Precipitation increases while temperature decreases with elevation and rainshadows occur on the eastern side of the Sierra Nevada. These climatic changes play an important role in the distribution of plant communities along this transect.

considered transitional between montane mixed coniferous forests of lower elevations where precipitation is less but the growing season longer and upper montane forest communities where precipitation may be greater but the growing season is shorter. White fir forests often occur along an ecological gradient resulting in a mixture of associates that may be more characteristic of communities at higher or lower elevations. Although in some areas *Abies concolor* is the dominant species, up to six species of conifers may be present in individual stands. Common associated trees are sugar pine and incense cedar. Giant sequoia is a common associate on moist, unglaciated flats in the central and southern Sierra Nevada. Ponderosa pine (at lower elevations) and Jeffrey pine (at higher elevations) are common associates on drier sites. Douglas-fir is an important associate in northern California and tends to gradually replace white fir in the northern Sierra, the Cascades and the Klamath Mountains.

At lower elevations, white fir forests grade into other forms of montane mixed coniferous forests. At higher elevations, white fir forests often merge with red fir forests over a broad ecotone. In southern California, where red fir is absent, white fir extends up to lodgepole pine or subalpine forest zones. White fir forests are poorly developed on the eastern slopes of the California mountains. Except for isolated groves of *Abies concolor* var. *concolor* (the Rocky Mountain race of white fir) in the Clark, Kingston, and New York Mountains, this species is absent from high elevational areas in the desert ranges of the state.

White fir forests are highly variable. Most commonly they are dense with a rather lush appearance and multi-layered structure. The upper canopy layer is usually composed of 80 percent or more white fir. The other 20 percent consists of some mixture of the six species of conifers mentioned previously. Mature white fir trees commonly are 50 to 63 m (150 to 200 ft) tall with trunk diameters of 1 to 2 m (3 to 6 ft) and average ages of 300 to 400 years. The understory vegetation varies in density and often is multilayered with a tree layer composed mostly of hardwoods (broadleaf trees) such as black oak, big leaf maple, mountain dogwood, and madrone and an understory shrub layer that is extremely variable both in species composition and percent cover. Shrub coverage values of 5 to 10 percent are most common, but values of 30 percent or more also occur. Common shrub species in the White fir forests include:

Arctostaphylos patula	greenleaf manzanita
Ceanothus integerrimus	deer-brush
Ceanothus cordulatus	mountain whitethorn
Ceanothus parvifolius	ceanothus
Corylus cornuta	hazelnut
Chrysolepis sempervirens	mountain chinquapin

Fig. 16-12. Forest regeneration in the Sierra Nevada of Tulare County. In the foreground shade-intolerant Jeffrey pines and species of *Ribes*, *Ceanothus*, and *Arctostaphylos* occupy a clearing. In the dense fir forest in the background the only young trees are the shade-tolerant red and white firs. Photo by David Keil.

Prunus emarginata	bitter cherry
Ribes spp.	gooseberries and currants
Sambucus spp.	elderberry

Several understory shrubs and trees typical of the forests in northern coastal California (mixed evergreen forest) are found in white fir forests of the northern Sierra, Cascade and Klamath Mountains. These include *Arbutus menziesii* (madrone), *Chrysolepis chrysophylla* (giant chinquapin), and *Taxus brevifolia* (yew). *Torreya californica* (California-nutmeg) occurs in scattered locations from the coast ranges to the southern Sierra Nevada.

Herbaceous cover is generally very sparse (less than 5 percent cover) in white fir forests except in very moist, sunny areas such as swales and drainages where the herbaceous cover is dense (around 100 per cent) and resembles a meadow in some cases. White fir forests share many herbaceous species with mixed coniferous forests and with red fir forests.

An important component of the understory vegetation is a large number of seedlings and saplings. Most of these young trees are white firs and a few are incense cedars. These trees are able to reproduce successfully in dense shade. Other trees such as sugar pine, ponderosa pine, and Jeffrey pine require more sunlight and small trees of these species are usually restricted to open areas or sites where a recent fire or other disturbance has opened the canopy (Fig. 16-12). Without periodic disturbances, a climax forest dominated by white fir would eventually occupy most upper elevational moist sites. The other conifers are seral species. In areas where fires have been suppressed, white fir is becoming increasingly the dominant species. The decrease in such trees as the sugar pine in recent years along with the denser, lusher growth of forests (as opposed to open park like stands) is the result of human control of fire. Shrub cover and diversity are extremely variable under stands depending on the fire history. *Chamaebatia foliolosa*, *Arctostaphylos viscida* and several other shrub species are favored by fire and are eliminated by shading in dense older stands. Historical records indicate that shrub density and cover in many fir and pine forests have decreased with fire suppression.

2. Red Fir Forests

Red fir (*Abies magnifica*, Fig. 16-7D) is largely restricted to the high mountains of California. It extends only a short distance into southern Oregon in the Cascades and extreme western Nevada near Lake Tahoe in the Sierra Nevada. The species comprises two varieties that differ in cone structure: the cone bracts of var. *magnifica* are hidden by the cone scales whereas those of var. *shastensis* (Shasta red fir) are exerted from the cone. *Abies magnifica* var. *magnifica* is widely distributed in the higher mountains of northern California and ranges as far south in the Sierra Nevada as Kern County. From about Lassen Peak in the Cascades and Snow Mountain (Lake County) in the North Coast Range north, red fir is often represented by var. *shastensis*. This variety also occurs in disjunct populations in the southern Sierra Nevada of Tulare and Kern Counties. It grades into and is very hard to distinguish from *Abies procera* (noble fir) in the Klamath Mountains. Noble fir completely replaces Shasta red fir north of about 44 degrees latitude in Oregon. Red firs are absent from the South Coast Ranges, the Transverse Ranges, the Peninsular Ranges, and the desert ranges.

Red fir forest communities are most widespread and best developed in northern California where glaciation was limited and snow fall is greatest. North of Lake Tahoe, deep, well-drained soils are dominated by red fir communities at high elevations even on steep slopes and exposed ridge tops. South of Lake Tahoe, favorable sites become more restricted and red fir forest communities tend to be confined to sheltered, gently sloping uplands and often form a mosaic with other forest types. Good examples of red fir forest communities are very uncommon in the

Fig. 16-13. Red fir forest in Sierra Nevada of Tulare County. Young red firs can grow in the shady environment of the forest interior. Photo by David Keil.

extreme southern part of the Sierra Nevada where precipitation is lower and temperatures higher. Red fir forest communities are usually poorly developed on the eastern slope of the Sierra-Cascade mountain chain. This is especially true south of Sonora Pass where red fir forests in transmontane California are confined to small stands in moist canyons above 2450 m (8000 ft).

In areas where it is well represented, red fir is overwhelmingly dominant in the coniferous forest zone just above the white fir forests (Fig. 16-13). It usually occurs on fairly deep, well-drained soils along an elevational belt from about 1600 to 2700 m (5000 to 9000 ft). Because red firs often occur in pure, dense stands that are distinct from adjacent communities, these forests are usually described as separate communities. Red fir forests often form such a dense canopy that no other conifers can successfully compete.

Because of the complexity of the montane environment, there is often a mosaic of forests within relatively small areas. For example, red fir is very successful in well developed soils that are neither too dry nor too wet. If the soils tend to be waterlogged, lodgepole pine and quaking aspen dominate. On soils that are rocky and poorly developed , with a low water holding capacity, Jeffrey pine, western white pine and western juniper are most successful. Lodgepole pine also occurs in rock crevices in areas next to these species.

The ecotones between adjacent forest communities may be abrupt or rather gradual. Where their elevational ranges overlap red fir and white fir are often equally mixed. (They do not hybridize.) At high elevations red firs occur with subalpine trees and in some cases krummholz forms of red fir extend up to the tree line.

The red fir forest zone is usually the area that receives the greatest precipitation of any montane area of California (Fig. 16-11). The precipitation is generally higher than in either the montane mixed coniferous forest zone of lower elevations or the lodgepole pine and subalpine forests at higher elevations. The average annual precipitation is 90 to 165 cm (35 to 63 in), about 80 percent of which falls as snow. As a result, some have referred to red fir forests as the "snow forest". Average snow depths range from 3 to 5 m (10 to 15 ft) and snow patches often do not melt until the middle of the summer in the dense shade of the forest. The depth of winter snow accumulation is indicated by the level on the tree trunk where the dense mantle of yellow-green lichen growth begins. *Letharia vulpina* and *Letharia columbiana*, species of fruticose staghorn lichens, grow on the trunks of red fir just above the snow line.

Soil moisture appears to be the most significant factor affecting the distribution of red fir forests. At lower elevations, the soils are too dry for red fir due to the combination of reduced precipitation and warmer temperatures, whereas at higher elevations the soils are either too wet or too dry. Well-developed soils in the subalpine areas are mostly restricted to glacial basins or drainages that have waterlogged soils with meadow vegetation or forests of lodgepole pine. Otherwise the soils are very poorly developed and hold very little moisture especially in the summer. The fact that red fir does extend up to the tree line in some areas indicates that the colder temperatures and shorter growing season are not nearly as important as the soil moisture in determining its upper boundaries.

Another important environmental feature is the common occurrence of lightning in red fir forest communities. Thunderstorms and lightning seem to be most prevalent in the montane forest zone where red fir forests occur. Because the red firs are usually the dominant as well as the tallest tree in this area, they are most vulnerable to lightning strikes. Evidence of lightning damage can be seen by the regular occurrence of flat topped trees where lightning has taken off the apex of the crown, burnt stumps and trunks, and scattered sections of trunks lying on the forest floor. Lightning fires are important in red fir forests because they burn the thick litter layer, expose the mineral soil and open up the forest allowing the establishment of young trees and understory shrubs and herbs.

Following fire, seral trees such as *Pinus contorta ssp. murrayana* (lodgepole pine) can often get established on the exposed mineral soils along with the red fir. After a period of time, red firs will overshadow the seral trees and dominate the site. Since red fir is tolerant of shade and is able to reproduce in its own shade, it is the climax species in these areas and dominates the favorable sites. In intermediate sites where the forest remains more open, other trees such as lodgepole pine, Jeffrey pine and western white pine are more successful and remain common components of the forest. In favorable sites where red fir forests form dense stands about the only small trees that can get established are the red firs themselves.

Because red firs are often the first trees to recolonize an area after fire, most of the trees of a stand are about the same size and same age. However, a red fir forest may be a mosaic of different-aged patches, the result of fires of different years at different sites. Additional seedlings may become established in the shade of the taller, older trees, but most of them eventually die. As a result, many red fir forest communities are even aged stands of red fir with little understory vegetation. The combination of dense shade, low temperatures, heavy snow, thick litter layer and competition for soil moisture makes it very difficult for plants to get established in the forest floor. Consequently, natural regeneration of red fir as well as the establishment of understory shrubs and herbs is restricted to small open areas in the forest created by the death of one or two trees or in larger open areas created by wind damage, insect epidemics, logging or fire. Without additional disturbance these areas will, in time, give way to the more typical even-aged stands of red fir with little understory vegetation.

The most common species of shrubs in red fir forest communities are listed below.

Lonicera conjugalis	double honeysuckle
Quercus vaccinifolia	huckleberry oak
Ribes viscosissimum	sticky currant
Symphoricarpos rotundifolius [*S. vaccinioides*]	mountain snowberry

Many of the shrubs listed for white fir forests often are found in the areas that have been opened up by fire or other disturbance but are shaded out fairly quickly once the forest gets reestablished. Herbaceous vegetation is usually sparse but is generally better developed than the shrub understory both in terms of coverage and species diversity. The most common species are listed below.

Aster breweri [*Heterotheca b.*]	Brewer's golden aster
Chimaphila umbellata	prince's-pine
Hieracium albiflorum	white-flowered hawkweed
Monardella odoratissima	mountain pennyroyal
Pyrola picta	white-veined wintergreen

In open areas within the forest and following fire, the herbaceous cover may be well developed before being shaded out by the red firs. In these areas the diversity of herbaceous species is greater, consisting of species more common to lower elevation montane coniferous forests as well as those more typical of red fir forests.

Red fir forests (and upper elevation montane mixed coniferous forests) have several species of mycotrophic herbs. These plants have no chlorophyll and do not manufacture their own food in photosynthesis, but instead rely on symbiotic relationships with soil fungi. *Sarcodes sanguinea* (snow plant), for example, has a mycorrhizal fungus associated with its roots that supplies the snow plant with nutrients taken from the breakdown of the forest humus or from carbohydrates produced by adjacent conifers to which the fungus is also attached. *Corallorhiza maculata* (spotted coralroot) and *Pterospora andromeda* (pinedrops) are also mycotrophic (sometimes inaccurately described as saprophytic) plants common in mixed coniferous and red fir forests. These mycotrophic species as well others not listed thrive in dense shade and thick litter probably because they do not rely on photosynthesis (and therefore sunlight) to provide carbohydrates.

3. Lodgepole Pine Forests

Above the red fir forests in northern California and above the montane mixed coniferous forests in southern California, forest communities are usually more open, and the trees are mostly shorter in stature. The forest just above the red fir forest is usually dominated by *Pinus contorta ssp. murrayana* [*P. murrayana*] (lodgepole pine).

Pinus contorta is widespread in western North America from Alaska and the central Yukon to Baja California and east through the Rocky Mountains to the Black Hills of South Dakota. Within this area it grows at a wider range of elevations than any other pine ranging from sea level along the coast (ssp. *contorta*, the shore pine) to 3600 m (12,000 ft) in the interior mountains. High elevation races are differentiated into ssp. *latifolia* (Rocky Mountain lodgepole pine) and ssp. *murrayana* (Sierran lodgepole pine, tamarack pine).

In California, the largest stands of *Pinus contorta* are in the central and northern Sierra Nevada where this species forms extensive forests above the red fir forest (Fig. 16-18A). Lodgepole pines are also common in the higher elevations of the Cascade and Klamath Mountains of northern California, but only the shore pine occurs to the south along the coast. In southern California, lodgepole pines are absent from the Transverse Ranges, but form rather extensive forests at high elevations in the Peninsular Ranges.

Within its broad range, *Pinus contorta* varies both ecologically and morphologically. *Pinus contorta ssp. contorta* (shore pine) is ecologically a very different taxon from the high elevation races of the species and is discussed in the coastal closed cone conifer forests (Chapter 12). Lodgepole pines (the high elevation races) are variable in cone morphology. The lodgepole pines in the Rocky Mountains often have closed cones that remain on the tree sealed by resins until fire burns the forest. Heat causes the cones to open and prolific seeding occurs following fires. This is also the case for the shore pine. However, *Pinus contorta ssp. murrayana*, the Sierran lodgepole pine, has cones that open on the branches although they may remain on the trees for a few years. Fires are important in the successful regeneration of lodgepole pines in California, but they are not essential for cones to open and seeds to be released. In the following discussion of California lodgepole pine forests only ssp. *murrayana* will be considered.

Pinus contorta grows in upper montane forests of California from about 1800 to 2450 m (6000 to 8000 ft) in the north and 2450 to 3600 m (8000 to 12000 ft) in the south. Lodgepole pines sometimes occur at lower elevations around bogs, meadows and lakes where they often form pure stands in saturated soils. Climatically, lodgepole pine forests are intermediate between red fir forests and subalpine forests in terms of both growing season and precipitation (Fig. 16-11). The average annual precipitation is 75 to 150 cm (31 to 62 in), most of which falls as snow during the winter, and the growing season is 2 to 3 months long.

Lodgepole pine is perhaps the most versatile of the western conifers. It is able to physiologically alter both its rate of moisture uptake and its transpiration rate depending on soil moisture conditions. At high elevations, soils tend to be either saturated for most of the growing season or very poorly developed and dry, yet lodgepole pine can grow in both types of soil (Fig. 16-14). Lodgepole pine is one of the only trees that can tolerate the waterlogged soils of bogs, meadows and lakes. In these areas it often occurs in pure stands (Fig. 16-14B). At high elevations these grade into red fir forest on surrounding drier soils.

At lower elevations stands of *Pinus contorta* are usually surrounded by mixed coniferous forest. The lodgepole pines grow around mountain meadows or snow-melt gullies which are found

Fig. 16-14. Lodgepole pine forests in contrasting environments. **A.** Open forest with trees growing in crevices of granite outcrop in Yosemite National Park. **B.** Dense forest in seasonally waterlogged soil surrounding meadow. Note the young pines invading the meadow. Photos by V. L. Holland.

in red fir or white fir-mixed conifer forests. On high protected slopes snowfields melt slowly allowing more or less permanent cold water seepage or drainage down slopes or gullies during the growing season.

Lodgepole pines can also grow on thin, dry soils in upper montane regions (Fig. 16-14A). However, lodgepole pines usually do not dominate sites that are intermediate between the wet and dry areas where red fir dominate. In glacially scoured areas, lodgepole pines colonize joint planes on which soil has accumulated and grow next to other drought tolerant species common to the subalpine such as western juniper, Jeffrey pine, white bark pine and limber pine. Some lodgepole pines extend up to tree line where krummholz forms grow in the rock crevices (Fig. 16-16A). Even though they have a wide range and overlap extensively with other communities, we are treating lodgepole pine forests separately because lodgepole pines form such extensive stands where they overwhelmingly dominate the forest.

Lightning strikes are common in lodgepole pine forest communities as are periodic lightning fires. These fires open up the forest and allow extensive lodgepole pine reproduction. Since they recolonize these areas after fire, lodgepole pines often form rather even-aged stands much like those discussed previously for red fir. In the absence of periodic fires lodgepole pine forests are successionally replaced by red fir forests in areas where soils are well developed and well drained. Lodgepole pines cannot reproduce successfully in the shade of the red firs and consequently can not compete with the red fir for the more favorable soil sites. In an effort to reintroduce fire in something approaching its natural role in these areas, the National Park Service has undertaken a program to allow lightning fires to run their course. However, fires are allowed to burn unchecked only in areas where fuel has not accumulated to dangerous proportions.

Under ideal conditions, lodgepole pines grow as tall as 30 m (100 ft) with trunk diameters of 0.6 to 1.3 m (2 to 5 ft). However, because they are not successful competitors, they are usually relegated to the less favorable sites where red fir cannot grow. In these sites they generally form slender, straight-trunked trees 15 to 25 m (50 to 80 ft) tall and 15 to 50 cm (6 to 20 in) in trunk diameter. In rocky sites exposed to wind and other subalpine features, they form dwarfed, bushy trees with a growth form similar to that of other subalpine species.

In recent years it appears that fire suppression may be resulting in an increase of lodgepole pine forests in some areas and reduction in others. In well drained soils, fire control has lead to a general degradation of lodgepole pine and its replace-

Fig. 16-15. Lodgepole pines killed by lodgepole needle miner in the Sierra Nevada of Yosemite National Park. Photo by V. L. Holland.

ment in some cases by red fir. On the other hand, invasion of many high elevation meadows by lodgepole pine may also be a result of fire suppression. Periodic fires prevent lodgepole pine trees that are invading the fringes of meadows from reaching maturity.

Sheep grazing has also reduced lodgepole pine reproduction in the past. With reduced grazing pressure in areas such as Yosemite National Park, lodgepole pines are successfully colonizing the fringes of the meadows.

The lodgepole needle-miner (*Coleotechnites milleri*) has killed large stands of lodgepole pine, such as the so called ghost forest in the Yosemite National Park (Fig. 16-15). Most of the time the needle-miner, which is the larval stage of a small moth, is kept under control by more than 40 parasitic insects and by several species of birds. The needle-miner has always been a part of the lodgepole pine forest ecosystem, but its periodic infestations only served to open up the forest, much like periodic fires, and recovery following the infestation was fairly rapid. In recent times, control of the needle-miner has been thrown off and the ecosystem imbalanced by human activities. Spraying with DDT killed more of the needle-miners' predators and parasites than needle-miners. As a result, the needle-miner continues to be a serious problem in lodgepole pine forests and one that is receiving considerable attention by forest managers.

Lodgepole pine is the overwhelming dominant in lodgepole pine communities and usually occurs in almost pure stands. However, at the lower limits of its elevational range it is often mixed with red fir. This ecotone has often been referred to as the red fir-lodgepole pine forest. At its upper elevational limits, lodgepole pine extends into and often becomes an important component of subalpine forests. Understory shrubs are not common in lodgepole pine forests and generally are of little importance. The following shrubs are sometimes locally important:

Arctostaphylos nevadensis	manzanita
Ceanothus cordulatus	mountain white-thorn, snowbush
Cercocarpus ledifolius	curl-leaf mountain-mahogany
Chrysolepis sempervirens	mountain chinquapin
Phyllodoce breweri	pink heather
Ribes montigenum	mountain gooseberry

The herbaceous understory is extremely variable in both composition and percent cover. In areas where lodgepole pine forests occur at the fringe of meadows and bogs, the herbaceous understory is lush and composed of species that are also common to the adjacent wet soils. At the other extreme on dry, glacial scoured areas, the herbaceous understory is sparse and composed of species also found in red fir forests at lower elevations and in subalpine forests at higher elevations.

4. Subalpine Forests

Subalpine communities are the uppermost coniferous forests. The only communities occurring above the subalpine forests are the treeless alpine communities (Chapter 17). These forests are most extensive in the high mountains of the central Sierra Nevada from Tulare County to Lake Tahoe. North of Lake Tahoe, scattered stands of subalpine forest occur in the northern Sierra Nevada, the Cascades and the Klamath Mountains. South of Mount Whitney, small areas of subalpine vegetation occur in the southern Sierra, the Transverse and Peninsular Ranges. These communities are more extensive on the western slopes than on the steeper eastern slopes. Desert subalpine forests occur at high elevations in the Inyo, White, Funeral, Grapevine, and Panamint Mountains of transmontane California.

Subalpine communities usually occur above 2000 m (6600 ft) in the Klamath Mountains, above 2450 m (8000 ft) in the northern Sierra and Cascade Mountains, and above 2900 (9500) in the southern Sierra and southern California mountains. These

Fig. 16-16. Krummholz growth form of trees near timberline. The wind-pruned, prostrate, shrublike conifers are indicative of the severe growing conditions near the upper-elevation limit of trees. **A.** Lodgepole pine (*Pinus contorta ssp. murrayana*) near timberline. **B.** Open subalpine forest Sonora Pass in the central Sierra Nevada. Photos by V. L. Holland.

forests (or in some cases woodlands) extend upward to timberline which ranges from about 2500 m (8200 ft) in the Klamath Range to 3500 m (11,500 ft) in the southern California mountains. These communities intermingle with lodgepole pine and red fir forests at lower elevations. Both lodgepole pine and red fir occur in subalpine forests in some areas but generally do not form a dominant component of these higher elevations forests.

Subalpine forests grow under the most severe environmental conditions of any forest vegetation in California (Fig. 16-11). Annual average precipitation ranges from about 70 to 125 cm (17 to 30 in) all of which falls as snow except for occasional summer thundershowers. Temperatures are very cold except for the short growing season of 7 to 9 weeks that occurs in the late summer. Even during this period frosts are not uncommon. Even though precipitation seems adequate, portions of the subalpine zone resemble a very cold desert because soil moisture is only available for a short period of time. During most of the year it is frozen and once the temperatures warm up enough to melt the snow, the water is only held in the soil for a short period because the soils are poorly developed, shallow and contain little organic matter. In basins and swales where soils are deeper, water often accumulates resulting in waterlogged soils that are unfavorable for tree growth. These areas are generally occupied by subalpine meadows that are occasionally invaded by lodgepole pines.

Wind is an important environmental feature in subalpine communities. It increases the aridity of the area and propels abrasive ice particles. Wind flagged trees are common in the subalpine zone. On the side of the tree facing the prevailing winds, leaves, branches and even bark are stripped away by wind-borne ice. The branches that grow away from the winds on the leeward side of the tree become permanently wind-trained and the tree develops a flag-like appearance. The trees near timberline are dwarfed and often grow almost prostrate along the rocks (Fig. 16-16). These krummholz (German for bent or twisted wood) tree forms are largely a result of frost damage, moisture stress and abrasion by windblown snow and ice. The height of krummholz trees is often directly correlated with snow depth. Snow acts as an insulating blanket that protects the trees. Any portion of a tree that extends above the snow line may be killed by severe winds and cold temperatures.

The species composition of the subalpine forests varies from south to north and from west to east, largely as a result of differences in available moisture. The most common and characteristic trees of the subalpine forests in California include:

Fig. 16-17. Trees of the subalpine. **A.** Mountain hemlock (*Tsuga mertensiana*) occurs from Alaska to the central Sierra Nevada. **B.** Mountain juniper (*Juniperus occidentalis* var. *australis*) occurs in dry subalpine forests in the Sierra Nevada. **C.** Bristlecone pine (*Pinus longaeva*) forms open subalpine forests in the White Mountains and several other desert ranges and extends eastward across the Great Basin. A & C, photos by David Keil; B, photo by V. L. Holland.

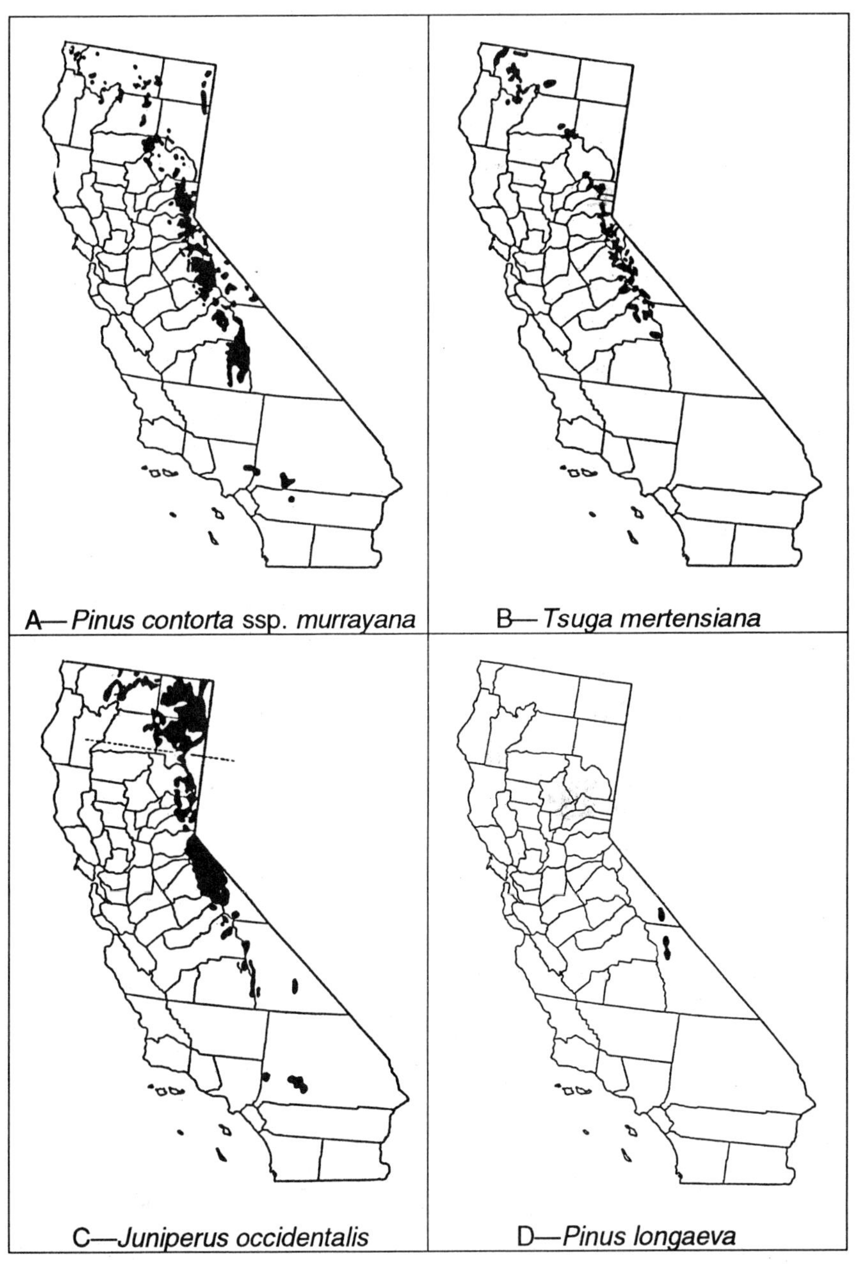

Fig. 16-18. Distribution of *Pinus contorta* ssp. *murrayana* (lodgepole pine), *Tsuga mertensiana* (mountain hemlock), *Juniperus occidentalis*, and *Pinus longaeva* (bristlecone pine) in California. The dotted line on the map of *Juniperus occidentalis* marks the transition between var. *occidentalis* (western juniper) to the north and var. *australis* (mountain juniper) to the south. Modified from Griffin and Critchfield, 1972)

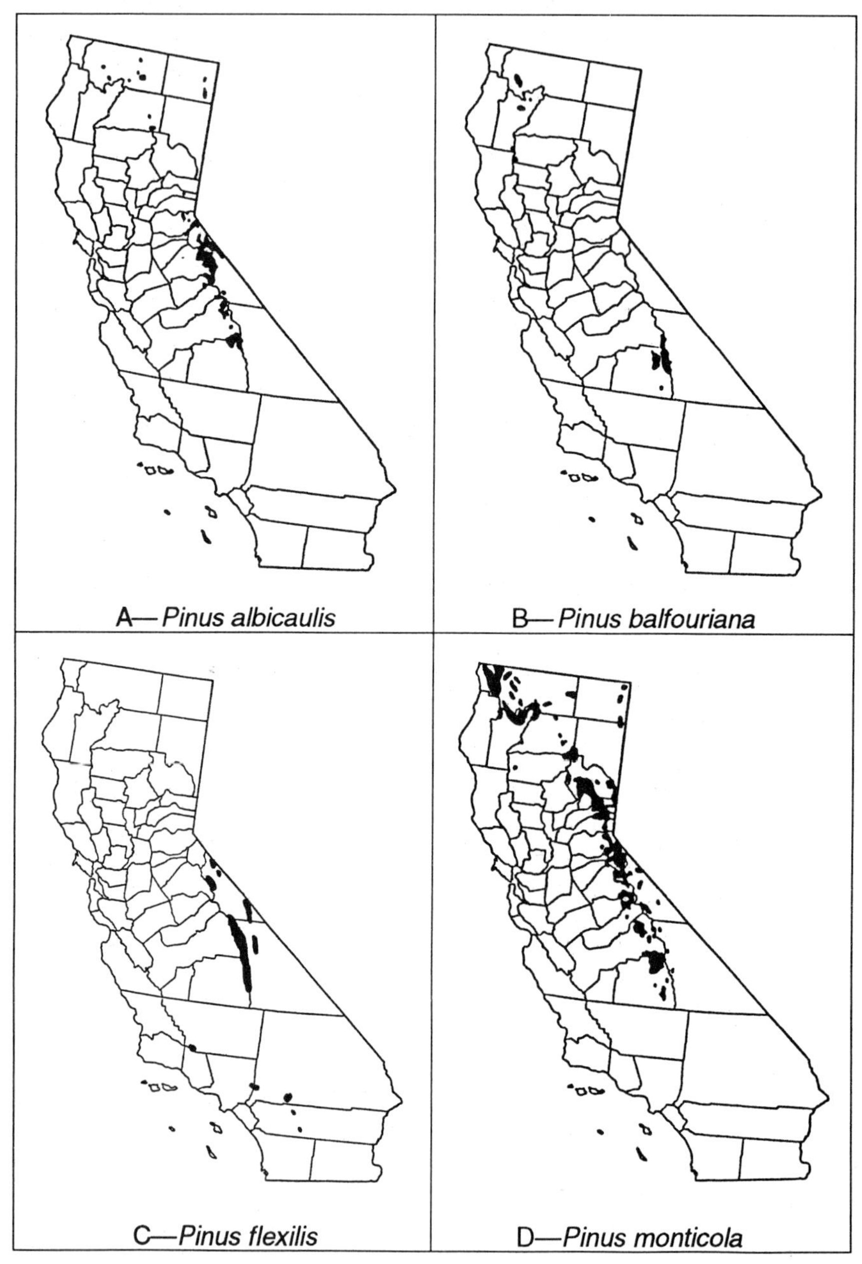

Fig. 16-19. Distribution of *Pinus albicaulis* (white bark pine), *Pinus balfouriana* (foxtail pine), *Pinus flexilis* (limber pine), and *Pinus monticola* (western white pine) in California. Modified from Griffin and Critchfield, 1972)

Juniperus occidentalis var. *australis* (Figs. 16-17, 16-18C)	mountain juniper
Pinus albicaulis (Fig. 16-19A)	whitebark pine
Pinus balfouriana (Fig. 16-19B)	foxtail pine
Pinus contorta ssp. murrayana (Figs. 16-16A, 16-18A)	lodgepole pine
Pinus flexilis (Fig. 16-19C)	limber pine
Pinus longaeva (Figs. 16-17C, 16-18D)	bristlecone pine
Pinus monticola (Fig. 16-19A)	western white pine
Tsuga mertensiana(Figs. 16-17A, 16-18B)	mountain hemlock

The subalpine regions in southern and eastern California tend to receive less precipitation and perhaps have more severe climatic conditions than those to the north and west. These trees sometimes occur in mixed stands and at other times they form almost pure stands. On the western slopes of the Sierra Nevada-Cascade axis and in the Klamath Mountains, subalpine forests are dominated by white bark pine, mountain hemlock and foxtail pine. Limber pine is usually the dominant in the drier phases of the subalpine zone.

Shrubby vegetation is usually sparse in subalpine zones. Shrubs and subshrubs such as *Salix* spp. (willows), *Vaccinium uliginosum* ssp. *occidentale* [*V. occidentale*] (western blueberry), *V. caespitosum* [*V. nivictum*] (dwarf bilberry), and *Kalmia polifolia* (mountain-laurel) are local in moist sites. Occasional shrubs include *Ribes cereum*; (wax currant), *Holodiscus microphyllus* (rock-spiraea) and *Artemisia tridentata* (Great Basin sagebrush). High elevation forms of montane chaparral (Fig. 16-20A) dominated by such species as *Quercus vaccinifolia* (huckleberry oak) and *Chrysolepis sempervirens* (mountain chinquapin) occur in some areas. Herbaceous vegetation ranges from very sparse in dry soils to dense, sedge-grass meadows with various wildflowers.

Probably the most characteristic tree of the subalpine forests from the central Sierra north is *Pinus albicaulis* (whitebark pine). It is common in subalpine forests from the central Sierra Nevada north (Fig. 16-19A). It occurs from central British Columbia east to Wyoming and reaches its southernmost limits in the central Sierra Nevada. It often forms pure stands of sprawling shrublike trees just below the alpine zone where conditions are too harsh to support any other conifers. At lower elevations where environmental conditions are somewhat more favorable, it grows in mixed stands with mountain hemlock, foxtail pine and lodgepole pine, reaching heights of up to 25 m (80 ft) although shorter trees are more common.

Pinus balfouriana (foxtail pine; Fig. 16-19B), a California endemic, has an unusual, disjunct distribution pattern. It occurs in the subalpine zone of the Klamath Mountains and in the southern

Fig. 16-20. Subalpine forests in the Sierra Nevada. **A.** Forest and high-elevation montane chaparral in Mammoth Lakes area of Mono County. **B.** Open forest with snowfields near Sonora Pass. Photos by V. L. Holland.

Sierra Nevada south of the South Fork of the San Joaquin River. These two populations, which are about 300 miles apart, are apparently remnants of a more extensive and continuous subalpine zone that existed in the Miocene and Pliocene epochs when summer precipitation was apparently more common. Summer thunderstorms have allowed the foxtail pine to remain in the two isolated areas where it exists today. Unlike the whitebark pine, foxtail pine remains erect regardless of environmental conditions and rarely occurs on the harsh, exposed slopes near tree line. Although it never becomes shrublike, it does get severely wind flagged in exposed areas. Foxtail pine occurs in extensive pure stands in some areas and also occurs in mixed stands with whitebark pine lodgepole pine, mountain hemlock and limber pine.

Tsuga mertensiana (mountain hemlock; Figs. 16-17A, 16-18B) is a widespread subalpine tree from Alaska to California. It is common from coastal areas of Alaska south through the mountains of British Columbia, Washington and Oregon and eastward into the northern Rockies. In the northernmost part of its range, it occurs at low elevations but it is a common component of subalpine forests in the Pacific Northwest.

Mountain hemlock is common in the subalpine zone in northern California. It grows on exposed ridges and slopes from its southernmost limits in Tulare County to southeastern Alaska and east into the northern Rocky Mountains of British Columbia and Idaho. Mountain hemlock is most common in cool, mesic sites where it often forms pure, dense stands with little understory. It occurs mostly as erect, single trucked trees but may form krummholz forms in exposed areas near timberline where it mixes with whitebark pine. At lower elevations and in sheltered sites, mountain hemlock mixes with lodgepole pine, western white pine and red fir.

Pinus flexilis (limber pine; Fig. 16-19C) is also widely distributed in western North America, but generally in much drier habitats than *T. mertensiana* and *P. albicaulis*. It occurs from Canada to northern New Mexico in the Rocky Mountains where it is common on dry, rocky ridges and peaks. Limber pine is also common on the tops of high elevation desert mountains from Utah and Nevada to eastern California. In the Inyo, White and Panamint Mountains of the desert regions of California, limber pine is commonly associated with *Pinus longaeva* (bristlecone pine). Like the bristlecone pine, limber pine is a slow growing, long lived tree. These rather small, gnarled trees have large taproots that anchor them and tap available soil moisture. They are characterized by their soft, light, flexible wood which gives them their common name.

Limber pine is perhaps the most characteristic tree in the dry subalpine regions of central and southern California. In southern California it grows in isolated stands in the Peninsular and Transverse Ranges where it reaches its westernmost limits on Mt. Pinos. In the Sierra Nevada it is mostly restricted to dry, harsh eastern slopes where it extends as far north as Inyo County. Limber pine is considered a rather poor competitor, and as a result, it is usually restricted to eroded, steep, rocky sites with well drained, infertile soils that front desert regions and receive little precipitation. North of Yosemite, limber pine is replaced by whitebark pine and generally does not mix with it. Although limber pine often grows in pure stands, it also frequently associates with foxtail pine and lodgepole pine in the Sierra Nevada. In southern California, the most common associate is lodgepole pine although Jeffrey pine and to a lesser extent white fir extend into the subalpine as associates.

Juniperus occidentalis var. *australis* (mountain juniper; Figs. 16-17B, 16-18C) is locally common in the subalpine forests from about Susanville (Lassen County) to southern California. It is rarely dominant and seldom extends up to timberline. It is closely related to and grades into var. *occidentalis* (western juniper) north of Susanville. Western juniper (var. *occidentalis*) is common in the Cascades and on the Modoc Plateau where it is the dominant in the northern juniper woodland communities (Chapter 18). Mountain juniper is generally restricted to steep, arid slopes where soils are very shallow and rocky. It is mostly an erect but highly twisted tree although shrubby, prostrate forms also occur especially in northern California. Mountain juniper is mostly a subalpine species, although it occurs as low as 1800 m (6000 ft) in montane mixed coniferous forests and sometimes forms a high-elevation juniper woodland in transmontane sites. It is the only subalpine tree able to occur over such a broad elevational range.

Pinus monticola (western white pine; Fig. 16-19D) is a fairly common component of subalpine as well as red fir forest communities in California. Its main distribution is in the Rocky Mountains from southeastern British Columbia to central Idaho. It also occurs in the Cascades and extends as far south as the southern Sierra Nevada in Tulare County. In California it is most abundant in the Sierra from Kings Canyon National Park to Lake Tahoe where it forms small rather open stands on dry, exposed granite slopes. It is usually found in association with red fir, lodgepole pine and mountain hemlock. It is occasionally a component of the upper elevational portions of the montane mixed coniferous forests.

Pinus longaeva (Great Basin bristlecone pine; Fig. 16-18D) extends from Utah into the Inyo, White, and Panamint Mountains

Fig. 16-21. Desert subalpine forest in the White Mountains of Mono County. The sparse forest is dominated by a mixture of bristlecone pines and limber pines. Photo by David Keil.

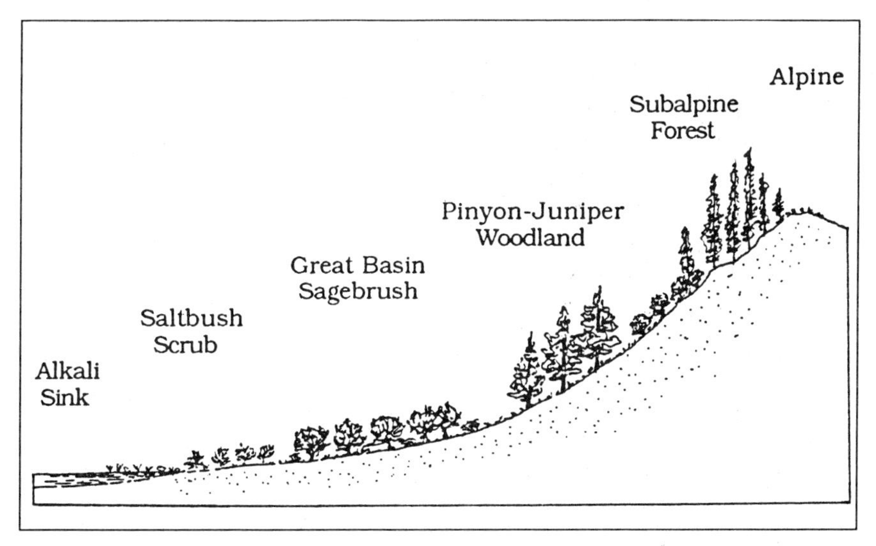

Fig. 16-22. Vegetation transect from up the western flank of the White Mountains. Upper slopes lack montane forests between the upper elevational limits of the piñon-juniper woodlands and the lower elevational limits of the limber pine-bristlecone pine subalpine woodland. Instead there is a high-elevation desert scrub dominated by Great Basin sagebrush.

of the desert regions of California where it mixes with limber pine to form the Great Basin subalpine forests (Fig. 16-21). At higher elevations and mostly on dolomite derived soils, it forms pure stands. It also occurs in pure stands in the Last Chance Mountains of the southern California desert where limber pine does not occur. Great Basin bristlecone pine is a true timberline tree in these arid desert mountains as is its close relative, the Rocky Mountain bristlecone pine (*Pinus aristata*) in the Rocky Mountains of Colorado, New Mexico and Arizona. Great Basin subalpine communities have perhaps the harshest environment of the California subalpine forests because of long, cold winters and low average annual precipitation (40 cm or 16 in) which falls almost entirely as winter snow.

Both bristlecone pines are gnarled, slow growing, long lived trees; however, the Great Basin bristlecone pines in the White Mountains (Fig. 16-17C) have attracted special interest because the oldest are believed to be among the world's oldest living organisms at over 4600 years old. On the most ancient trees (those over 4000 years old), there is commonly only a narrow strip of living bark with only a small group of twigs attached to a single live branch.

Montane Coniferous Forests—References

Agee, J. K., and H. H. Biswell. 1969. Seedling survival in a giant sequoia forest. Calif. Agric. 23:18–19.

Arkley, R. J. 1981. Soil moisture use by mixed conifer forest in a summer-dry climate. Soil Sci. Soc. Am. J. 45:423–427.

Aune, P. S. (tech. coord.). 1994. Giant sequoias: their place in the ecosystem and society. U.S.D.A. Forest Service, PSW, Albany, CA. 170 pp.

Axelrod, D. I. 1976. History of the coniferous forests, California and Nevada. Univ. Calif. Publ. Bot. 70:1–62.

———. 1986. Cenozoic history of some western American pines. Ann. Missouri Botanical Garden 73:565–641.

———. 1988. An interpretation of high montane conifers in western Tertiary floras. Paleobiology 14:301–306.

Bailey, D. K. 1970. Phytogeography and taxonomy of *Pinus* subsection Balfourianae. Ann. Missouri Bot. Gard. 57:210–249.

Barbour, M. G. 1984. Can a red fir forest be restored? Fremontia 11(4):18–19.

———. 1988. California upland forests. Pp. 131–164 *in* M. G. Barbour and W. D. Billings (eds.), North American Terrestrial Vegetation. Cambridge Univ. Press, Cambridge.

———, N. H. Berg, T. G. F. Kittel, and M. E. Kunz. 1991. Snowpack and the distribution of a major vegetation ecotone in the Sierra Nevada of California. J. Biogeography 18:141–149.

———, and R. A. Woodward. 1985. The Shasta red fir forest of California. Can. J. For. Res. 15:570–576.

Beasley, R. S., and J. O. Klemmedson. 1976. Water stress in bristlecone pine and associated plants. Commun. Soil Sci. Plant Anal. 7:609–618.

———, and ———. 1980. Ecological relationships of bristlecone pine. Amer. Midl. Naturalist 104:242–252.

Benedict, N. B. 1982. Mountain meadows: stability and change. Madroño 29:148–153.

Biswell, H. H. 1961. The big trees and fires. National Parks Magazine 35:11–14.

———, H. Buchanan, and R. P. Gibbins. 1966. Ecology of the vegetation of a second-growth big tree forest. Ecology 46:630–634.

Bonnicksen, T. M., and E. C. Stone. 1981. The giant sequoia-mixed conifer forest community characterized through pattern analysis as a mosaic of aggregations. Forest Ecol. Manage. 3:307–328.

———, and ———. 1982. Reconstruction of a presettlement giant sequoia—mixed conifer forest community using the aggregation approach. Ecology 63:1134–1148.

Borchert, M. 1985. Serotiny and cone-habit variation in populations of *Pinus coulteri* (Pinaceae) in the southern Coast Ranges of California. Madroño 32:29–48.

———, and M. Hibberd. 1984. Gradient analysis of a north slope montane forest in the western Transverse Range of southern California. Madroño 31:129–139.

Brown, D. E. 1982. Great Basin conifer woodland. Pp. 52–57 *in* D. E. Brown (ed.). Biotic communities of the American Southwest—United States and Mexico. Desert Plants 4:1–341.

Burke, M. T. The vegetation of the Rae Lakes basin, Southern Sierra Nevada. Madroño 29:164–176.

Clausen, J. 1965. Microclimatic and vegetational contrasts within a subalpine valley. Proc. Natl. Acad. Sci. 53:1315–1319.

Clausen, J. 1965. Population studies of alpine and subalpine races of conifers and willows in the California high Sierra Nevada. Evolution 19:52–68.

Conard, S., and S. R. Radosevich. 1982. Post-fire succession in white fir(*Abies concolor*) vegetation of the northern Sierra Nevada. Madroño 29:42–56.

Cook, L. F. 1942. The giant sequoias of California. U.S. Dept. Interior, Natl. Park Svce., Washington, D.C. 28 pp.

Cooper, C. F. 1960. Changes in vegetation, structure, and growth of southwestern pine forests since white settlement. Ecol. Monogr. 30:129–164.

———. 1961. Patterns in ponderosa pine forests. Ecology 42:493–499.

Critchfield, W. B. 1957. Geographic variation in *Pinus contorta*. Maria Moors Cabot Found. Publ. 3:1–118. Harvard Univ., Cambridge.

———, and G. L. Allenbaugh. 1969. The distribution of Pinaceae in or near northern Nevada. Madroño 20:12–26.

———, and E. L. Little. 1966. Geographical distribution of the pines of the world. U.S.D.A. Forest Service Misc. Publ. 991. 97 pp.

Edmonds, R. L. 1982. Analysis of coniferous forest ecosystems in the western United States. US/IBP Synthesis Series 14. Hutchinson Ross Publ. Co., Stroudsburg, PA.

Glock, W. S. 1937. Observations on the western juniper. Madroño 4:21–28.

Gordon, D. T. 1970. Natural regeneration of white and red fir ... influence of several factors. Forest Service Res. Pap. PSW-58, Pacific SW Forest and Range Exp. Stn, Berkeley, Calif.

Griffin, J. R. 1964. Isolated *Pinus ponderosa* forests on sandy soils near Santa Cruz, California. Ecology 45:410–412.

———. 1975. Plants of the highest Santa Lucia and Diablo Range peaks, California. USDA Forest Serv. Res. Paper PSW-110.

———. 1982. Pine seedlings, native ground cover, and *Lolium multiflorum* on the Marble-Cone burn, Santa Lucia Range, California. Madroño 29:177–188.

———, and W. B. Critchfield. 1972. The distribution of forest trees in California. Pacific S.W. Forest and Range Exp. Stn., Berkeley, Calif. 114 pp. (U.S.D.A. Serv. Res. Paper PSW 82).

Haller, J. R. 1959. Factors affecting the distribution of ponderosa and Jeffrey pines in California. Madroño 15:65–71.

———. 1962. Variation and hybridization in ponderosa and Jeffrey pines. Univ. Calif. Publ. Bot. 34:123–165.

Hanes, T. L. 1976. Vegetation types of the San Gabriel Mountains. Pp. 65–76 *in* J. Latting (ed.). Plant communities of southern California. Calif. Native Pl. Soc. Spec. Publ. 2.

Hartesveldt, R. J. 1964. Fire ecology of the giant sequoias: controlled fires may be one solution to survival of the species. Natural History 73:12–19.

———, and H. T. Harvey. 1967. The fire ecology of sequoia regeneration. Proc. Tall Timbers Fire Ecol. Conf. 7:65–77.

Harvey, H. T., H. S. Shellhammer, and R. E. Stecker. 1980. Giant sequoia ecology. Fire and reproduction. U.S. Dept. Interior Natl. Park Serv., Sci. Monogr. Ser. No. 12. 182 pp.

———, and ———. 1991. Survivorship and growth of giant sequoia (*Sequoiadendron giganteum* (Lindl.) Buchh.) seedlings after fire. Madroño 38:14–20.

Heath, J. P. 1967. Primary conifer succession, Lassen Volcanic National Park. Ecology 48:270–275.

Helms, J. A. 1987. Invasion of *Pinus contorta ssp. murrayana* (Pinaceae) into mountain meadows at Yosemite National Park, California. Madroño 34:91–97.

———, and R. D. Ratliff. 1987. Germination and establishment of *Pinus contorta ssp. murrayana* (Pinaceae) in mountain meadows of Yosemite National Park, California. Madroño 34:77–90.

Henrickson, J., and B. Prigge. 1975. White fir in the mountains of eastern Mojave desert of California. Madroño 23:164–168.

Hitch, C. J. 1982. Dendrochronology and serendipity. Amer. Sci. 70:300–305.

Holmgren, N. H. 1972. Plant geography of the intermountain region. Pp. 77–161 *in* A. Cronquist, A. H. Holmgren, N. H. Holmgren and J. L. Reveal. Intermountain Flora. Vol. 1. New York Bot. Gard. and Hafner Publ. Co., N.Y.

Johnston, V. R. 1970. Sierra Nevada. Houghton, Mifflin Co., Boston.

———. 1994. California forests and woodlands. A natural history. Univ. California Press, Berkeley & Los Angeles.

Keeley. 1988. Bibliographies on chaparral and the fire ecology of other Mediterranean systems, 2nd ed. Calif. Water Resource Cntr., Univ. Calif. Davis, Rep. No. 69. 328 pp.

Kilgore, B. M. 1971. The role of fire in managing red fir forests. Trans. North Amer. Wild. Res. Conf. 36:405–416.

———. 1971. The role of fire in a giant sequoia-mixed conifer forest. Pp. 93–116 *in* Research in the parks. Trans. Natl. Park Serv. Symp. Ser. 1, Natl. Park Serv., Washington, D.C.

———. 1973. Impact of prescribed burning on a sequoia-mixed conifer forest. Proc. Tall Timbers Fire Ecol. Conf. 12:345–375.

———. 1973. The ecological role of fire in Sierran conifer forests: its application to national park management. J. Quaternary Res. 3:396–513.

———, and G. F. Briggs. 1972. Restoring fire to high elevation forests in California. J. For. 70:226–271.

———, and D. Taylor. 1979. Fire history of a sequoia-mixed conifer forest. Ecology 60:129–142.

Klikoff, L. L. 1965. Microenvironmental influence on vegetational pattern near timberline in the central Sierra Nevada. Ecol. Monogr. 35:188–211.

Klyver, F. D. 1931. Major plant communities in a transect of the Sierra Nevada Mountains of California. Ecology 12:1–17.

LaMarche, V. C. 1969. Environment in relation to age of bristlecone pine. Ecology 50:53–59.

———. 1973. Holocene climatic variations inferred from treeline fluctuations in the White Mountains, California. Quaternary Res. 3:632–660.

Lathrop, E. W., and B. D. Martin. 1982. Response of understory vegetation to prescribed burning in yellow pine forests of Cuyamaca Rancho State Park, California. Aliso 10:329–343.

Minnich, R. A. 1976. Vegetation of the San Bernardino Mountains. Pp. 99–124 *in* J. Latting (ed.). Plant communities of southern California. Calif. Native Plant Soc. Spec. Publ. 2.

———. 1977. The geography of fire and big-cone Douglas-fir, Coulter pine and western conifer forests in the eastern Transverse Ranges. Pp. 443–450 *in* H. A. Mooney and E. C. Conrad (tech. coordinators). Proceedings of the symposium on the environmental consequences of fire and fuel management in Mediterranean ecosystems. U. S. D. A. Forest Service Gen. Tech. Rep. WO-3.

———. 1987. The distribution of forest trees in northern Baja California. Madroño 34:98–127.

Mooney, H. A. 1973. Plant communities and vegetation. Pp. 7–17 *in* R. M. Lloyd and R. S. Mitchell. A flora of the White Mountains, California and Nevada. Univ. Calif. Pr., Berkeley.

———, G. St. Andre, and R. D. Wright. 1962. Alpine and subalpine vegetation patterns in the White Mts. of California. Amer. Midl. Naturalist 68:257–273.

Oosting, H. J., and W. D. Billings. 1943. The red fir forest of the Sierra Nevada: *Abietum magnificae*. Ecol. Monogr. 13:259–274.

Pase, C. P. 1982. Sierran montane conifer forest. Pp. 49–51 *in* D. E. Brown (ed.). Biotic communities of the American Southwest—United States and Mexico. Desert Plants 4:1–341.

———. 1982. Sierran subalpine conifer forest. Pp. 40–41 *in* D. E. Brown (ed.). Biotic communities of the American Southwest—United States and Mexico. Desert Plants 4:1–341.

Parker, A. J. 1982. Comparative structural/functional features in conifer forests of Yosemite and Glacier National Parks, USA. Amer. Midl. Naturalist 107:55–68.

———. 1982. Environmental and compositional ordinations of conifer forests in Yosemite National Park, California. Madroño 29:109–118.

———. 1986. Persistence of lodgepole pine forests in the central Sierra Nevada. Ecology 67:1560–1567.

Parsons, D. J. 1972. The southern extensions of *Tsuga mertensiana* (mountain hemlock) in the Sierra Nevada. Madroño 21:536–539.

———. 1981. Fire in a subalpine meadow. Fremontia 9(2):16–18.

Patterson, M. T., and P. W. Rundel. 1995. Stand characteristics of ozone-stressed populations of *Pinus jeffreyi* (Pinaceae): extent, development, and physiological consequences of visible injury. Amer. J. Bot. 82:150–158.

Potter, D. 1994. Guide to forested communities of the upper montane in the central and northern Sierra Nevada. U.S.D.A. Forest Service, PSW, Albany, CA.

———. 1995. Ecological classification in upper montane forests of the central and southern Sierra Nevada. U.S.D.A. Forest Service Gen Tech. Rep., PSW, Albany, CA.

Reveal, J. L. 1964. Single leaf pinyon and Utah juniper woodlands of western Nevada. J. For. 42:276–278.

Riegel, G. M., D. A. Thornburgh, and J. O. Sawyer. 1990. Forest habitat types in the South Warner Mountains, Modoc County, California. Madroño 37:88–112.

Rundel, P. W. 1971. Community structure and stability in the giant sequoia groves of the Sierra Nevada, California. Amer. Midl. Naturalist 85:478–492.

———. 1972. Habitat restriction in giant sequoia: the environmental control of grove boundaries. Amer. Midl. Naturalist 87:81–99.

———, D. J. Parsons, and D. T. Gordon. 1977. Montane and subalpine vegetation of the Sierra Nevada and Cascade Ranges. Pp. 559–599 *in* M. G. Barbour and J. Major (eds.). Terrestrial vegetation of California. John Wiley and Sons, N.Y.

Sawyer, J. O., and D. A. Thornburgh. 1977. Montane and subalpine vegetation of the Klamath Mountains. Pp. 699–732 *in* M. G. Barbour and J. Major (eds.). Terrestrial vegetation of California. John Wiley and Sons, N.Y.

Schierenbeck, K. A., and D. B. Jensen. 1994. Vegetation of the Upper Raider and Hornback Creek basins, South Warner Mountains: northwestern limit of *Abies concolor* var. *lowiana*. *Madroño 41:53–64.*

Schulman, E. 1958. Bristlecone pine, oldest known living thing. Nat. Geogr. 113:355–372.

Shirley, J. C. 1942. The redwoods of coast and Sierra. Univ. Calif. Pr., Berkeley. 84 pp.

Show, S. B., and E. I. Kotok. 1924. The role of fire in the California pine forests. U.S.D.A. Bull. 1294. 80 pp.

Sigg, J. 1983. The foxtail pine of the Sierra. Fremontia 11(1):3–8.

Smiley, F. J. 1915. The alpine and subalpine vegetation of the Lake Tahoe region. Bot. Gaz. 2:265–286.

Stark, N. 1965. Natural regeneration of Sierra Nevada mixed conifers after logging. J. Forestry 63:456–461.

———. 1968. Seed ecology of *Sequoiadendron giganteum*. Madroño 19:267–277.

Storer, T. I., and R. L. Usinger. 1963. Sierra Nevada natural history. Univ. Calif. Pr., Berkeley.

Sudworth, G. B. 1908. Forest trees of the Pacific slope. U.S.D.A. Forest Service. Washington, D.C. 441 pp.

Talley, S. N., and J. R. Griffen. 1980. Fire ecology of a montane pine forest, Junipero Serra Peak, California. Madroño 27:49–60.

Thorne, R. F. 1977. Montane and subalpine forests of the Transverse and peninsular Ranges. Pp. 537–557 *in* M. G. Barbour and J. Major (eds.). Terrestrial vegetation of California. John Wiley and Sons, N.Y.

———. 1982. The desert and other transmontane plant communities of southern California. Aliso 10:219–257.

Tranquillini, W. 1979. Physiological ecology of the alpine timberline. Springer-Verlag, N. Y. 137 pp.

Vale, T. R. 1981. Tree invasion of montane meadows in Oregon. Amer. Midl. Naturalist 105:61–69.

Vankat, J. L. 1977. Fire and man in Sequoia National Park. Ann. Assoc. Amer. Geogr. 67:17–27.

———. 1982. A gradient perspective on the vegetation of Sequoia National Park, California. Madroño 29:200–214.

———, and J. Major. 1978. Vegetation changes in Sequoia National Park, California. J. Biogeogr. 5:377–402.

Vale, T. R. 1981. Ages of invasive trees in Dana Meadows, Yosemite National Park, California. Madroño 28:45–47.

Vasek, F. C., and R. F. Thorne. 1977. Transmontane coniferous vegetation. Pp. 797–832 *in* M. G. Barbour and J. Major (eds.). Terrestrial vegetation of California. John Wiley and Sons, N.Y.

Vogl, R. J. 1976. An introduction to the plant communities of the Santa Ana and San Jacinto Mountains. Pp. 77–98 *in* J. Latting (ed.). Plant communities of southern California. Calif. Native Plant Soc. Spec. Publ. 2.

———, and B. C. Miller. 1968. The vegetational composition of the south slope of Mt. Pinos, California. Madroño 19:225–288.

Wagener, W. W. 1961. Past fire incidence in Sierra Nevada forests. J. Forestry 51:739–748.

Waring, R. H. 1969. Forest plants of the eastern Siskiyous: their environmental and vegetational distribution. Northwest Sci. 43:1–17.

Weaver, H. 1943. Fire as an ecological and silvicultural factor in the ponderosa pine region of the Pacific slope. J. For. 41:7–15.

———. 1967. Fire and its relationship to ponderosa pine. Proc. Tall Timbers Fire Ecol. Conf. 7:127–149.

Whitney, S. 1979. A Sierra Club naturalist's guide to the Sierra Nevada. Sierra Club Books, San Francisco.

Whittaker, R. H. 1960. Vegetation of the Siskyou Mountains, Oregon and California. Ecol. Monogr. 30:279–338.

Wilken, G. C. 1967. History and fire record of a timberland brush field in the Sierra Nevada of California. Ecology 48:302–304.

Willard, D. 1994. Giant sequoia groves of the Sierra Nevada. A reference guide. Published by the author, Berkeley, CA.

Wright, R. D., and H. A. Mooney. 1965. Substrate-oriented distribution of bristlecone pine in the White Mountains of California. Amer. Midl. Naturalist 73:257–284.

Yeaton, R. I. 1981. Seedling characteristics and elevational distributions of pines (Pinaceae) in the Sierra Nevada of central California: a hypothesis. Madroño 28:67–77.

———, R. W. Yeaton, and J. E. Horenstein. 1980. The altitudinal replacement of foothill pine by ponderosa pine on the western slopes of the Sierra Nevada. Bull. Torrey Bot. Club 107:487–495.

Chapter 17

Alpine Communities

Only the highest mountains in California have alpine communities. These assemblages of plants, often called alpine tundra or alpine fell fields, are treeless communities dominated by low-growing herbaceous plants and dwarf shrubs. They occur at elevations where the climatic conditions are too cold to support trees. With the exception of Mount Shasta in the Cascade Range and the high crest of the Sierra Nevada, California's mountains today have rather small areas of alpine vegetation (Fig. 17-1). These zones are isolated on high peaks much like islands in an archipelago. At present less than one percent of the state is above the treeline.

The lower limit of the alpine zone is the irregular timberline that marks the zone above which trees are unable to survive. At or near timberline, trees of the subalpine zone are often gnarled, twisted and highly wind-pruned, forming a krummholz growth form (Chapter 16). The elevation of timberline is dependent upon temperature conditions, slope aspect and snow accumulation. The elevation of timberline is determined primarily by local microclimate. Timberline tends to occur at lower elevations on north-facing slopes or sites where snow accumulation is high. Timberline (Fig. 17-2) ranges in elevation from as low as 2000 m (6600 ft) in the Klamath Mountains of northwestern California to 2700 m (8900 ft) in the southern Cascades, 3350 m (11,000 ft) in the southern Sierra Nevada and 3500 m (11,500 ft) in the Peninsular ranges and in the White Mountains of the eastern desert region.

Past geological and climatic changes have caused the elevation of treeline to vary greatly from its present level. The high peaks of the Cascades, the Sierra Nevada and other high mountains of California have risen to their present elevation only during the Pleistocene Epoch. Prior to that time the areas available for alpine vegetation may have been very localized and restricted to only a few high peaks. During the great ice ages of the Pleistocene, glacial ice covered the peaks of most of the high mountains of California, and alpine vegetation was more widespread in California's mountains. It occurred as much as

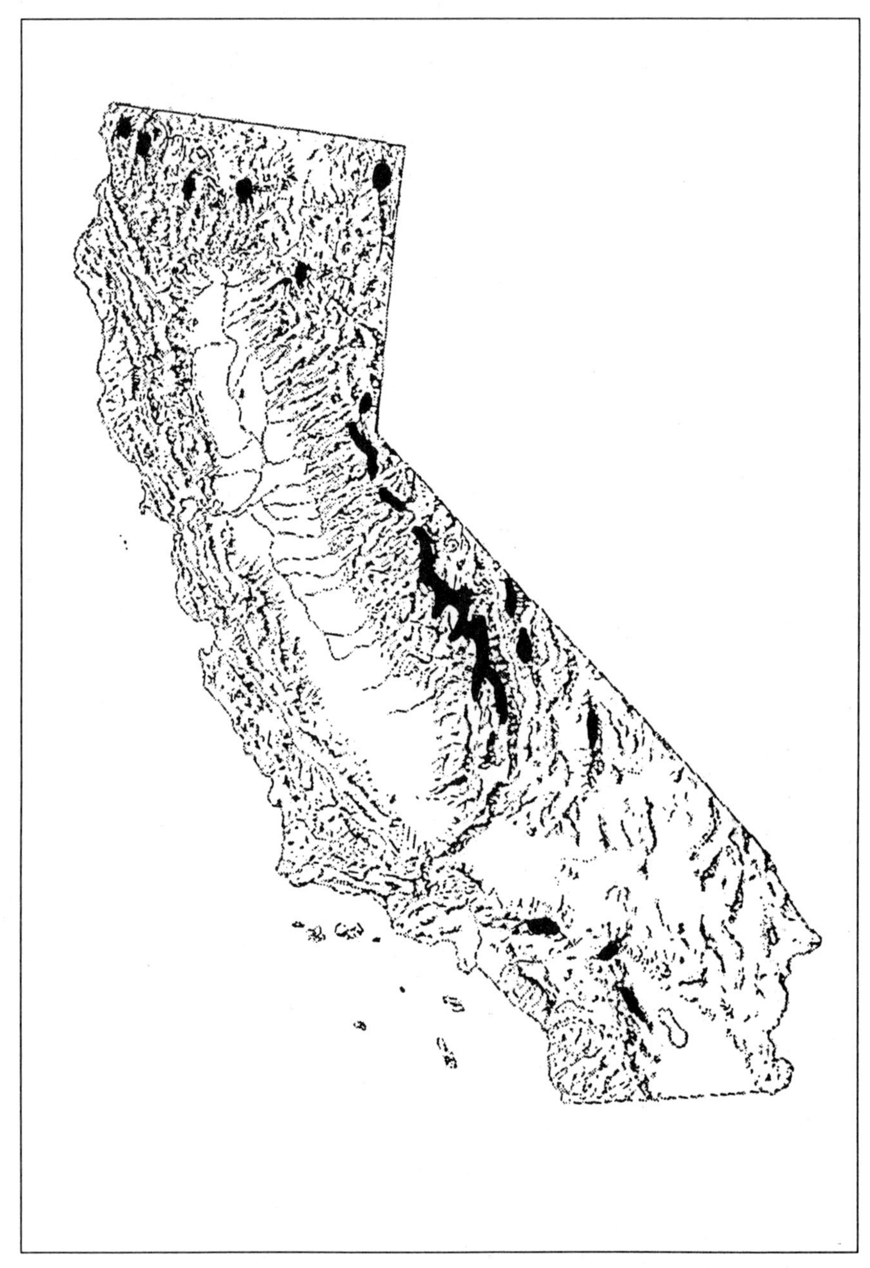

Fig. 17-1. Distribution of alpine communities in California.

Fig. 17-2. Timberline and alpine zone with snowfields near Sonora Pass in the Sierra Nevada of Mono County. Photo by V. L. Holland.

1300 m (4300 ft) lower on mountain slopes than it does today and may have extended from one mountain range to another. About 12,000 years ago during times of much colder climates, about 10 percent of the state was alpine or ice-covered.

Alpine communities are similar in various ways to the arctic tundra of the far north. The environments of both areas are characterized by long, cold winters, a short growing season and frequent frosts. The soils are often shallow and poorly developed. Although alpine and arctic areas are similar in various ways, there are also significant differences. Alpine communities occur at high elevations with much relief, are exposed to a thin atmosphere that allows intense insolation but holds little heat, have regular diurnal rhythms of sunlight and darkness and corresponding diurnal temperature fluctuations, and are exposed to very high winds. Arctic communities occur at lower elevations and often have little relief; the atmosphere is comparatively thick, and the suns rays pass through it at a glancing angle of inclination; there are prolonged periods of 24-hour daylight or 24-hour darkness or twilight with little day-night temperature fluctuation. Winds on the average are not as intense. Plant adaptations to the harsh environmental conditions are similar in both regions and some species occur in both alpine and arctic communities. Somewhat less than 20 percent of California's alpine species also occur in the arctic tundra.

Environmental conditions above timberline are very harsh. Winter is prolonged and freezing weather can occur at any time of the year. Because of the thin air at high elevations, there is much day to night temperature fluctuation with frost common at night even in mid-summer. The growing season may be as short as 4 to 8 weeks. Areas of snow accumulation have the shortest growing seasons. Snow melts slowly at high elevations, and plant growth is not possible until the snow-pack has melted. Snow occasionally falls in the alpine zone even in mid-summer, blanketing the plants as they grow and flower. Plants that grow in such areas must be able to photosynthesis at low temperatures, to grow rapidly, and to complete their reproductive cycle during the brief summer period. On cloudless days insolation may be intense, and even on cloudy days ultraviolet radiation is at a high level.

Soils in the alpine zone are often shallow and rocky. On exposed slopes wind and water erosion of fine-grained soil particles results in a coarse gravelly or rocky soil with little organic matter and essentially no soil differentiation. Alternate freezing and thawing churns the soil. On open sites snow accumulation may be very slight because of wind action, and the water that does melt from snow is swiftly lost to runoff. On other, more sheltered sites, snow may accumulate in deep drifts and snow banks may not melt until late in the summer. Rocky alpine areas are consequently rather xeric environments, and plants growing there often have adaptations to drought similar to those of desert plants.

The plants that occur in the alpine tundra have been derived from various sources. Some have evolved from species of nearby lower-elevation areas. Various herbaceous species do not recognize the elevational boundary that characterizes the alpine treeline and grow in the subalpine as well as in the alpine communities. The alpine flora of high California mountains has some species in common with the alpine floras of other high mountains in western North America. As mentioned above, some of these species are shared with the Arctic tundra of the far North. However, a large percentage of California's alpine tundra flora shows affinities not to other high elevation regions but rather to the desert regions of the west, particularly the Great Basin desert. The drier areas of alpine vegetation on the rain-shadow slopes of the Sierras and Cascades and on the desert mountains have the highest representation of these desert-derived species. These species may have adapted to alpine conditions during the lowering of elevational zones in the Pleistocene.

There are over 600 plant species that have been reported growing in California's alpine community. About two-thirds of these (400 species) occur in the alpine and also in surrounding plant communities, especially the subalpine. About one-third (200) of the species are restricted to alpine areas. However, even some of the 200 species considered to be restricted have been found in microhabitats of the surrounding subalpine such as the edge of large snow banks and along shaded, moist cliffs. These areas, while elevationally in the subalpine, actually have environmental conditions more like the alpine.

Alpine flora and vegetation is complex and extremely variable from site to site. Recent analysis of California's alpine resulted in over 30 site specific assemblages of plants. However, for simplicity sake we have lumped these into 3 broad categories: Alpine meadow communities, Rocky alpine communities and Desert Alpine Communities.

1. Alpine Meadows

Alpine meadows occur in the high elevation areas of the Klamath Mts. and in subalpine and alpine areas of the Cascades and northern Sierra Nevada. These communities have high snow accumulation and much moisture available during the summer from both snow-melt and summer precipitation. Soils are often well-vegetated, and meadow or bog communities commonly develop. Sedges and grasses are dominant, often with various associated forbs. These wetland communities are discussed in greater detail in Chapter 21 sections C and D with other meadow and bog communities.

In the less extensive alpine and subalpine meadows of the southern Sierra Nevada, the soil dries out earlier in the season, and grasses and sedges play a lesser role. Forbs are well-represented, though. These communities are transitional to the xerophytic rocky alpine and desert alpine communities discussed below.

Overall, the vegetation of the alpine meadows is dominated by sod-forming grasses and sedges such as *Carex subnigricans* (alpine sedge), *Carex vernacula* (common sedge), and *Calamagrostis breweri* (Brewer's shorthair grass). However, there are also numerous broadleafed herbs that can produce a showy wildflower display during the late summer flowering season. Some of the most characteristic examples are:

Potentilla drummondii	Drummond's cinquefoil
Antennaria media [*A. alpina* var. *m.*]	alpine pussytoes
Lewisia pygmaea	dwarf lewisia
Erigeron algidus [*E. petiolaris*]	Sierra daisy

Solidago multiradiata	alpine goldenrod
Dodecatheon alpinum	alpine shooting star
Ranunculus eschscholtzii	alpine buttercup
Lupinus lepidus	dwarf lupine

Shrubs are not usually dominant in alpine meadows, but several species are fairly common. These include several species of *Salix* (willows), *Kalmia polifolia* var. *microphylla* (alpine-laurel), *Vaccinium caespitosum* [*V. nivictum*] (dwarf bilberry), *Cassiope mertensiana* (white heather), and *Phyllodoce breweri* (red mountain heather).

2. Rocky Alpine Communities

Rocky alpine communities (alpine fell fields communities) are located on high-elevation ridges in the Cascades, the Sierra Nevada, and Transverse Ranges. Snow accumulation is usually high, but runoff is rapid and summer precipitation is low or absent. Soils are usually shallow, poorly developed, and well-drained. Plant cover is often spotty with much exposed bare soil or rock scree. Summer temperatures are usually higher, and summer cloud cover is lower in these areas. Overall temperature fluctuation on both a diurnal and seasonal basis is often greater in the xerophytic type communities than in the alpine meadows.

Plants that grow in well-drained alpine soils are mostly perennial herbs. The underground portions of these plants often exceed the above-ground parts. Perennials are generally better represented at high elevations than are annuals because the photosynthates stored in their subterranean roots and stems enables them to quickly produce new stems, leaves and flowers. If they fail to reproduce during a particularly harsh growing season, they remain a part of the community. Annuals must be able to establish themselves, photosynthesize, and reproduce during the brief alpine summer. Reproductive failure over a few unfavorable growing seasons can eliminate an annual from the community.

Because of the immature rocky soil, rocky alpine communities have the appearance of a rock garden with masses of perennial herbs scattered among the rocks (Fig. 17-3). These perennial plants are usually low-growing herbs or subshrubs with reduced, often densely compacted, pubescent leaves, and frequently form dense mats. Often several species grow together to form a single "cushion". The cushion collects particles of soil and decomposing organic litter and shields the soil from wind and water erosion. Mosses and lichens also may contribute to the structure of a cushion.

Fig. 17-3. Cushion plants in alpine tundra. In rocky soils slow-growing perennials form tight mats, sometimes composed of several species. Photo by David Keil.

The time available for reproduction is very limited. Some of the plants produce flower-buds a year before the flowers open. During the flowering season many of these plants produce large colorful flowers that attract pollinating insects, even though most are capable of self-fertilization, a feature that ensures high seed set. The large size of the flowers relative to the vegetative portion of the plant is often striking. Along with the diverse group of dicots, there often are dwarfed cespitose or rhizomatous grasses and sedges in the community.

Species composition in the rocky alpine is quite variable and is related to the variation in the type of rocky substrate. For example, some areas are flat and gravelly whereas others are steeper and covered by coarse boulders with large crevices. Some areas are exposed, and snow does not accumulate as much as it does in other areas such as the rock crevices mentioned above. As a result, the rocky alpine can be divided into many microassociations. Overall, the most common plants in these rocky alpine areas are low growing herbaceous perennials that form "cushion plants", bunch grasses, and prostrate shrubs. The following list includes some of the most characteristic herbaceous plants:

Festuca brachyphylla	alpine fescue
Draba densifolia	dense-leafed draba
Astragalus kentrophyta	spiny-leafed milkvetch
Eriogonum ovalifolium	oval-leafed eriogonum
Phlox condensata	carpet phlox
Castilleja nana	alpine paintbrush
Raillardella argentea	silver raillardella
Arenaria nuttallii	Nuttall's sandwort
Primula suffrutescens	Sierra primrose
Oxyria digyna	alpine sorrel
Aquilegia pubescens	alpine columbine

Some common low shrubs are *Ribes cereum* (wax currant), *Leptodactylon pungens* (granite gilia), and *Ericameria discoidea* [*Haplopappus macronema*] (whitestem goldenbush).

3. Desert Alpine Communities

Desert alpine communities (alpine steppe communities; Fig. 17-4) are located on the eastern (rain-shadow) slope of the southern Sierra Nevada and on the highest of the desert mountain ranges (White, Sweetwater, and Warner Mts.). Snow accumulation is low in these areas, and snowmelt and runoff take place early in the year. Summer drought is prolonged. Soils are shallow and very poorly developed with scant humus accumulation. Plant cover is very sparse, and much bare soil and rock is exposed. Seasonal and diurnal temperature fluctuations are high, and during the summer the daytime temperature is often high. Plants of these alpine areas have a very similar appearance and share some species with the rocky alpine areas discussed previously. Characteristic plants include several species each of *Eriogonum* (buckwheat), *Astragalus* (locoweed, milk-vetch), *Oxytropis* (oxytrope), *Draba* (draba), *Arabis* (rockcresses), *Arenaria* (sandwort), *Erigeron* (fleabane daisy), and *Potentilla* (cinquefoil). Bunch grasses and sedges form scattered dense clumps.

Human Impacts On Alpine Communities

Human impact in alpine tundra takes several forms. In the past, grazing of livestock has occurred during the relatively brief period of high productivity in the summer., and grazing still occurs in some alpine areas, including the White Mountains. Grazing affects plants of alpine areas rather severely because of their dependence on a brief period of photosynthetic activity. The hooves of livestock also can cause severe disruption of the root systems of the plants and the soil that they bind. Recovery of vegetation in alpine areas may take hundreds of years.

More recently in some areas the impact has come mainly from human foot traffic. Hiking tends to wear away the plant cover,

Fig. 17-4. Desert alpine area in White Mountains of Inyo County. Photo by David Keil.

dislodge the soil, and initiate erosion. Because of the slow overall growth of the alpine plants (seasonal growth may be rapid, but annual accumulation is low), successional restabilization of disturbed sites is slow. Serious erosion has been detected in areas of heavy foot traffic. In certain areas the erosion has been exacerbated by the use of off-road vehicles in sensitive alpine environments.

Alpine Communities —References

Billings, W. D. 1974. Adaptations and origins of alpine plants. Arc. and Alp. Res. 6:129–142.

———. 1975. Arctic and alpine vegetation: plant adaptations to cold summer climates. Pp. 403–443 *in* J. D. Ives and R. G. Barry (eds.). Arctic and alpine environments. Methuen Pr., London.

———. 1978. Alpine phytogeography across the Great Basin. Great Basin Naturalist Mem. 2:105–117.

———. 1979. High mountain ecosystems. Evolution, structure, operation and maintenance. Pp. 97–125 *in* P. J. Webber (ed.). High altitude geoecology. AAAS Selected Symposium 12. Westview Pr., Boulder, Colorado.

———. 1981. Plants in high places. Natural History 90(10):82–88.

———. 1988. Alpine vegetation. Pp. 391–420 *in* M. G. Barbour and W. D. Billings (eds.)., North American Terrestrial Vegetation. Cambridge Univ. Press, Cambridge.

———, and H. A. Mooney. 1968. The ecology of arctic and alpine plants. Biol. Rev. 43:481–529.

Bliss, L. C. 1971. Arctic and alpine plant life cycles. Ann. Rev. Ecol. Syst. 2:405–438.

Burke, M. T. 1982. The vegetation of the Rae Lakes basin, southern Sierra Nevada. Madroño 29:164–176.

Chabot, B. F., and W. D. Billings. 1972. Origins and ecology of the Sierran alpine flora and vegetation. Ecol. Monogr. 42:163–199.

Clausen, J. 1965. Population studies of alpine and subalpine races of conifers and willows in the California high Sierra Nevada. Evolution 19:52–68.

Daubenmire, R. F. 1941. Some ecologic features of the subterranean organs of alpine plants. Ecology 22:370–378.

Hanes, T. L. 1976. Vegetation types of the San Gabriel Mountains. Pp. 65–76 *in* J. Latting (ed.). Plant communities of southern California. Calif. Native Pl. Soc. Spec. Publ. 2.

Howell, J. T. 1951. The arctic-alpine flora of three peaks in the Sierra Nevada. Leafl. West. Bot. 6:141–156.

Hunter, K. B., and R. E. Johnson. 1983. Alpine flora of the Sweetwater Mountains, Mono County, California. Madroño 30(4, suppl.): 89–105.

Jackson, L. E. 1982. Distribution of ephemeral herbaceous plants near treeline in the Sierra Nevada, California, U.S.A. Arc. and Alp. Res. 14:33–42.

Klikoff, L. L. 1965. Microenvironmental influence on vegetational pattern near timberline in the central Sierra Nevada. Ecol. Monogr. 35:188–211.

Major, J., and D. W. Taylor. 1977. Alpine. Pp. 601–675 *in* M. G. Barbour and J. Major (eds.). Terrestrial vegetation of California. John Wiley and Sons, N.Y.

———, and S. A. Bamberg. 1967. Comparison of some North American and Eurasian alpine ecosystems. Pp. 89–118 *in* H. E. Wright and W. H. Osborn (eds.). Arctic and alpine environments. Indiana Univ. Pr., Bloomington.

———, and S. A. Bamberg. 1967. Some cordilleran plants disjunct in the Sierra Nevada of California and their bearing on Pleistocene ecological conditions. Pp. 171–178 *in* H. E. Wright and W. H. Osborn (eds.). Arctic and alpine environments. Indiana Univ. Pr., Bloomington.

Mitchell, R. S., V. C. LaMarche, and R. M. Lloyd. 1966. Alpine vegetation and active frost features of Pellisier Flats, White Mts., California. Amer. Midl. Naturalist 75:516–525.

Mooney, H. A. 1973. Plant communities and vegetation. Pp. 7–17 in R. M. Lloyd and R. S. Mitchell. A flora of the White Mountains, California and Nevada. Univ. Calif. Pr., Berkeley.

———, G. St. Andre, and R. D. Wright. 1962. Alpine and subalpine vegetation patterns in the White Mts. of California. Amer. Midl. Naturalist 68:257–273.

———, W. D. Billings, and R. D. Hillier. 1965. Transpiration rates of alpine plants in the Sierra Nevada of California. Amer. Midl. Naturalist 74:374–376.

Pase, C. P. 1982. Alpine tundra. Pp. 27–33 *in* D. E. Brown (ed.). Biotic communities of the American Southwest—United States and Mexico. Desert Plants 4:1–341.

Spira, T. P. 1987. Alpine annual plant species in the White Mountains of eastern California. Madroño 34:315–323.

Stebbins, G. L. 1982. Floristic affinities of the high Sierra Nevada. Madroño 29:189–199.

Thorne, R. F. 1982. The desert and other transmontane plant communities of southern California. Aliso 10:219–257.

Wardle, P. 1971. An explanation for alpine timberlines. New Zealand J. Bot. 9:371–402.

———. 1974. Alpine timberlines. Pp. 371–402 *in* J. D. Ives and R. G. Barry (eds.). Arctic and alpine environments. Methuen Pr., London.

Went, F. W. 1948. Some parallels between desert and alpine flora in California. Madroño 9:241–249.

Whitney, S. 1979. A Sierra Club naturalist's guide to the Sierra Nevada. Sierra Club Books, San Francisco. 526 pp.

Zwinger, A. H., and B. E. Willard. 1972. Land above the trees: a guide to American alpine tundra. Harper and Row, N.Y. 489 pp.

Chapter 18

Desert Scrub Communities

Desert scrub communities are shrublands located in areas of markedly low precipitation. These communities are generally very open, and the component species are highly adapted to survival under harsh climatic conditions with one or more drought adaptations. Large expanses of bare ground are a feature of desert areas. However, deserts are by no means environmentally uniform. They vary in temperature conditions, amount, quality and seasonality of precipitation, elevation, topography, soils, and many other factors. Because all these factors vary in California's deserts, the desert communities in the state are quite diverse.

The major factor controlling the distribution of deserts is, of course, the low amount of precipitation. Desert areas in California generally receive 20 cm (8 in) or less annual precipitation, and in some areas the total is 5 cm (2 in) or less. The precipitation is irregular, and in some years little or no rain or snow may fall at all. In fact, an entire year's precipitation can come in a single storm.

Several major factors contribute to the low precipitation, including rain shadows, high-pressure atmospheric cells, and cool ocean currents. The numerous north-south trending mountain ranges in California tend to intercept much precipitation on their western (windward) slopes and cast low-precipitation rain shadows on their eastern (leeward) slopes as well as on areas further to the east. Larger mountain masses cast larger rain shadows; consequently, the rain shadow of the Sierra-Cascade axis is much more extensive than that of the Coast Ranges. The driest areas in the state are in the transmontane part of California, and most of the state's deserts are located in these areas. Smaller deserts are located in the rain shadows of the South Coast Ranges and western Transverse Ranges. Overall, desert vegetation occupies about 30–34 percent of California's total land area.

On a worldwide basis, the climate around 30° N and S latitudes (the Horse Latitudes) tends to be dominated by persistent subtropical high-pressure atmospheric cells. Dry air flows downward toward the sea, warming as it descends, and restricts precipitation by shunting storms to the north or to the south. During much of the year, high atmospheric pressure dominates the climate of the lowland deserts of southern California and adjacent regions, keeping storm fronts from moving onshore in these areas.

The cold California Current also contributes to the dryness of southern California. Air masses moving across this ocean current are cooled, and their moisture-holding capacity is reduced. As the air moves onto the land, it is warmed and draws moisture from the soil and plants. In Baja California this is so pronounced that deserts occur along the coast. In California the effects are less prominent, resulting in Mediterranean vegetation along the coast and desert vegetation inland. The overall effect is a progressive decrease in precipitation from northern to southern California.

The desert communities of California occur at elevations below the coniferous forests and usually below the desert woodlands. The deserts often occupy elevational zones that support much more mesic communities on the windward sides of major mountain ranges. Deserts vary from west to east, from north to south, and from upland to lowland sites. On the basis of elevation, the desert communities can be classified as either high deserts, mostly above 1300 m (4300 ft) elevation, or as low deserts, below sea level in some transmontane valleys to about 1300 m. These deserts differ in several important environmental features. The high deserts generally have: (1) a shorter growing season, (2) lower summer temperatures, (3) colder winter temperatures with frequent frosts, (4) greater diurnal temperature fluctuations, (5) higher annual precipitation with snow as a frequent source of precipitation, and (6) a more northerly distribution. High desert communities include Great Basin sagebrush scrub, blackbush scrub, and parts of the saltbush scrub. Low desert communities include creosote bush scrub, desert sand dune communities, desert dry wash scrub, parts of the saltbush scrub, alkali sink, and west Central Valley scrub.

Soils in California desert regions are usually azonal aridosols or entisols. They range from very rocky soils, to gravel, sand, or silt. In lowland areas these soils may be heavily impregnated with salts and may have a strongly alkaline pH. Fine-grained alluvial soils may be underlain by an impervious claypan or calcareous hardpan (caliche). Extreme salinity and alkalinity occur together in and around the dry beds of ancient lakes. Litter accumulation is very slight, and the organic content of desert soils is low. Desert soils often are not very permeable to water and a considerable fraction of the limited precipitation may be lost to runoff. Storms in desert regions often result in flash floods because so little water penetrates the soil. Water that enters the soil may not penetrate below the upper few cm and may be quickly lost to evaporation.

Desert communities in California are mostly dominated by shrubby species. These shrubs have various morphological and physiological adaptations that allow them to survive prolonged exposure to drought conditions. With low water availability as a limiting factor, desert shrubs have independently evolved an assortment of adaptations that reduce transpirational water loss. Most have small leaves with correspondingly small surface area. Some have such small leaves that most or all of their photosynthesis is carried out in green stems rather than in leaves (e.g., palo verde, cacti). The leaves and stems are often coated with whitish pubescence, resinous exudates, or waxy powders that increase the reflectivity of the surfaces and decrease their light and heat absorption (e.g., saltbushes, Fig. 18-4). Dense pubescence also decreases the evaporative potential of air moving near the leaf surface by interfering with molecular diffusion. Many of the shrubs are wholly or partially drought-deciduous with leaf loss or stem dieback reducing the evaporative surface when moisture stress is greatest (e.g., ocotillo). Some, such as *Encelia farinosa*, produce large, thin, greenish leaves when water supplies are ample, and smaller, thickish, densely white-pubescent leaves as temperatures rise in the spring.

Succulents such as cacti, yuccas, and agaves retain moisture in their swollen stems or leaves. Most succulents have the CAM (Crassulacean Acid Metabolism) photosynthetic syndrome that allows them to absorb CO_2 at night and keep their stomates closed in the day when transpiration loss is generally at its peak. Others, such as some Chenopodiaceae, have C-4 photosynthesis, a combined morphological and physiological system that increases photosynthetic rates at high temperatures and decreases the rate of transpiration under stressful situations. Some species that retain their foliage during drought periods have leaves that are internally reinforced with sclerenchymatous tissues, preventing structural collapse during periods of physiological wilting.

Root systems of desert shrubs are often much more extensive than are the above-ground portions of the plants. Roots may be widespread in the shallow upper layers of the soil or may extend very deep into the subsoil. Shallow roots enable a plant to utilize moisture from rainfall that does not penetrate more than 10 cm into the soil. Deeply rooted shrubs usually occur in areas such as washes where deep porous sediments allow moisture to percolate into the soil. The root systems of some shrubs such as mesquite may penetrate all the way to the water table, sometimes

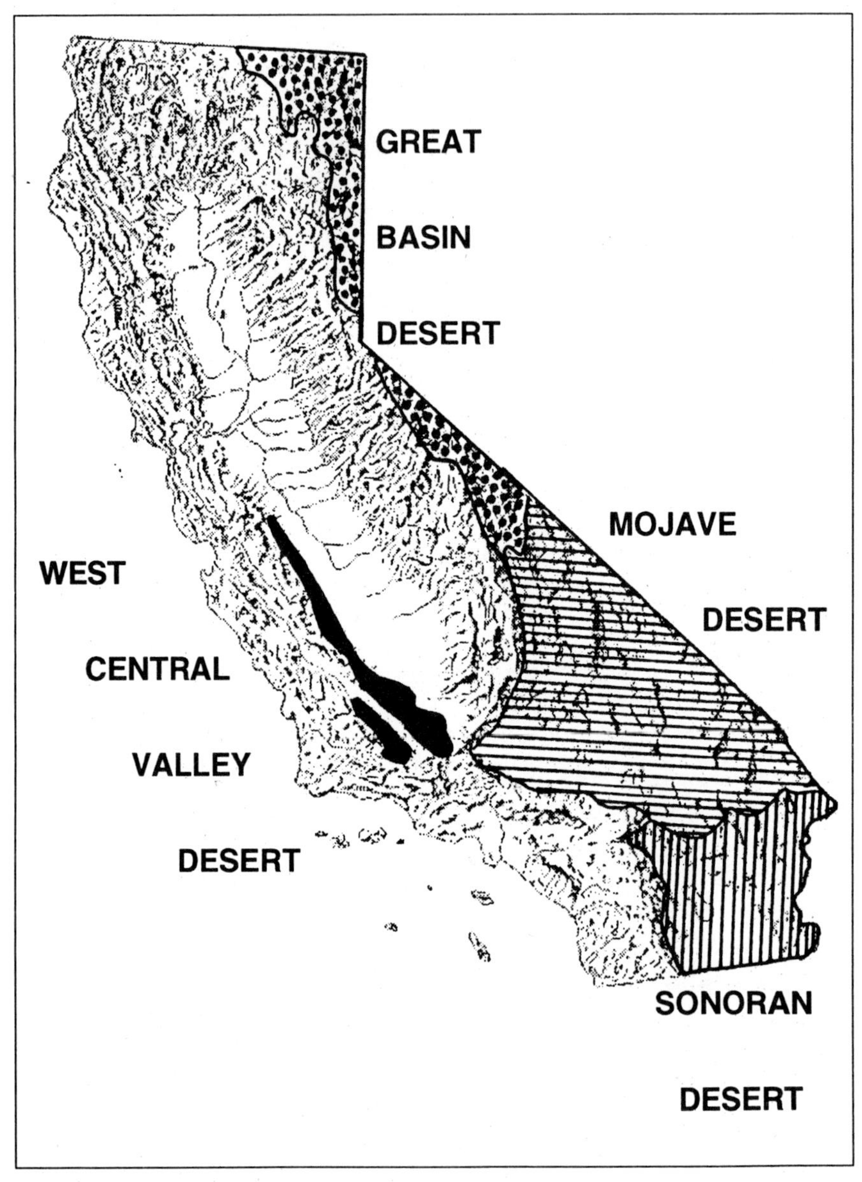

Fig. 18-1. Distribution of Great Basin, Mojave, Sonoran, and West Central Valley Deserts in California. The Great Basin, Mojave, and Sonoran Deserts all extend well beyond the boundaries of California.

several m below the soil surface. Plants that tap the water table are often known as phreatophytes (well-plants). Most phreatophytes occur as members of desert dry wash communities or desert riparian communities.

A significant part of the desert flora is the crop of ephemeral species. Ephemerals are either perennials with deeply seated bulbs, corms or rhizomes, or annual species that complete their entire life cycle in a period ranging from a few weeks to a few months. Ephemerals generally grow only when a sufficient amount of precipitation has fallen to permit hydration of their seeds, to dampen the upper layer of soil, and to permit seedling establishment. The life cycle of ephemerals is therefore very tightly correlated with seasonal precipitation.

Ephemerals are often described as "drought evaders" because they grow only when drought conditions have been temporarily alleviated by precipitation. Some ephemerals are rather mesophytic in appearance, but most have one or more drought adaptations. In seasons that follow favorable precipitation, the desert ephemerals can form beautiful floral displays. Many ephemeral species have showy, disproportionately large flowers or inflorescence. Distribution of precipitation varies from year to year, and some years are warmer or cooler than others. The composition of the ephemeral wildflower display reflects this seasonal variation, and year to year differences can be striking. The factors that cause species to grow and flower in one year and not in another remain to be investigated.

1. Great Basin Sagebrush Scrub

Great Basin sagebrush scrub communities are among the most widespread of the North American desert communities, occupying much of the Great Basin region of Utah, Nevada and adjacent portions of surrounding states. In California (Fig. 1) these communities occur in relatively deep but well-drained, non-alkaline soils from the eastern base of the Cascades across the Modoc Plateau and south along the eastern flanks of the Sierra Nevada to the Transverse Ranges. They also extend into cismontane California as far north as the upper drainage of the Cuyama River. Isolated patches of Great Basin sagebrush scrub occur in Siskiyou and San Diego Counties but are poorly developed in these areas. They also can be found on upland slopes of the higher desert mountain ranges. These communities normally occur between 1300 and 2300 m (4250–7550 ft) elevation but may extend to 3600 m (11,800 ft) in such areas as the White Mountains.

A

B

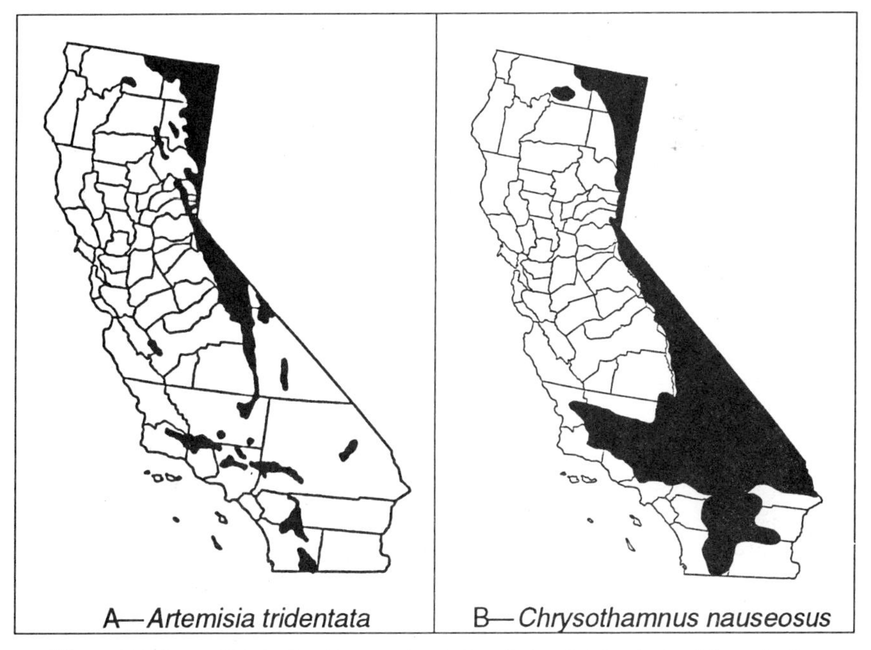

Fig. 18-3. Distribution of *Artemisia tridentata* (Great Basin sagebrush) and *Chrysothamnus nauseosus* (rabbitbush) in California.

The dominant shrub in most areas occupied by these communities is Great Basin sagebrush (*Artemisia tridentata*; Figs. 18-2, 18-3A) which sometimes forms pure stands but often grows together with a mixture of other shrubs. Common associates are:

Atriplex spp.	saltbush
Chrysothamnus nauseosus (Fig. 18-3B)	rabbitbush
Chrysothamnus viscidiflorus	rabbitbush
Ephedra viridis	Mormon-tea
Grayia spinosa	hopsage
Purshia tridentata	bitterbrush
Tetradymia canescens	gray horsebrush

Other sagebrush species such as *A. arbuscula*, *A. nova*, and *A. cana* may be locally dominant. Edaphic conditions are in large part the factors determining local species composition.

In addition to the shrubby dominants, Great Basin sagebrush communities usually have a significant herbaceous component.

Fig. 18-2. A. Great Basin sagebrush (*Artemisia tridentata*). The grayish leaves of this species give a gray-green color to the vegetation it dominates. Photo by David Keil. **B.** Great Basin sagebrush scrub in the Mammoth Lakes area of Mono County. *Artemisia tridentata* forms a monotonous gray-green mantle over extensive areas of eastern California and adjacent states. Photo by V. L. Holland.

The herbaceous cover is generally dominated by a mixture of perennial bunch grasses such as:

Achnatherum hymenoides [*Oryzopsis h.*]	ricegrass
Achnatherum spp. [*Stipa*]	needlegrass
Bromus carinatus	California brome
Elymus elymoides [*Sitanion hystrix*]	squirreltail
Leymus cinereus [*Elymus c.*]	Great Basin wild-rye
Pseudoroegneria spicata [*Agropyron s.*]	bluebunch wheatgrass

In areas with sparse or no shrub cover, these grasses may form desert grasslands (Chapter 11). Annual grasses and various annual and perennial forbs also occur in these communities. In many areas an introduced annual grass, *Bromus tectorum* (cheatgrass), has become the dominant herbaceous species.

The overall appearance of sagebrush communities is usually a monotonous gray-green expanse of shrubs with bare ground and scattered herbs among the shrubs (Fig. 18-2B). Local species diversity is often very low. These communities often extend for miles with little change in appearance or species composition.

Great Basin sagebrush scrub communities form ecotones with several other communities. At high elevations on the eastern slopes of the Sierra Nevada and in the desert mountain ranges, components of the Great Basin sagebrush scrub communities often form the understory layers of such tree-dominated communities as piñon-juniper woodlands, montane mixed coniferous forests, and even subalpine forests. At lower elevations the sagebrush scrub communities intergrade with saltbush scrub, blackbush scrub, Joshua tree woodlands, and occasionally creosote bush scrub communities. Where soil salinity and alkalinity are high, sagebrush scrub communities are largely replaced by saltbush scrub.

Areas of Great Basin sagebrush scrub are often used as grazing land for cattle and sheep. Overgrazing historically has tended to eliminate the more palatable species such as the perennial grasses and to increase dominance of shrubs, particularly such unpalatable plants as *Artemisia* and *Chrysothamnus* spp. Various range-management practices have been implemented to enhance growth of grasses and other forage species. Burning, which can increase the growth of both grasses and forbs, has the drawback of killing *Purshia tridentata*, which is an important forage species. Range managers sometimes seed *Purshia tridentata* into burned areas to re-establish its cover.

However, the introduced *Bromus tectorum* sometimes invades and outcompetes the desirable native forage species. This grass is both low in forage value and very flammable when dry. Once established, *B. tectorum* often grows very densely, and it burns frequently. These fires may be hot enough to kill native bunch grasses and eliminate desirable shrubs.

2. Saltbush Scrub

Saltbush scrub communities occur mostly on soils in which the soluble salt content is high (Figs. 18-6, 18-8). Usually these soils also have a high soil pH. Sites where such soils occur are mostly broad desert plains and beds of ancient lakes. Soils of these areas often have an impervious claypan or caliche layer. Not all saltbush scrub communities occur in highly saline sites, however. Upland slopes with stony soils may also support saltbush scrub, as in the mountains around Death Valley in Inyo County.

Saltbush scrub communities are mostly dominated by shrubby members of the Chenopodiaceae. Common chenopods include:

Atriplex canescens (Figs. 18-4A, 18-5A)	four-wing saltbush
Atriplex confertifolia (Figs. 18-4C, 18-5C)	shadscale
Atriplex polycarpa (Figs. 18-4B, 18-5B)	allscale
Grayia spinosa (Fig. 18-7A)	hopsage
Krascheninnikovia lanata [*Ceratoides l.*, *Eurotia l.*] (Fig. 18-5D)	winterfat
Sarcobatus vermiculatus	greasewood

Other shrubs common in such communities are:

Artemisia spinescens	bud-sage
Chrysothamnus viscidiflorus	rabbitbush
Ephedra spp.	Mormon-tea
Psorothamnus polyadenius [*Dalea p.*]	indigo-bush
Tetradymia spp.	cottonthorn, horsebrush

The most common grasses are the perennials, *Achnatherum hymenoides* and *Elymus elymoides*, and the introduced annual, *Bromus tectorum*. Various other herbs and forbs also occur in these communities, including the introduced *Salsola tragus* [*S. iberica*, *S. kali*] (Russian-thistle) which is particularly common in some sites.

Saltbush scrub communities have a wide elevational range and occur in both transmontane and cismontane California. At low elevations and in strongly saline soils, such species as *Atriplex polycarpa*, *A. canescens* (four-wing saltbush) and *A. hymenelytra* (desert-holly) predominate. In the West Central Valley Desert

Fig. 18-4. Saltbushes are common gray-leaved shrubs in desert areas with saline soils. Species differ in foliage features and in the form of the bracts that surround the dry fruits. **A.** *Atriplex canescens* (four-wing saltbush). **B.** *Atriplex confertifolia* (shadscale). **C.** *Atriplex polycarpa* (allscale). Photos by David Keil.

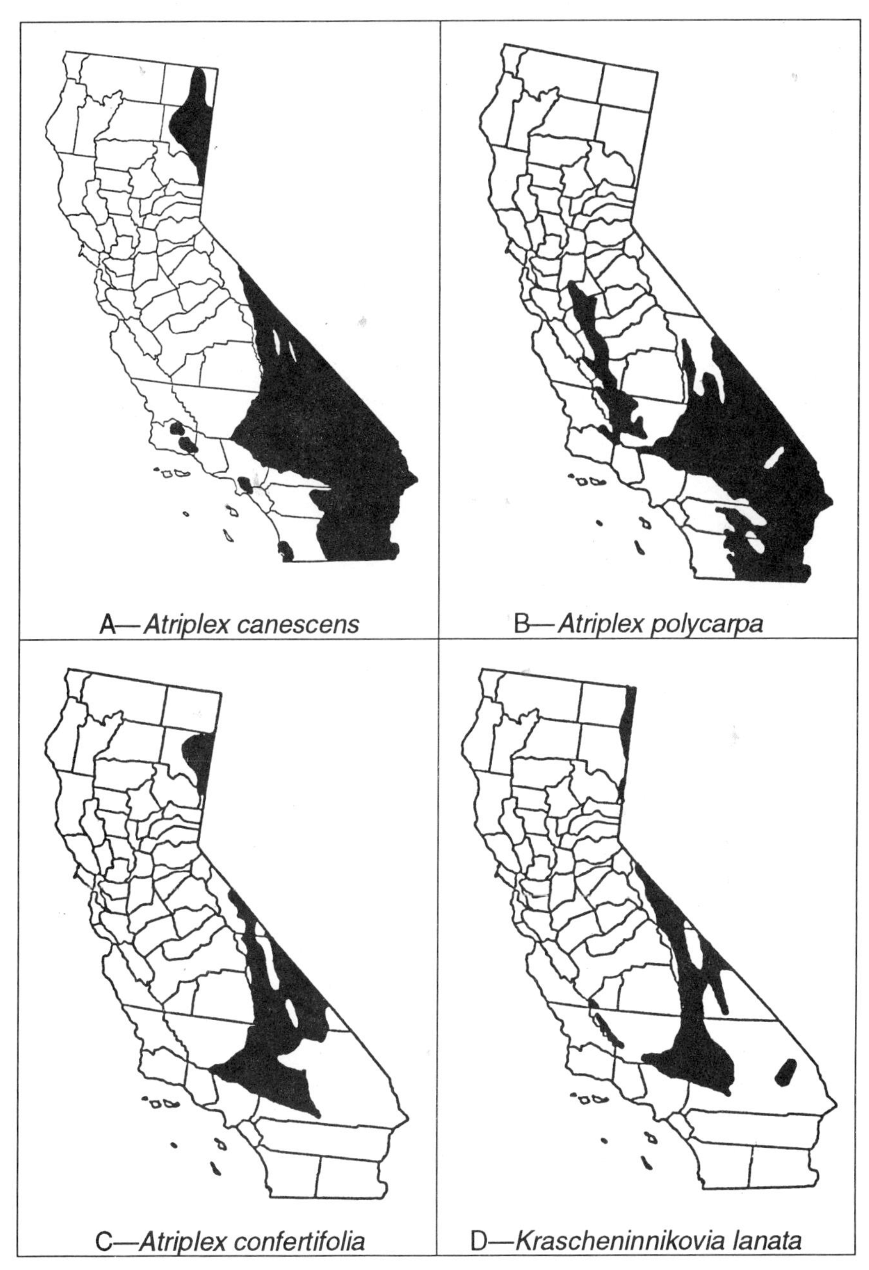

Fig. 18-5. Distribution of *Atriplex canescens* (four-wing saltbush), *Atriplex confertifolia* (shadscale), *Atriplex polycarpa* (allscale), and *Krascheninnikovia lanata* (winterfat) in California.

Fig. 18-6. Alluvial fan (bajada) at base of El Paso Mountains in eastern Kern County. Creosote bush scrub occupies the bajada and mountain slopes and saltbush scrub the dry valley. The transition from

Atriplex spinifera (spine-scale), *A. polycarpa*, and *Ephedra californica* (Mormon-tea) often are dominant.

Saltbush scrub communities are ecotonally related to several other communities (Fig. 18-6). In areas of progressively increasing soil salinity bordering the playa of dried up lakes, saltbush scrub grades into alkali sink communities (Fig. 18-8). Where soil salinity decreases significantly, saltbush scrub communities grade into Great Basin sagebrush scrub, creosote bush scrub or valley grasslands. The ecotone between saltbush scrub and blackbush scrub is often very indistinct, and these communities may intergrade to a considerable extent.

3. Blackbush Scrub

Blackbush scrub communities are elevationally intermediate between the Great Basin sagebrush scrub communities of the high desert and the creosote bush scrub communities of the low desert. They occur sometimes as an understory in Joshua tree or piñon-juniper woodland areas and intergrade over a broad ecotone with saltbush scrub communities. In California, blackbush scrub communities are restricted to transmontane areas; they extend eastward to the southern part of the Great Basin desert in southern Utah and northern Arizona.

creosote bush scrub to saltbush scrub marks an increase in soil salinity and pH from the well-drained bajada to the soils of the valley. Photos by V. L. Holland.

Blackbush scrub communities are aptly named. The dominant shrub, *Coleogyne ramosissima* (blackbush; Fig. 18-7B) is a low-growing, intricately branched gray-green bush. En masse these shrubs give the landscape a blackish appearance. The species that occur with blackbush are as common members of one or another of the communities with which blackbush scrub shares an ecotonal relationship. Common shrubby associates include:

Ephedra spp.	Mormon-tea
Eriogonum fasciculatum	California buckwheat
Grayia spinosa	hopsage
Hymenoclea salsola	burrobrush
Krascheninnikovia lanata	winterfat
Lycium cooperi	wolfberry
Psorothamnus polyadenius	indigo-bush
Tetradymia spp.	cottonthorn, horsebrush
Thamnosma montana	turpentine broom

4. Creosote Bush Scrub

Creosote bush scrub communities cover extensive areas in the hot deserts of southeastern California. Creosote bush (*Larrea tridentata*, Figs. 18-7C, 18-9, 18-10A, C) is one of the most

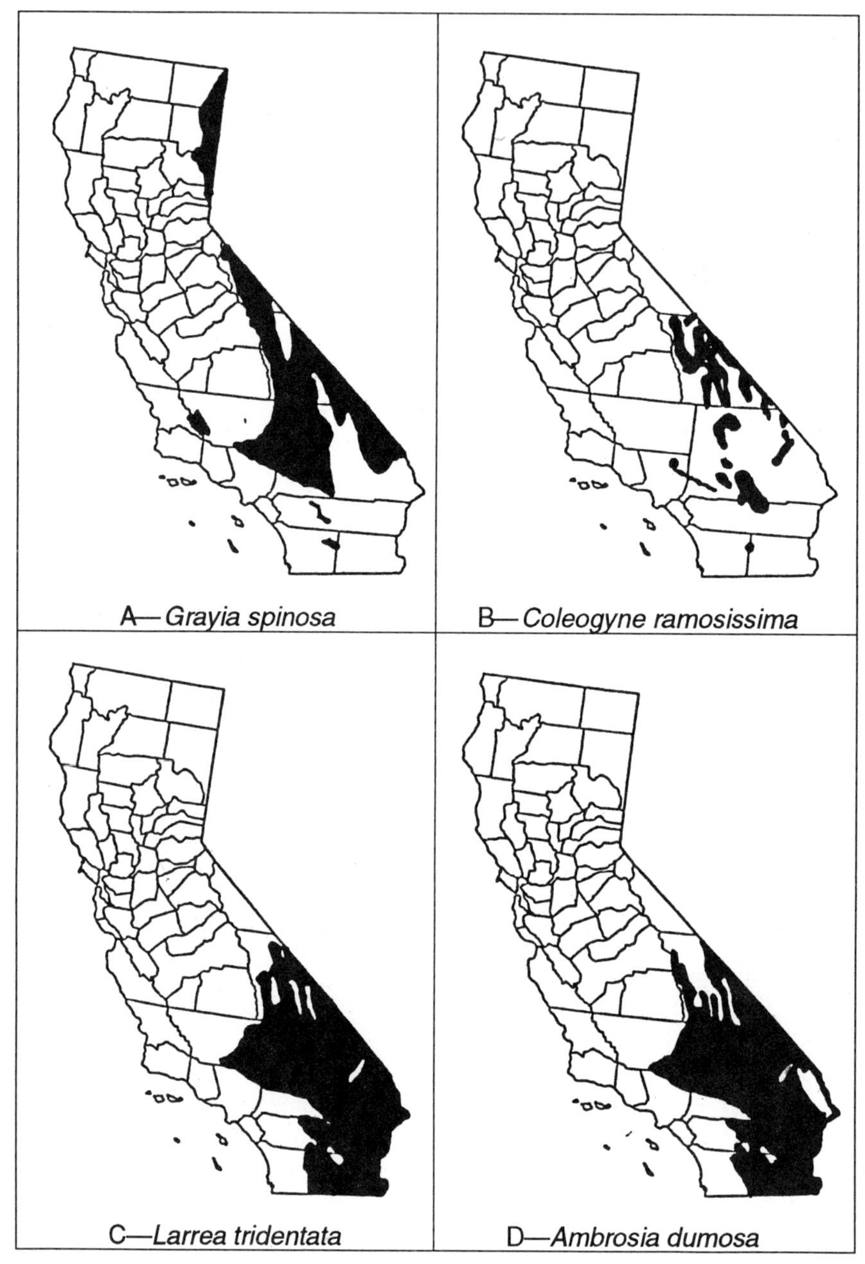

Fig. 18-7. Distribution of *Grayia spinosa* (hopsage), *Coleogyne ramosissima* (black bush), *Larrea tridentata* (creosote bush), and *Ambrosia dumosa* (white bursage) in California.

Piñon Pine-Juniper Woodland

↓

Joshua Tree Woodland

↓

Creosote Bush Scrub

Environmental gradient to Creosote Bush Scrub
- Decrease in elevation
- Increase in summer and winter temperatures
- Increase in frost free days
- Decrease in precipitation (rainfall and snow)

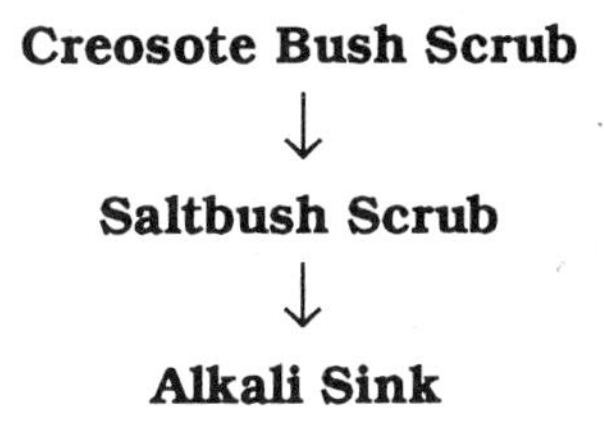

Environmental gradient to Alkali Sink
- Soil salinity, pH, and water increases
- Soil texture becomes heavier (clayey)
- Soil aeration decreases
- Soil drainage decreases

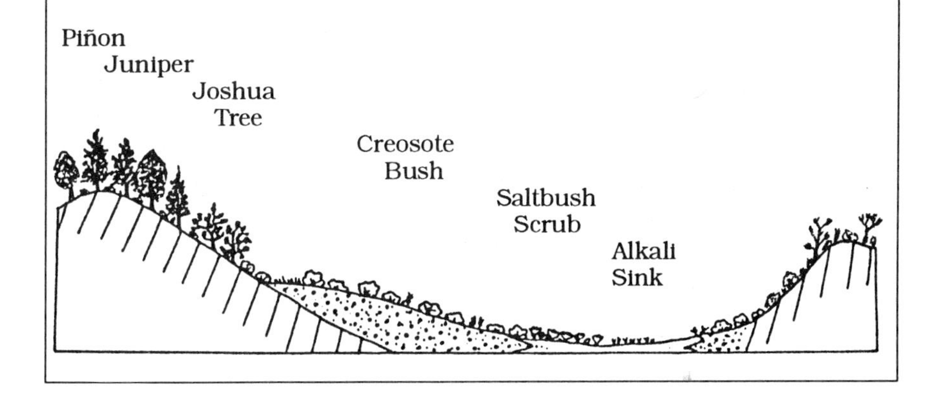

Fig. 18-8. Clinal variation among desert communities. **A.** Along a transect from piñon-juniper woodland to creosote bush scrub the vegetation reflects climatic gradients. **B.** The community composition along a transect from creosote bush scrub to alkali sink communities reflects a change in soil conditions.

widespread shrubs in North America. This species, sometimes considered to be conspecific with a South American species, *Larrea divaricata*, occurs as a dominant shrub in most of the hot desert areas of North America. Three major hot desert areas are often delimited: the Chihuahuan Desert, the Mojave Desert and the Sonoran Desert. The Chihuahuan Desert, characterized by summer precipitation and little or no winter precipitation, is centered in north-central Mexico and extends northward to western Texas and southern New Mexico. Creosote bush is dominant in much of the Chihuahuan Desert, but as this desert does not occur in California, it will not be further considered here.

The Mojave Desert (Fig. 18-1) is centered in southern California and extends into southern Nevada and northwestern Arizona. This desert region occurs at moderate elevations mostly between 900 and 1300 m (3000–4200 ft) and is characterized by mostly winter precipitation with infrequent summer rainfall. The Mojave desert is the driest of the North American deserts.

The Sonoran Desert (Fig. 18-1) is a mostly lowland desert extending from southern California to southwestern and central Arizona, and south to southern Baja California Sur and western Sonora. In the Sonoran Desert region, precipitation occurs during two rainy seasons. Winter rain falls during the general winter storms that sweep inland from the Pacific. During the late summer months, tropical monsoon storms sometimes move northward and result in localized, brief, and often intense thunderstorms. The portion of the Sonoran Desert that occurs in southern California is called the Colorado Desert. It extends from the Colorado River area westward to the southern flank of the eastern Transverse Ranges and to the desert flanks of the Peninsular Ranges. Within this region summer precipitation decreases from east to west and from south to north. The Colorado Desert area receives the least summer precipitation of any region of the Sonoran Desert.

The hot desert environment is at times very severe. During the summer, atmospheric temperatures can exceed 46° C (115° F) in the shade, and the temperatures of the soil surface can reach as high as 71° C (160° F). Insolation is intense with both direct solar radiation and reflection from bare soil surfaces. The potential evapotranspiration vastly exceeds the limited available moisture, and relative humidity is commonly below 20%. Although summer precipitation does occur in some areas, it is of irregular distribution, and in some years is absent. Consequently, perennials must be able to survive very prolonged drought under the harshest of conditions.

Winters in desert areas dominated by creosote bush range from cool to warm. In Mojave Desert areas snow falls occasionally, but

Fig. 18-9. Pure stand of creosote bush (*Larrea tridentata*) in Mojave Desert north of Mojave in eastern Kern County. Photo by David Keil.

it does not persist. In the Colorado Desert, snowfall is uncommon or completely absent. Frost is more frequent in the Mojave Desert than in the Colorado Desert. Frost is a factor that limits the range of some desert species. Creosote bush can tolerate moderate frost although it suffers stem dieback under severe freezing conditions. Daytime winter temperatures are usually mild, ranging from 7 to 24° C (45–75° F) during the coolest winter periods. Soil moisture is retained much longer after winter storms than after summer storms because winter temperatures are so mild.

Communities dominated by creosote bush are quite varied. Few species share the broad ecological tolerances of *Larrea tridentata*; consequently, local conditions determine the associated species although *Larrea* may remain dominant. Lowland valleys and outwash slopes are often dominated by communities of low diversity. In some areas *Larrea* occurs in extensive pure stands of widely spaced individuals with large expanses of bare soil (Fig. 18-9) Many such areas support an association of 1-2 meter tall *Larrea* bushes and much shorter bushes of *Ambrosia dumosa* [*Franseria d.*] (white bursage; Figs. 18-7D, 18-10B, C). Other frequent associates include *Hymenoclea salsola* (burrobrush), *Atriplex canescens* and *Atriplex polycarpa*. Steep slopes and bajadas (outwash slopes) with numerous small arroyos and rock outcrops offer a diversity of microhabitats. Increased elevation also often increases habitat diversity. In such situations the number of shrubby species may

Fig. 18-10. A. Creosote bush (*Larrea tridentata*). The dark green bilobed leaves are covered with a resinous exudate and at least some of them persist through the heat of the summer. **B.** White bursage (*Ambrosia dumosa*). Most of the small, gray-green leaves are summer deciduous. **C.** Creosote bush-white bursage association. Creosote bush and white bursage commonly grow together over large areas in both the Mojave and Colorado Deserts. Photos by David Keil.

be quite high and dominance very difficult to ascertain. The following lists of species that occur with *Larrea*, or that locally replace it, are by no means complete. In Mojave Desert areas, common associates of creosote bush include:

Acamptopappus sphaerocephalus	goldenhead
Ambrosia dumosa	white bursage
Coleogyne ramosissima	blackbush
Encelia actoni	Acton encelia
Encelia farinosa	brittlebush
Encelia frutescens	rayless encelia
Encelia virginensis	Virgin River encelia
Ephedra spp.	Mormon-tea
Ericameria cooperi [*Haplopappus c.*]	Cooper's goldenbush
Eriogonum inflatum	bladderstem
Grayia spinosa	hopsage
Hymenoclea salsola	burrobrush
Krameria spp.	ratany
Krascheninnikovia lanata	winterfat
Lepidium fremontii	shrubby peppercress
Opuntia basilaris (Fig. 18-11)	beavertail cactus
Opuntia echinocarpa	staghorn cholla
Pleuraphis rigida [*Hilaria r.*]	big galleta grass
Salazaria mexicana	bladder-sage
Salvia dorrii	desert purple sage
Yucca brevifolia (Fig. 18-13)	Joshua tree
Yucca schidigera	Mojave yucca

In Colorado desert areas *Larrea* often associates with such shrubs and succulents as:

Agave deserti (Fig. 18-12)	century plant
Ambrosia dumosa	white bursage
Echinocereus engelmannii	hedgehog cactus
Encelia farinosa	brittlebush
Ephedra spp.	Mormon-tea
Eriogonum fasciculatum	California-buckwheat
Fagonia spp.	fagonia
Ferocactus acanthodes	barrel cactus
Fouquieria splendens	ocotillo
Krameria grayi	ratany
Lycium spp.	wolfberry
Opuntia bigelovii	teddybear cholla
Opuntia echinocarpa	staghorn cholla
Simmondsia chinensis	jojoba

The *Larrea*-dominated communities of the Mojave Desert and Colorado Desert grade into one another and share numerous species. Cacti are better represented in the Colorado Desert both in total number of individuals and in diversity. Creosote bush scrub communities are ecotonally related to a wide variety of other California plant communities. These communities intergrade with such desert communities as saltbush scrub,

Fig. 18-11. Beavertail cactus (*Opuntia basilaris*) is a common species in Creosote bush scrub communities in the Mojave Desert. Photo by David Keil.

Fig. 18-12. *Agave deserti* (century plant) in Sonoran desert creosote bush scrub community in Anza Borrego State Park, San Diego County. Photo by E. Craig Cunningham.

Fig. 18-13. Creosote bush-white bursage association in Mojave Desert at Red Rock Canyon State Park in eastern Kern County. A few small Joshua trees are scattered through the community. Photo by David Keil.

blackbush scrub, and alkali sink scrub, and are traversed by various manifestations of desert dry wash communities. Creosote bush and its associates sometimes form the understory in desert woodlands dominated by junipers or Joshua trees. In areas where the interior forms of southern coastal scrub approach the desert fringes, considerable overlap of the component species may occur. Species from semidesert chaparral communities sometimes also grow with plants of the creosote bush scrub.

In areas that receive summer rainfall, creosote bush scrub communities have two crops of ephemerals per year. One crop develops after winter storms. Winter rains are usually cool, and the moderately long spring growing season is usually characterized by equable temperatures. Some of the winter ephemerals are mesomorphic in their vegetative structure and are unable to withstand severe drought stress. Numerous annuals and perennial ephemerals go through their reproductive cycles during the spring. In seasons with adequate moisture a great diversity of annuals may flower. and present colorful wildflower displays.

The winter-spring ephemerals have considerable taxonomic diversity, representing a large variety of families and genera. A few of the most common are listed below:

Asteraceae

Chaenactis spp.	pincushion flower
Eriophyllum spp.	woolly daisy
Filago spp.	herba impia
Geraea canescens	desert-sunflower
Monoptilon spp.	desert star

Malacothrix glabrata	desert dandelion
Stylocline spp.	nest straw
Brassicaceae	
Lepidium spp.	peppercress
Boraginaceae	
Amsinckia spp.	fiddleneck
Cryptantha spp.	cryptantha
Pectocarya spp.	comb-seed
Fabaceae	
Lotus spp.	deervetch
Lupinus spp.	lupine
Hydrophyllaceae	
Nama spp.	purple mat
Phacelia spp.	phacelia
Lamiaceae	
Salvia columbariae	chia
Onagraceae	
Camissonia spp.	evening-primrose
Oenothera spp.	evening-primrose
Poaceae	
Bromus spp.	brome (introduced)
Schismus spp.	Mediterranean grass (introduced)
Polemoniaceae	
Eriastrum spp.	woollystar
Gilia spp.	gilia
Langloisia setosissima	sunbonnet
Linanthus spp.	linanthus
Loeseliastrum spp.	desert calico
Polygonaceae	
Chorizanthe spp.	spineflower
Eriogonum spp.	annual buckwheat
Scrophulariaceae	
Mimulus spp.	monkeyflower
Mojavea spp.	desert snapdragon

The annuals may produce very showy floral displays in some years, but in seasons with low precipitation, few seeds germinate and annuals that do grow often are stunted.

The conditions under which summer ephemerals develop are markedly different. Summer rains usually come in the form of thunderstorms. Temperatures are high at the time the rain is falling and during the subsequent growing season. A very different crop of annuals germinates following summer rains. These plants grow in the heat of the desert summer and must be able to reach reproductive maturity in the brief period before the moisture evaporates from the soil.

Physiologically the differences between the winter and summer annuals are striking. Germination responses of the two annual crops are temperature related. When soil temperatures are cool, only winter ephemerals germinate. Conversely, summer ephemerals germinate only if temperatures are high. Almost all winter ephemerals use the C-3 photosynthetic pathway, which is more efficient in cool temperatures than the C-4 pathway. On the other hand, almost all summer ephemerals are C-4 plants.

The summer ephemerals represent several different families in which evolution of C-4 photosynthesis has occurred independently, but the overall family diversity is much lower for summer-flowering than for winter-flowering annuals. Far fewer species comprise the summer annual flora than the spring flora, and the species composition is almost entirely different. Most of the families that are represented in the spring flora are absent in the summer. Among the most common summer-flowering ephemerals are:

Amaranthaceae	
Amaranthus fimbriatus	fringed amaranth
Asteraceae	
Pectis papposa	chinchweed
Euphorbiaceae	
Chamaesyce spp. [*Euphorbia* spp.]	prostrate spurge
Molluginaceae	
Mollugo cerviana	threadstem carpetweed (introduced)
Nyctaginaceae	
Allionia incarnata	windmills
Boerhaavia spp.	boerhaavia
Poaceae	
Aristida adscensionis	six-weeks three-awn
Bouteloua aristidoides	needle grammagramma
Bouteloua barbata	six-weeks gramma

5. Desert Sand Dunes

Desert sand dune communities occur where erosional processes in desert areas have yielded significant quantities of sand. Where there is insufficient vegetation or physical obstacles to impede wind erosion, sand can be blown for considerable distances. Desert sand dunes have developed in various parts of southern California. However, sandy deserts form a rather small portion of the total area of deserts in California. A characteristic of desert sand dunes is their ability to absorb water. Most desert soils are so impervious that much of the annual rainfall is ultimately lost to runoff or to surface evaporation. Desert sands, on the other hand, are porous, and water generally soaks in

Fig. 18-14. Desert sand dune at Death Valley National Park. On the sand dunes creosote bush (*Larrea tridentata*) grows to a greater stature than on other substrates nearby. The partially exposed stems were buried before the shifting sands exposed them. Tap-rooted annuals dot the sand. Photo by David Keil.

rather than running off. Additionally, because water can penetrate deeply into the sand, it mostly is below the upper layers that are warmed by the sun. As a result, sand dunes in desert areas retain moisture for longer periods than do most other desert soils.

Plants growing on desert sand dunes, like those of coastal dunes, must be able to exploit the relatively deep water supply and to withstand burial. Relatively few species occur on most desert dunes, but those that do grow there form an important plant cover on the dunes. Dominant species on desert sand dunes include *Larrea tridentata* (Fig. 19-19), *Prosopis glandulosa* (mesquite) and sometimes *Atriplex canescens*, *Psorothamnus schottii* [*Dalea s.*] or other shrubs. These species also occur in non-dune sites. However, on dunes they often reach considerably greater stature than they do in non-dune sites. This is a reflection of the availability of deep moisture supplies below the dry surface of the dunes. Perennial grasses that grow on desert sand dunes include *Achnatherum hymenoides*, *Pleuraphis rigida*, and the endemic Eureka Valley dune grass, *Swallenia alexandrae* which occurs only in Inyo County.

Herbaceous dune plants include:

Abronia villosa	sand-verbena
Baileya pauciradiata	small-headed desert-marigold
Cryptantha sp.	cryptantha
Dicoria canescens	dicoria
Geraea canescens	desert sunflower
Oenothera deltoides	evening-primrose
Palafoxia arida [*P. linearis*]	Spanish-needles
Tiquilia plicata [*Coldenia pl.*]	dune-mat
Tiquilia palmeri [*Coldenia pa.*]	dune-mat

These herbaceous plants are mostly spring-flowering ephemerals. A summer dune flora usually does not develop unless heavy rainfall occurs, and the species that do germinate are often those of the adjacent areas of creosote bush scrub. An exception is *Dicoria canescens* which flowers from late summer to early winter. In seasons with exceptionally heavy rainfall, *Geraea canescens* and *Palafoxia arida* may persist into the summer months.

6. Desert Dry Wash Communities

Crisscrossing the hot desert areas of California are innumerable arroyos, ravines, canyons, gullies, and dry washes. These normally dry channels are carved into the sides of rocky mountain slopes and broad desert plains alike. Although the channels only occasionally carry water, they quickly become raging torrents during desert storms. Just as quickly the flow can ebb, and the wash channels are soon dry again. The depth and width of these channels vary considerably. The small ones are sometimes a meter or less wide. The broadest are braided streams with anastomosing channels. The shallowest and steepest channels and those cut into bedrock have very little, if any, subsurface flow. They provide outlets for floodwater but retain very little of it. Large channels with deeply bedded sediments, however, may have subsurface water movement even though surface flow is absent. The alluvial fans (bajadas) that spread from the base of desert mountains often are dissected by numerous well-developed wash channels.

Desert dry wash communities develop along channels that are deep enough to have some subsurface water retention or flow. Dry wash communities are characterized by a mixture of shrubs from surrounding desert scrub communities and shrubs and trees that require a moisture supply greater than that available in areas away from the washes. Usually the plants of desert dry washes are taller than those of surrounding desert slopes or plains, and commonly they occur closer together as well.

Fig. 18-15. Desert dry wash community in Joshua Tree National Park. **A.** The large shrub to the right is *Chilopsis linearis* (desert-willow). The smaller shrubs are mostly *Hymenoclea salsola* (burrobrush). The wash channel is periodically swept clear of vegetation during flash floods. **B.** The gray-green shrub in the foreground is desert-lavender (*Hyptis emoryi*). The small tree in the background is palo verde (*Cercidium floridum*). Photos by David Keil.

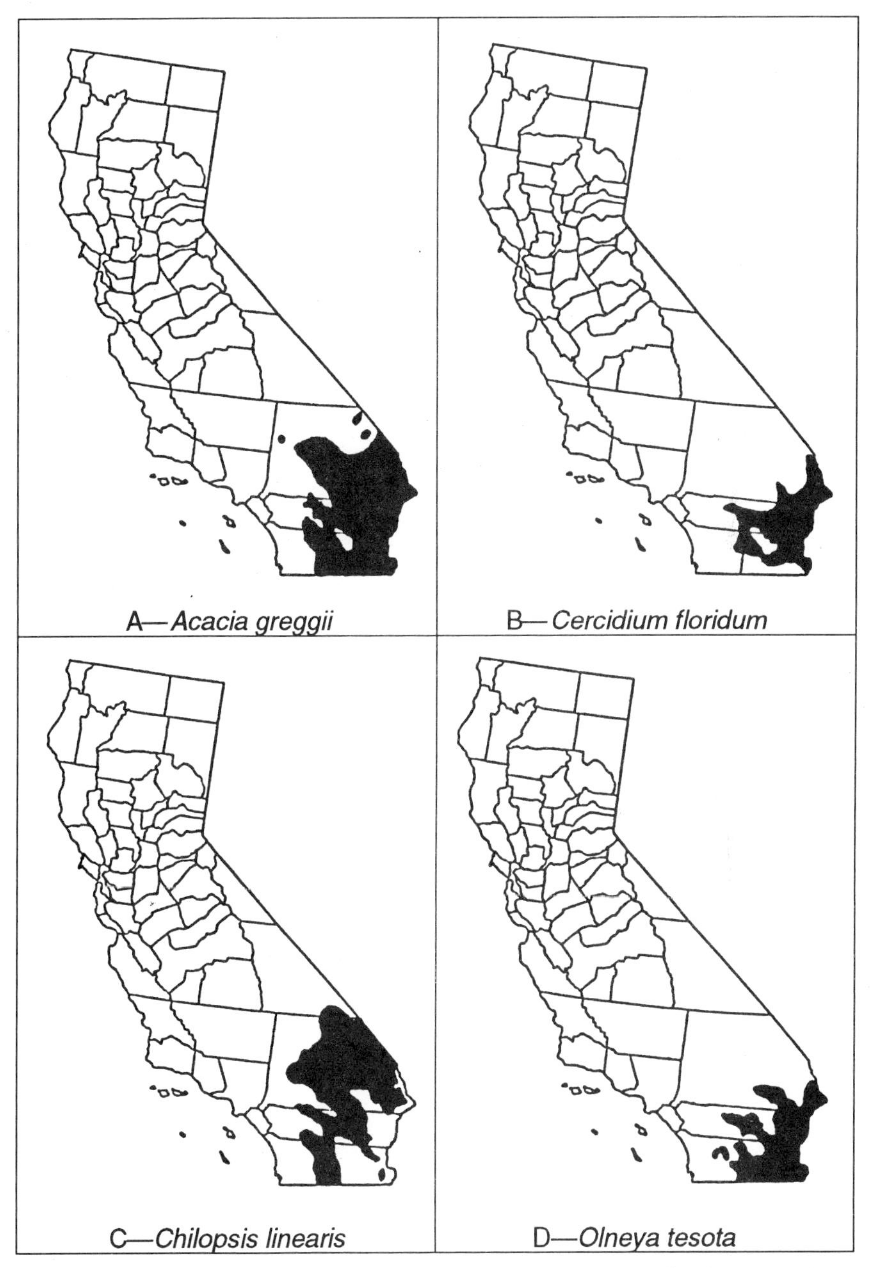

Fig. 18-16. Distribution of *Acacia greggii* (catclaw), *Cercidium floridum* (palo verde), *Chilopsis linearis* (desert-willow), and *Olneya tesota* (ironwood) in California.

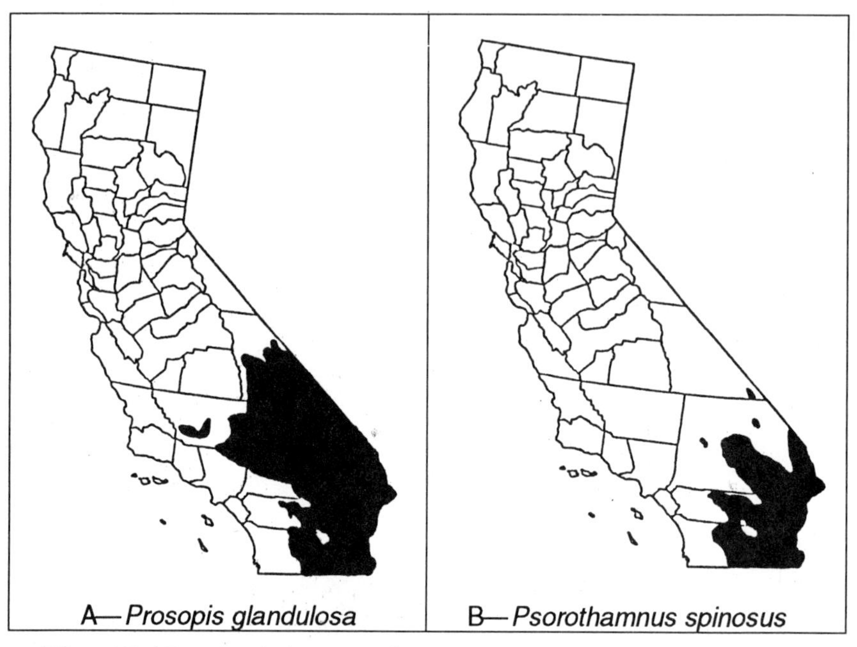

Fig. 18-17. Distribution of *Prosopis glandulosa* (mesquite) and *Psorothamnus spinosus* (smoke-tree) in California.

Often the only trees in a desert area are located along the wash channels. From above, these communities appear as dark green bands across the desert, and the contrast with the more open scrub is readily apparent.

Along dry washes, common trees or large shrubs include:

Acacia greggii (Fig. 18-15A)	catclaw
Cercidium floridum (Figs. 18-14A, 18-15B)	palo verde
Chilopsis linearis (Figs. 18-14A, 18-15C)	desert-willow
Olneya tesota (Fig. 18-15D)	ironwood
Prosopis glandulosa (Fig. 18-16A)	mesquite
Psorothamnus spinosus [*Dalea*](Fig. 18-16B)	smoke-tree

In addition, there are various other common shrubs including:

Baccharis sarothroides	broom baccharis
Bebbia juncea	sweet-bush
Hymenoclea salsola (Fig. 18-14A)	burrobrush
Hyptis emoryi (Fig. 18-14B)	desert-lavender
Isomeris arborea	bladderpod
Justicia californica [*Beloperone c.*]	chuparosa
Larrea tridentata	creosote bush
Lycium andersonii	wolfberry
Pluchea sericea	arrow-weed
Simmondsia chinensis	jojoba

Introduced species of the genus *Tamarix* (salt-cedar) also are frequent in these communities.

The majority of shrubs and trees occur on banks or islands of the wash, and few are actually found in the flood channel. The major reason for this distribution is the force of the infrequent flash floods. The sediment load for a wash during a flash flood is high, and large rocks can be moved by the stream. The force of the water is sufficient to break or uproot shrubs or trees anchored in the loose sediments of the stream channel. The movement of water-borne sediment has positive value to some plants that grow along desert washes. For several species, including *Olneya tesota*, *Cercidium floridum* and *Acacia greggii*, the seed coat must be broken before germination can occur. During flash floods, seeds are tumbled through churning sand and gravel, and this provides the necessary scarification.

Desert dry wash communities also include assorted ephemerals, most of which also occur in adjacent areas of desert scrub. The ephemerals often occur in the open sunny areas of the stream channel. Often the annuals achieve greater stature in washes than in adjacent areas with a lower water supply.

7. Alkali Sink Communities

Much of transmontane California and parts of the Central Valley are characterized by interior drainage. Streams that flow into these areas have no outlet to the ocean. Instead ,they empty
into basins where the only outlet for water is through evaporation (Figs. 18-17, 18-18). California and other western states are dotted with the dry or drying beds of ancient lakes. Wave-cut benches on now dry hillsides far above the dry lakebeds attest to the fact that some of these lakes were once extensive bodies of water. In fact, the drying of these lakes is recent in geological time. During the pluvial periods of the late Pleistocene, precipitation was somewhat higher in the desert regions than it is today, and because of lower overall temperature, the rate of evaporation was much lower. Massive mountain glaciers scoured the flanks and montane valleys in many of the high mountains of western North America. Melt water from these glaciers, plus rainwater and snow melt, served to fill the basins with water. About twelve thousand years ago, the climate became warmer and drier. With the retreat of the glaciers the water supply to the lakes was greatly decreased. The rate of annual evaporation exceeded the rate of inflow, and little by little the lakes dried up.

The water that filled the lakes was not pure. Contained within it were small amounts of dissolved impurities—minerals and salts of various kinds. As the water evaporated, the remaining solution

Fig. 18-18. Alkali sink communities. **A.** Rushes and sedges dominate a spring-fed interior salt marsh (alkali meadow) in Mono County. As water evaporates from the landlocked lake it leaves dissolved salts behind. The marsh vegetation gives way in the background to more salt-tolerant species and unvegetated salt flats. **B.** Dry alkali sink at Koehn Dry Lake in eastern Kern County. The lake bed is occasionally flooded by desert storms. The very sparse vegetation is dominated by *Atriplex* species and other members of the Chenopodiaceae. Photos by David Keil.

Fig. 18-19. Alkali sink communities are often dominated by succulent shrubs of the Chenopodiaceae. **A.** Iodine bush (*Allenrolfea occidentalis*). **B.** *Suaeda moquinii* (seep weed). Both are common in saline soils in cismontane as well as in transmontane California. **C.** Alkali sink community bordering Soda Lake in eastern San Luis Obispo County. The vegetation in the foreground is dominated by spiny saltbush (*Atriplex spinifera*) and iodine bush (*Allenrolfea occidentalis*). Photos by David Keil.

became progressively more concentrated. One by one the impurities reached a point of saturation and precipitated out. When all the water was gone, the beds of the former lakes were broad salt pans (playas). Subsequent rainfall and flash floods have swept additional dissolved salts into the lake beds and reworked the sediments around the margins. The result after thousands of years is a series of concentric rings of progressively more saline soils from the desert slopes above the lake beds to the salty centers of the playas (Fig. 18-8). Soils in the middle of a salt flat are usually too salty for any higher plant to grow. In the zones of decreasing salinity, some salt-tolerant species are able to survive. Water from desert storms sometimes floods the lake beds, but it is generally very shallow and ephemeral.

The assemblages of halophytic plants that occupy the borders of playas and shallow salty lakes are known as alkali sink communities. These communities have much in common with coastal salt marsh communities (Chapter 6). Both consist of halophytic plants and both occupy areas with zonation along salinity gradients. Both also have areas with considerable fresh-water influx and other areas with high salinity and often xeric conditions. These communities differ in that the climate of the coastal salt marshes is quite moderate because of the nearby ocean, whereas that of the alkali sink communities is often extremely hot. The soils of the coastal salt marsh are generally neutral or weakly acidic, whereas those of the alkali sink generally have a high basic pH because of the high carbonate content of the soil. Despite their differences, coastal salt marsh communities and alkali sink communities share several species and have quite a few genera in common. Shared species include:

Anemopsis californica	yerba mansa
Distichlis spicata	saltgrass
Frankenia salina [*F. grandifolia*]	frankenia
Salicornia subterminalis	pickleweed
Scirpus americanus [*S. olneyi*]	Olney threesquare

Shared genera include *Atriplex* (saltbushes), *Juncus* (rushes), *Suaeda* (seepweeds, sea-blites), and *Typha* (cattails).

Alkali sink communities are usually dominated by species of the family Chenopodiaceae (sometimes referred to as cheno-podiaceous species or simply chenopods) in dry sites (Figs. 18-17B, 18-18C) and by saltgrass, sedges and rushes in areas with an influx of fresh water (Fig. 18-17A). These areas are sometimes called alkaline meadows. Common plants of the alkali sink are listed below. Common chenopods include:

Allenrolfea occidentalis (Fig. 18-18A)	iodine bush
Atriplex hymenelytra	desert-holly
Atriplex polycarpa	allscale

Nitrophila occidentalis	nitrophila
Sarcobatus vermiculatus	greasewood
Suaeda moquinii (Fig. 18-18B) [*S. fruticosa*, *S. torreyana*]	seepweed

Other common species in moist to dry sites include:

Distichlis spicata	saltgrass
Frankenia salina	frankenia
Pluchea sericea	arrow-weed

Interior salt marshes form where freshwater streams or springs empty into an interior drainage basin (Fig. 18-17A). In such areas *Juncus cooperi* (rush), *Scirpus americanus*, *Typha domingensis* (cattail), *Phragmites australis* [*F. communis*] (common reed), and *Distichlis spicata* may occur together or in a vegetational mosaic and then grade into the drier zones dominated by *Allenrolfea*. Where sufficient fresh water is present, *Prosopis glandulosa* may be locally common. Eurasian weedy plants such as *Tamarix* spp. (salt-cedar), *Salsola tragus* (Russian-thistle) and *Bassia hyssopifolia* (bassia) also have become members of alkali sink communities in some areas.

8. West Central Valley Desert Scrub

The rain shadow of the California coast ranges is less extensive than that of the Sierra Nevada and other high mountains. Nevertheless, in the southern part of the San Joaquin Valley and adjacent dry valleys (Fig. 18-1), conditions are dry enough that the original plant cover in much of the area was a desert scrub vegetation. Some of this vegetation has been invaded by introduced grasses and now is at least in some areas intermediate in character between desert scrub and valley grassland communities. In parts of the San Joaquin Valley, however, the desertic nature of the communities is still apparent. West Central Valley desert scrub communities contain a mixture of cismontane and transmontane species. The codominants vary from area to area, and overall the diversity of shrubby species is low. Soil pH ranges from near neutral to strongly alkaline, and alkali sink communities occur in some areas.

On neutral to moderately alkaline soils the dominant plants of the West Central Valley desert scrub include:

Atriplex canescens	four-wing saltbush
Atriplex polycarpa	allscale
Atriplex spinifera	spinescale
Krascheninnikovia lanata	winterfat
Eastwoodia elegans	yellow mock-aster
Ephedra californica	Mormon-tea
Ericameria linearifolia [*Haplopappus l.*]	interior goldenbush

Eriogonum fasciculatum	California-buckwheat
Gutierrezia californica [*G. bracteata*]	snakeweed
Hymenoclea salsola	burrobrush
Isomeris arborea	bladderpod
Isocoma acradenia [*Haplopappus a.*]	alkali goldenbush
Lycium andersonii	wolfberry

Along the eastern flanks of the western Transverse Ranges are some well-developed Great Basin sagebrush communities. Ephemeral species of this area are largely cismontane and several are endemics. In alkali sink areas species such as the following predominate:

Allenrolfea occidentalis	iodine bush
Atriplex spinifera	spiny saltbush
Cressa truxillensis	alkali weed
Distichlis spicata	saltgrass
Isocoma acradenia	alkali goldenbush
Salsola tragus	Russian-thistle, tumbleweed
Suaeda moquinii	seepweed

In the southern part of the Central Valley there are areas of typical saltbush scrub. The West Central Valley desert scrub communities grade into valley grassland in various areas and into piñon-juniper woodland, juniper woodland or juniper-blue oak woodlands in other areas. Dry washes crossing these desert areas may contain desert shrubs, or woodland tree species, or such shrubs as *Baccharis salicifolia* [*B. glutinosa*, *B. viminea*] (seep-willow), *Pluchea sericea* (arrow-weed), *Chrysothamnus nauseosus* (rabbitbush), *Atriplex canescens* (four-wing saltbush), and *Salix* spp. (willow).

Human Impact On Desert Communities

Human impact on desert communities has taken, and continues to take, many forms. Areas of desert land have been mined for an assortment of minerals (Fig. 18-19A). In locations where the desert soils are suitable for cultivation and water is available for irrigation, desert land has become extremely productive agricultural land. Desert lands are marginal at best for grazing, but this has not prevented people from grazing sheep and cattle, sometimes at greater levels than the carrying capacity of the land. Feral burros are particularly troublesome in desert areas because of the damage they do to the vegetation. Wind erosion is one major consequence of overgrazing by domestic or feral livestock in desert regions.

Modern technology has allowed increasing numbers of people to live in or to visit the desert. Modern forms of transportation have made access to even the remotest areas of the desert practic-

Fig. 18-20. Human impacts in desert communities have taken many forms. **A.** Mines and mining towns such as Bodie, a ghost town in Mono County leave behind permanent scars on the landscape. Natural succession is slow in the Great Basin desert and other desert areas. **B.** Off-road vehicular activity in desert scrub communities crushes vegetation, causes soil erosion, and provides disturbance corridors for invasive weeds. This area north of Mojave in Kern County has been closed to ORV activity because of the severe damage, but many other areas are unprotected. Photos by V. L. Holland.

able. Air conditioning and other modern conveniences make living in desert areas far more comfortable than it ever was in the past. Human activities have resulted in many changes in desert plant communities. Cities and towns have expanded with housing developments, roads and businesses replacing desert plants. The military has developed enormous desert bases with airfields, practice bombing ranges, training areas for military maneuvers and numerous other facilities.

The increase in human occupancy and use of the desert has placed an inevitable strain on the critical limiting factor of the desert, its water supply. Water must come from either wells or from reservoirs. Removal of ground water has had deleterious effects on the vegetation of some desert areas. If the water table drops below the level to which the roots of phreatophytic species can grow, these plants may die. Extensive use of water for irrigation has resulted in the salinization of some agricultural soils. Some such areas that once were under cultivation now support successional communities of halophytic weeds. The construction of dams along the Colorado River and the network of irrigation canals have reduced the carrying capacity of this river and have allowed the desert riparian communities along its banks to become denser than in the past when periodic flash floods scoured the streamside vegetation. The Salton Sea, now a large body of water in the middle of the Colorado desert, owes its existence to human carelessness. This brackish lake was filled when irrigation canals from the Colorado River burst through their banks and diverted huge quantities of water from the river into a formerly dry lake bed.

Human activities in the desert areas of California include physical removal of plants and other direct damages. Cacti and other succulent plants have been extensively exploited for use in desert landscaping. The removal of mature plants from the community has resulted in degradation of the more accessible desert areas and such damage continues to occur. Off-road vehicular damage occurs in desert areas(Fig. 18-19B), just as it does in coastal dune areas (Fig. 8-10). The use and misuse of desert lands is a major political issue that will continue to be a source of controversy in California's future. As a result of passage of the Desert Protection Bill by the Congress in late 1994, a large portion of the California desert was placed under the protection of the National Parks Service.

Desert Scrub Communities—References

Adams, S., B. R. Strain, and M. S. Adams. 1970. Water-repellent soils, fire and annual plant cover in a desert scrub community of southeastern California. Ecology 51:696–700.

Anderson, D. J. 1971. Pattern in desert perennials. J. Ecol. 59:555–560.

Anonymous. 1984. Crassulacean acid metabolism. Desert Plants 5:192.

Antevs, E. 1938. Rainfall and tree growth in the Great Basin. Carnegie Inst. Wash. Publ. 469. 97 pp.

Applegate, E. I. 1938. Plants of the Lava Beds National Monument. Amer. Midl. Naturalist 19:334–368.

Axelrod, D. I. 1950. Evolution of desert vegetation in western North America. Carnegie Inst. Wash. Publ. 590:215–260.

———. Age and origin of Sonoran Desert vegetation. Occ. Pap. Calif. Acad. Sci. 132:1–74.

Barbour, M. G. 1969. Age and space distribution of the desert shrub *Larrea divaricata*. Ecology 50:679–685.

———. 1973. Desert dogma reexamined: Root/shoot production and plant spacing. Amer. Midl. Naturalist 89:41–50.

———, D. V. Diaz, and R. W. Breidenbach. 1974. Contributions to the biology of *Larrea* species. Ecology 55:1199–1215.

Batanouny, K. H. 1983. Human impact on desert vegetation. Pp. 139–149 *in* W. Holzner, M. J. A. Werger, and I. Ikusima (eds.). human impact on vegetation. Dr. W. Junk Publ., The Hague, Boston & London.

Beatley, J. C. 1967. Survival of winter annuals in the northern Mojave desert. Ecology 48:745–750.

———. 1974. Effects of rainfall and temperature on the distribution and behavior of *Larrea tridentata* (creosote bush) in the Mojave desert of Nevada. Ecology 55:245–261.

———. 1974. Phenological events and their environmental triggers in Mojave desert ecosystems. Ecology 55:856–863.

———. 1975. Climates and vegetation pattern across the Mojave/ Great Basin Desert transition of southern Nevada. Amer. Midl. Naturalist 93:53–70.

Benson, L. 1969. The native cacti of California. Stanford Univ. Press, Stanford. 243 pp.

———, and R. A. Darrow. 1981. Trees and shrubs of the southwestern deserts. 3rd ed. Univ. Arizona Press, Tucson. 416 pp.

Billings, W. D. 1949. The shadscale vegetation of Nevada and eastern California in relation to climate and soil. Amer. Midl. Naturalist 42:87–109.

———. 1951. Vegetational zonation in the Great Basin in western North America. Pp. 101–122. *in* Comp. rend. du colloque sur les bases ecologiques de regeneration de la vegetation des zones arides, Union Internat. Soc. Biol., Paris.

Blake, W. P. 1914. The Cahuilla basin and desert of Colorado in the Salton Sea, a study of the geography, geology, floristics, and the ecology of a desert basin. Carnegie Inst. Wash. Publ. 193.

Bloss, H. E. 1986. Studies of symbiotic microflora and their role in the ecology of desert plants. Desert Plants 7:119–127.

Bowers, J. E. 1980–81. Catastrophic freezes in the Sonoran Desert. Desert Plants 2:232–236.

Bradley, W. G. 1970. The vegetation of Saratoga Springs, Death Valley National Monument. Southw. Naturalist 15:111–129.

Britton, C. M., and R. G. 1985. Effects of fire on sagebrush and bitterbrush. Pp. 22–26 in K. Sanders and J. Durham (eds.). Rangeland fire effects. A symposium. U.S.D.I., Bureau of Land Management. Idaho State Office, Boise.

Brown, D. E. and R. A. Minnich. 1986. Fire and changes in creosote bush scrub of the western Sonoran Desert, California. Amer. Midl. Naturalist 116:411-422.

Brown, G. W. (ed.). 1968. Desert biology. Academic Press, N.Y.

Bunting, S. C. 1985. Fire in sagebrush-grass ecosystems: successional changes. Pp. 7–11 *in* K. Sanders and J. Durham (eds.). Rangeland fire effects. A symposium. U.S.D.I., Bureau of Land Management. Idaho State Office, Boise.

Burk, J. H. 1977. Sonoran desert vegetation. Pp. 869–889 *in* M. G. Barbour and J. Major (eds.). Terrestrial vegetation of California. John Wiley and Sons, N.Y.

———. 1982. Phenology, germination, and survival of desert ephemerals in Deep Canyon, Riverside County, California. Madroño 29:154–163.

Castagnoli, S. P., G. C. de Nevers, and R. D. Stone. 1983. Vegetation and flora. Pp. 43–104 *in* R. D. Stone and V. A. Sumida (eds.), The Kingston Range of California: A resource survey. Publ. 10, Environmental Field Program, Univ. California, Santa Cruz.

Cockerell, T. D. A. 1945. The Colorado Desert of California; its origin and biota. Trans. Kansas Acad. Sci. 48:1–39.

Cody, M. L. 1978. Distribution ecology of *Haplopappus* and *Chrysothamnus* in the Mojave desert. I. Niche position and niche shifts on north-facing granitic slopes. Amer. J. Bot. 65:1107–1116.

Crosswhite, F. S., and C. D. Crosswhite. 1984. A classification of life forms of the Sonoran Desert, with emphasis on the seed plants and their survival strategies. Desert Plants 5:131–161, 186–190.

Daubenmire, R. 1970. Steppe vegetation of Washington. Wash. Agric. Exp. Stn. Tech. Bull. 62. 131 pp.

Davidson, E., and M. Fox. 1974. Effects of off-road motorcycle activity on Mojave desert vegetation and soil. Madroño 22:381–390.

Devine, R. 1993. The cheatgrass problem. Atlantic 271(5):43–48.

Garcia-Moya, E., and C. M. McKell. 1970. Contribution of shrubs to the nitrogen economy of a desert wash plant community. Ecology 51:81–88.

Glendinning, R. M. 1949. Desert contrasts illustrated by the Coachella. Geogr. Rev. 39:220–228.

Goodin, J. R., and A. Mozafar. 1972. Physiology of salinity stress. Pp. 255–259 *in* C. M. McKell, J. P. Blaisdell, and J. R. Goodin (eds.). Wildlife shrubs—their biology and utilization. U.S.D.A. Forest Service Gen. Tech. Rep. Int. 1. Washington, D.C.

Goodman, P. J. Physiological and ecotypic adaptations of plants to salt desert conditions in Utah. J. Ecol. 61:473–494.

Gulmon, S. L., and H. A. Mooney. 1977. Spatial and temporal relationships between two desert shrubs, *Atriplex hymenelytra* and *Tidestromia oblongifolia* in Death Valley, California. J. Ecol. 65:831–838.

Hanes, T. L. 1976. Vegetation types of the San Gabriel Mountains. Pp. 65–76 *in* J. Latting (ed.). Plant communities of southern California. Calif. Native Pl. Soc. Spec. Publ. 2.

Holmgren, N. H. 1972. Plant geography of the intermountain region. Pp. 77–161 *in* A. Cronquist, A. H. Holmgren, N. H. Holmgren and J. L. Reveal. Intermountain Flora. Vol. 1. New York Bot. Gard. and Hafner Publ. Co., N.Y.

Hunt, C. B. 1966. Plant ecology of Death Valley, California. U.S. Dept. Interior, Geol. Survey Prof. Paper 509. 68 pp.

Ives, R. L. 1949. Climate of the Sonoran Desert region. Ann. Assoc. Amer. Geogr. 39:143–187.

Jackson, D. D. 1975. Sagebrush country. Time-Life Books, N.Y. Jaeger, E. C. 1957. The North American deserts. Stanford Univ. Press, Stanford. 308 pp.

Jaeger, E. C. 1965. The California deserts, from Death Valley to the Mexican border. Stanford Univ. Press, Stanford. 308 pp.

Johnson, A. W. 1968. The evolution of desert vegetation in western North America. pp. 590–606 *in* G. W. Brown (ed.). Desert biology. Academic Press, N.Y.

Johnson, H. B. 1976. Vegetation and plant communities of Southern California deserts—a functional view. Pp. 125–164 *in* J. Latting (ed.). Plant communities of southern California. Calif. Native Plant Soc. Spec. Publ. 2.

Kemp, P. R., and G. L. Cunningham. 1981. Light, temperature and salinity effects on growth, leaf anatomy and photosynthesis of *Distichlis spicata* (L.) Greene. Amer. J. Bot. 68:507–516.

King, T. J., and S. J. R. Woodell. 1973. The causes of regular pattern in desert perennials. J. Ecol. 61:761-765.

Klemmedson, J. O., and J. G. Smith. 1964. Cheatgrass (*Bromus tectorum* L.). Bot. Rev. 30:226-262.

Little, E. L. 1976. Atlas of United States trees. Vol. 3. Minor western hardwoods. U.S.D.A. Forest Service Misc. Publ. 1314. 13 pp. + 210 maps.

Mack, R. N. 1981. Invasion of *Bromus tectorum* into western North America: an ecological chronicle. Agro-Ecosystems 7:145-165.

———. 1986. Alien plant invasion into the Intermountain West. Pp. 191-213 *in* H. A. Mooney and J. A. Drake (eds.), Ecology of biological invasions of North America and Hawaii. Springer Verlag, New York.

MacMahon, J. A. 1979. North American deserts: their floral and faunal components. Pp. 21-82 *in* D. W. Goodall, R. A Perry, and K. M. W. Howes (eds.). Arid-land ecosystems: structure, functioning and management. Vol. 1. Cambridge Univ. Press, Cambridge.

———. 1988. Warm deserts. Pp. 231-264 *in* M. G. Barbour and W. D. Billings (eds.), North American Terrestrial Vegetation. Cambridge Univ. Press, Cambridge.

Marks, J. B. 1950. Vegetation and soil relations in the lower Colorado desert. Ecology 31:171-193.

McArthur, E. D., and A. P. Plummer. 1978. Biogeography and management of native western shrubs: a case study, section *Tridentatae* of *Artemisia*. Great Basin Naturalist Mem. 2:229-243.

McGinnies, W. G., B. J. Goldman, and P. Paylore (eds.). 1968. Deserts of the world. An appraisal of research into their physical and biological environments. Univ. Arizona Press, Tucson. 788 pp.

Mooney, H. A. 1973. Plant communities and vegetation. Pp. 7-17 *in* R. M. Lloyd and R. S. Mitchell. A flora of the White Mountains, California and Nevada. Univ. Calif. Press, Berkeley.

Muller, C. H. 1953. The association of desert annuals with shrubs. Amer. J. Bot. 40:53-60.

Mulroy, T. W., and P. W. Rundel. 1977. Annual plants: adaptations to desert environments. Bioscience 27:109-114.

Munz, P. A. 1974. A flora of southern California. Univ. Calif. Press, Berkeley. 1086 pp.

Nilsen, E. T., M. R. Sharifi and P. W. Rundel. 1981. Summer water relations of the desert phreatophyte *Prosopis glandulosa* in the Sonoran Desert of southern California. Oecologia 50:271-276.

———, ———, and ———. 1984. Comparative water relations of phreatophytes in the Sonoran Desert of California. Ecology 65:767-778.

Nord, E. C. 1965. Autecology of bitterbrush in California. Ecol. Monogr. 35:307-334.

Nowak, C. L., R. S. Nowak, R. J. Tausch, and P. E. Wigand. 1994. Tree and shrub dynamics in northwestern Great Basin woodland and shrub steppe during the Late-Pleistocene and Holocene. Amer. J. Bot. 81:265-277.

O'Leary, J. F., and R. A. Minnich. 1981. Postfire recovery of creosote bush scrub vegetation in the western Colorado desert. Madroño 28:61-66.

Parish, S. B. 1930. Vegetation of the Mojave and Colorado deserts of southern California. Ecology 11:481-503.

Pavlik, B. M. 1980. Patterns of water potential and photosynthesis of desert sand dune plants, Eureka Valley, California. Oecologia 46:147-154.

Phillips, E. A., K. K. Page, and S. D. Knapp. 1980. Vegetational characteristics of two stands of Joshua tree woodland. Madroño 27:43-47.

Rempel, P. J. 1936. The crescentic dunes of the Salton Sea and the relation to the vegetation. Ecology 17:347-358.

Robison, T. W. 1958. Phreatophytes. U.S. Geol. Surv. Water Supply Pap. 1423. 84 pp.

Runyon, E. H. 1934. The organization of the creosote bush with respect to drought. Ecology 15:128–138.

Sanders, , K., J. Durham, et al. 1985. Rangeland fire effects: a symposium. U.S.D.I., Bureau of Land Management, Idaho State Office, Boise. 124 pp.

Sankary, M. N., and M. G. Barbour. 1972. Autecology of *Atriplex polycarpa* from California. Ecology 53:1155–1162.

Shantz, H. L. 1925. Plant communities in Utah and Nevada. Contr. U.S. Natl. Herb. 25:15–23.

Shreve, F. 1942. The desert vegetation of North America. Bot. Rev. 8:195–246.

———, and I. M. Wiggins. 1963. Vegetation and flora of the Sonoran Desert. 2 vols. Stanford Univ. Press, Stanford.

Solbrig, O. T., M. A. Barbour, J. Cross, G. Goldstein, C. H. Lowe, J. Morello, and T. W. Yang. 1977. The strategies and community patterns of desert plants. Pp. 67–106 *in* G. H. Orians and O. T. Solbrig (eds.). Convergent evolution in warm deserts. Dowden, Hutchinson Ross, Stroudsburg, Pa.

Sternberg, L. 1976. Growth forms of *Larrea tridentata*. Madroño 23:408–417.

Thorne, R. F. 1982. The desert and other transmontane plant communities of southern California. Aliso 10:219–257.

Turner, R. M. 1982. Great Basin desertscrub. Pp. 145–155 *in* D. E. Brown (ed.). Biotic communities of the American Southwest—United States and Mexico. Desert Plants 4:1-341.

———. 1982. Mojave desert scrub. Pp. 157–168 *in* D. E. Brown (ed.). Biotic communities of the American Southwest—United States and Mexico. Desert Plants 4:1–341.

———, and D. E. Brown. 1982. Sonoran desertscrub. Pp. 181–221 *in* D. E. Brown (ed.). Biotic communities of the American Southwest—United States and Mexico. Desert Plants 4:1–341.

Ungar, I. A. 1974. Inland halophytes of the United States. Pp. 235–305 *in* R. J. Reimold and W. H. Queen (eds.). Ecology of halophytes. Academic Press, N.Y.

U.S. Department of the Interior. 1980A. The California Desert Conservation Area: plan alternatives and environmental impact statement: draft. U.S. Dept. Interior, Bureau of Land Manage., Calif. State Office, Sacramento. 436 pp.

———. 1980B. The California Desert Conservation Area: final environmental impact statement and proposed plan. U. S. Dept. Interior, Bureau of Land Manage. Calif. State Office, Sacramento. 245 pp. in 2 sections.

Vasek, F. C. 1980. Ancient creosote bush rings in the Mojave desert. Fremontia 7(4):10–13.

———. 1980. Creosote bush: long-lived clones in the Mojave Desert. Amer. J. Bot. 67:246–255.

———. 1983. Plant succession in the Mojave desert. Crossosoma 9(1):1–23.

———, and L. J. Lund. 1980. Soil characteristics associated with a primary plant succession on a Mojave Desert dry lake. Ecology 61:1013–1018.

———, and M. G. Barbour. 1977. Mojave desert scrub vegetation. Pp. 835–867 *in* M. G. Barbour and J. Major (eds.). Terrestrial vegetation of California. John Wiley and Sons, N.Y.

Wells, P. V., and J. H. Hunziker. 1976. Origin of the creosote bush (*Larrea*) deserts of southwestern North America. Ann. Missouri Bot. Gard. 63:843–861.

———, and L. M. Shields. 1964. Distribution of *Larrea divaricata* in relation to temperature inversion at Yucca Flat, southern Nevada. Southwestern Naturalist 9:51–55.

Went, F. W. 1942. The dependence of certain annual plants on shrubs in southern California deserts. Bull. Torrey Bot. Club 69:100–114.

———. 1955. Ecology of desert plants. Sci. Amer. 192:68–75.

West, N. E. 1988. Intermountain deserts, shrub steppes, and woodlands. Pp. 209–230 *in* M. G. Barbour and W. D. Billings (eds.), North American Terrestrial Vegetation. Cambridge Univ. Press, Cambridge.

Wilson, R. C. 1972. *Abronia*. I. Distribution, ecology and habit of 9 species of *Abronia* found in California. Aliso 7:421–437.

Woodell, S. R. J., H. A. Mooney, and A. J. Hill. 1969. The behaviour of *Larrea divaricata* (creosote bush) in response to rainfall in California. J. Ecol. 57:37–44.

Wright, H. A. 1985. Effects of fire on grasses and forbs in sagebrush-grass communities. Pp. 12–21 *in* K. Sanders and J. Durham (eds.). Rangeland fire effects. A symposium. U.S.D.I., Bureau of Land Management. Idaho State Office, Boise.

Yeaton, R. I., and M. L. Cody. 1979. The distribution of cacti along environmental gradients in the Sonoran and Mojave deserts. J. Ecol. 67:529–541.

Yoder, V., M. G. Barbour, R, S, Boyd, and R. A. Woodward. 1983. Vegetation of the Alabama Hills region, Inyo County, California. Madroño 30:118–126.

Young, J. A., R. A. Evans, and J. Major. 1977. Sagebrush steppe. Pp. 763–796 *in* M. G. Barbour and J. Major (eds.). Terrestrial vegetation of California. John Wiley and Sons, N.Y.

———, ———, and P. T. Tueller. 1975. Great Basin plant communities—pristine and grazed. Pp. 186–215, *in* R. Elston (ed.). Holocene climates in the Great Basin. Occas. Paper, Nevada. Archeol. Survey, Reno.

Zamora, B., and P. T. Tueller. 1973. *Artemisia arbuscula, A. longiloba* and *A. nova* habitat types in northern Nevada. Great Basin Naturalist 19:33:225–242.

Chapter 19

Desert Woodland Communities

Desert woodlands are communities dominated by a mixture of desert shrubs and small xerophytic trees. These communities occur below the montane coniferous forest zones and above the various desert scrub communities (Fig. 19-1). They are mostly located in transmontane California, but extend into cismontane areas along the interior slopes of the Transverse Ranges and South Coast Ranges (Figs. 19-2, 19-8). They occupy about three percent of California's land area. Desert woodlands occur mostly between 750 and 2400 m (2500–7900 ft) elevation on both mountain slopes and level terrain. In rain shadow areas these communities extend throughout the length of California. On desert mountain ranges that are too low or too dry to support montane coniferous forests, desert woodlands often occupy the upper elevation zones. We have recognized two major types of desert woodlands: Piñon Pine and Juniper Woodlands and Joshua Tree Woodlands. The Piñon Pine and Juniper Woodlands have further been grouped into several plant associations based on geographic location and species composition.

Desert woodlands have warm to hot dry summers and cool to cold winters with most precipitation falling in winter. Two major environmental factors interact to determine the distribution of the small trees that characterize the desert woodlands: soil moisture and temperature extremes. Tree species generally have a greater moisture requirement than do herbs or shrubs. In dry areas tree distribution is often limited by the availability of soil moisture. In desert regions trees are usually encountered either along watercourses or in high elevation areas. Desert woodlands receive between 15 and 50 cm precipitation per year. With increased elevation there is a corresponding temperature decrease and much of the precipitation falls as snow. Each tree species of the desert woodlands has both a range of drought tolerance and a range of tolerances to temperature extremes. Where ranges of individual species overlap, a mixed woodland community occurs. In some sites conditions may be too dry or too cold for all but one

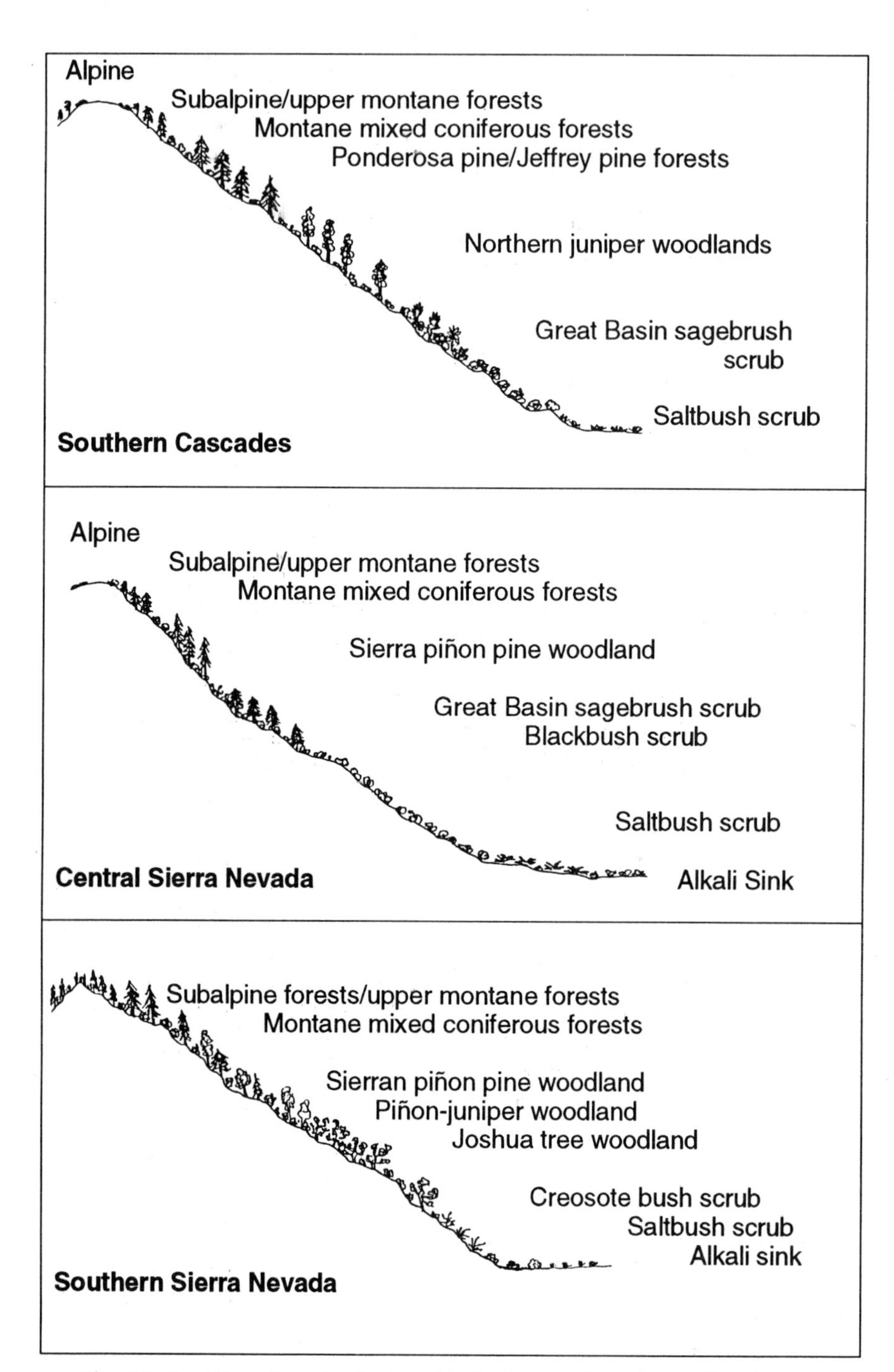

Fig. 19-1. Elevational relationship between desert woodland communities and other vegetation types in transmontane areas of California.

tree species and in these areas pure stands occur. The ranges of tolerance of many of the desert shrubs are different from those of the tree species. The understory layers in desert woodlands dominated by the same tree often vary from one area to another. On the other hand, understory composition may remain uniform even as tree species change.

1. Piñon Pine and Juniper Woodlands

Pygmy coniferous woodlands dominated by piñon pines or junipers or by a mixture of these trees are widespread in the dry mountains of western North America. They occur at elevations of 1200–2400 m (4000–8000 ft) in rain shadow areas in various parts of California (Fig. 19-2)13. These woodlands are best represented in transmontane California, but they also occur in cismontane areas in the Transverse Ranges of Kern, Los Angeles, Ventura and Santa Barbara Counties. Juniper woodlands extend north along the inner South Coast Ranges where they intergrade in places with the blue oak phase of the foothill woodland communities (Chapter 16). Sometimes curl-leaf mountain-mahogany (*Cercocarpus ledifolius*) is also an important component of these woodlands. Piñon pine and juniper woodlands are quite common at high elevations in desert ranges such as the White Mountains, Inyo Mountains, Panamint Mountains, etc. These woodlands occur along the steep eastern slope of the Sierra Nevada, on the eastern slopes of the Cascades and on the Modoc Plateau.

Piñon pine and juniper woodlands usually consist of scattered trees 3 to 15 m tall. Density of the woodlands varies considerably depending on local climatic and edaphic conditions. Junipers are often more xerophytic than piñon pines and sometimes grow in pure stands at lower elevations where the climate is warmer and drier. At high elevations piñon pines tend to increase in dominance and junipers may drop out entirely. Juniper woodlands in southern California often grade into Joshua tree woodlands or desert scrub communities at low elevations. Piñon pine woodlands often grade into the lower fringes of Jeffrey pine forests.

Junipers and piñon pines are not uniformly distributed in California, and in some areas one or the other is present alone. In northeastern California along the east slope of the Cascade Range and on the Modoc Plateau, piñon pines are absent and junipers grow in pure stands. Along much of the eastern side of the Sierra Nevada, junipers are absent and piñons are the sole dominants. Because of the geographical differentiation of these woodland communities, several associations can be recognized.

Piñon pine and juniper woodlands are widely distributed in the western mountains of North America. These woodlands occur

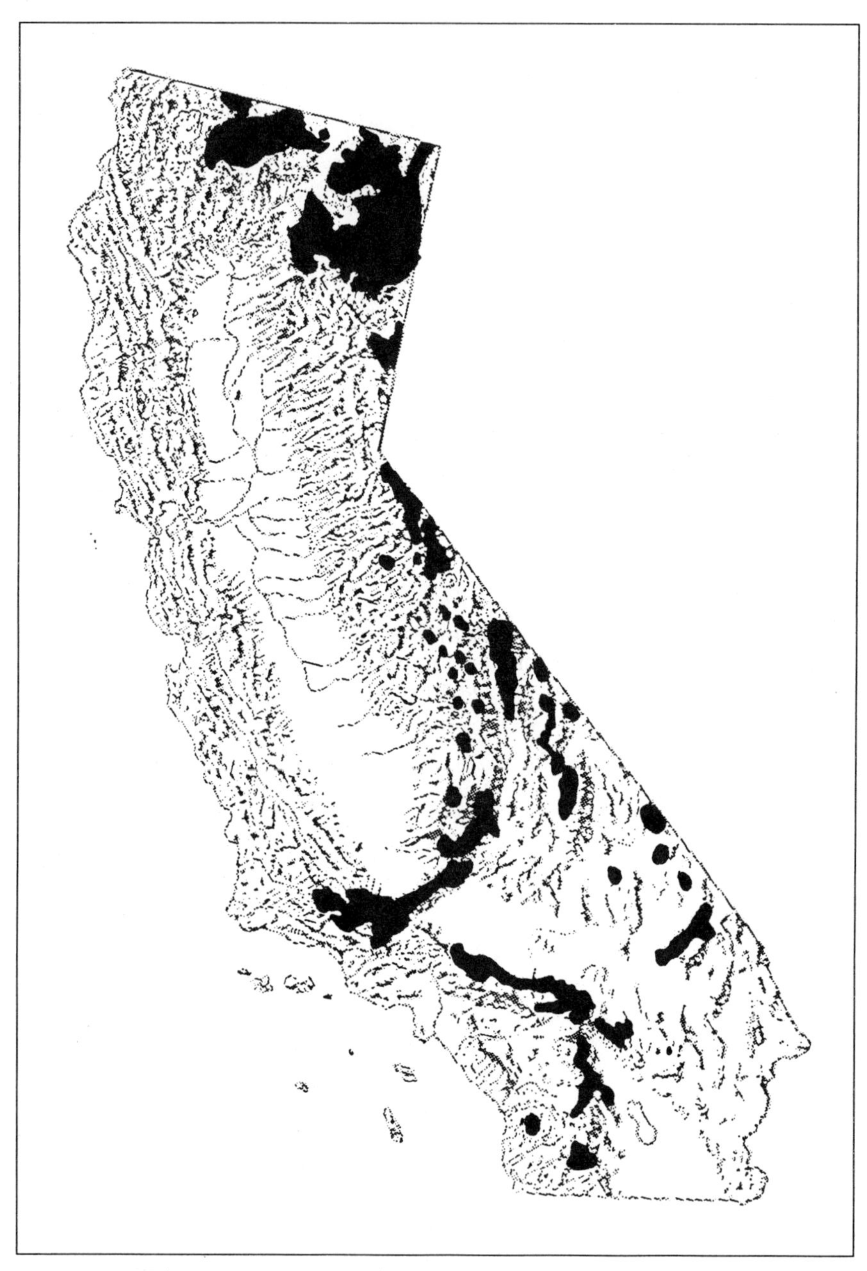

Fig. 19-2. Distribution of piñon pine-juniper and northern juniper woodland communities in California.

Fig. 19-3. Piñon-juniper woodland in vicinity of Walker Pass, eastern Kern County. The understory is dominated by Great Basin sagebrush (*Artemisia tridentata*). Photo by David Keil.

from the Rocky Mountains to the Sierra Nevada and are common on the numerous fault block mountains of the basin and range province. Several species each of piñon pine and juniper occur in the western mountains and several of these occur in California. The most common piñon pine in California woodlands is *Pinus monophylla* (one-needle piñon, Figs. 19-4A, 19-6) which is a characteristic species of the Great Basin region. *Pinus edulis*, two-needle piñon, is widespread in the southern Rocky Mountains and Arizona, and enters California only in the upper elevations of the New York Mountains in the eastern part of the Mojave desert region. *Pinus quadrifolia* (four-needle piñon) occurs mostly in the mountains of Baja California and extends into California only in San Diego and southern Riverside Counties.

Three juniper species occur in desert woodlands in California. *Juniperus californica* (California juniper, Figs. 19-4B. 19-5) occurs mainly in the mountains of southern California. *Juniperus occidentalis* (Fig. 19-4D) consists of two races. The first of these, var. *occidentalis* (western juniper; Fig. 19-7), ranges from northeastern California to eastern Oregon and adjacent parts of Washington, Idaho and Nevada. It is common on the arid volcanic plateaus in the Rain shadow of the Cascades. The second race, var. *australis* (mountain juniper), is a mostly high elevation race that occurs mainly in the Sierra Nevada with disjunct populations in the North Coast Range, the San Gabriel

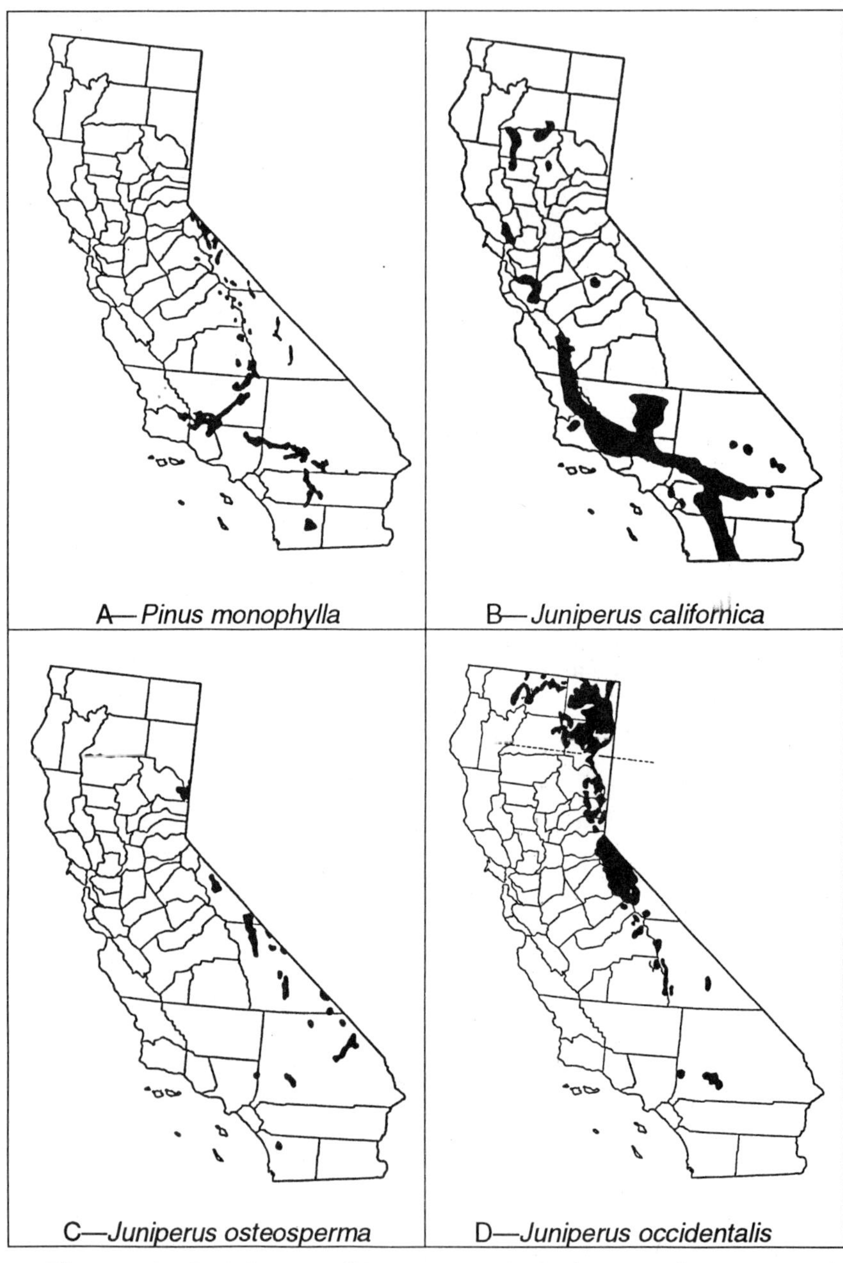

Fig. 19-4. Distribution of *Pinus monophylla* (one-needle piñon pine), *Juniperus californica* (California juniper), *Juniperus osteosperma* (Utah juniper), and *Juniperus occidentalis* in California. *Juniperus occidentalis* comprises two varieties. Western juniper (var. *occidentalis*) grows mostly north of the zone marked by the dotted line; mountain juniper (var. *australis*) occurs mostly to south, usually at higher elevations.

Fig. 19-5. *Juniperus californica* and *Yucca whipplei.* California juniper typically grows as a large shrub. Photo by David Keil.

and the San Bernardino Mountains. The latter taxon often occurs in montane mixed coniferous and subalpine forests above the elevation range of the desert woodlands, but in some locations in the southern Sierra Nevada it forms woodlands alone or together with *Pinus monophylla* at the upper elevational extent of the desert woodlands. *Juniperus osteosperma* (Utah juniper, Fig. 19-4C), is centered in the Great Basin region. It extends into the arid Basin Ranges of eastern California and also occurs in the San Gabriel and San Bernardino Mountains of the Transverse Range.

A. Piñon-Juniper Woodlands

These communities are common in the desert mountains of California, on the flanks of the Transverse Ranges and some of the Peninsular Ranges. In the Sierra Nevada, piñon-juniper woodlands occur only in the southern part of the range (Fig. 19-3). Dominant tree (or large shrub) species are *Pinus monophylla, Juniperus occidentalis, Juniperus californica* and sometimes *Cercocarpus ledifolius* or *Quercus johntuckeri* (desert scrub oak). *Pinus edulis* partially replaces *Pinus monophylla* in the New York Mountains where it is joined by *Quercus turbinella* [desert scrub oak]. *Juniperus osteosperma* is restricted for the most part to the transmontane desert mountains and is replaced by *J. californica* and *J. occidentalis* in other areas. *Quercus johntuckeri* and *Q.* -

Fig. 19-6. Sierran piñon pine woodland in eastern Sierra Nevada near Kennedy Meadows, Tulare County. The dominant understory shrub is *Artemisia tridentata* (Great Basin sagebrush). Photo by David Keil.

corneliusmulleri [desert scrub oaks] are conspicuous parts of these communities in the Transverse and Peninsular Ranges.

Associated species vary from region to region in California. In piñon-juniper woodlands bordering Great Basin sagebrush scrub communities (Chapter 18), the understory shrubs are commonly the dominants of the adjacent communities, such as:

Artemisia tridentata	Great Basin sagebrush
Chrysothamnus nauseosus	rabbitbrush
Chrysothamnus viscidiflorus	rabbitbrush
Grayia spinosa	hopsage
Sarcobatus vermiculatus	greasewood

Other common members of these communities are:

Cercocarpus ledifolius	curl-leaf mountain mahogany
Ephedra viridis	Mormon-tea
Fallugia paradoxa	Apache plume
Purshia mexicana [*Cowainia m.*]	cliff-rose
Purshia tridentata	bitterbrush, antelope brush

Piñon-juniper woodlands also grade into Joshua tree woodlands and creosote bush scrub communities. In such areas the understory species may include those listed below as well as many others.

Coleogyne ramosissima	blackbush
Ephedra spp.	Mormon-tea
Gutierrezia sp.	snakeweed
Hymenoclea salsola	burrobrush
Larrea tridentata	creosote bush
Lycium sp.	wolfberry

In the Transverse and Peninsular Ranges chaparral shrubs such as *Arctostaphylos* spp. (manzanitas), *Cercocarpus betuloides* (mountain mahogany), and *Ceanothus greggii* desert ceanothus) sometimes grow in piñon-juniper woodlands.

B. Sierran Piñon Pine Woodlands

Piñon pine is often the sole dominant in Sierran desert woodlands (Fig. 19-6). *Juniperus osteosperma* occurs only to a limited extent in the Sierra Nevada, and *J. californica* only grows in the extreme southern end of the range. *Juniperus occidentalis* is represented in the Sierra by its high-elevation race, var. *australis*, which is usually elevationally separated from piñon pine, growing in mixed coniferous or subalpine communities. The understory shrubs in the Sierran piñon pine woodlands are usually species from the Great Basin sagebrush scrub communities. Common associates include *Artemisia tridentata, Chrysothamnus nauseosus, Purshia tridentata*, *Prunus andersonii* (desert peach), and perennial grasses.

The distribution of piñon pines in the Sierra Nevada is mostly restricted to transmontane areas. These trees and some associated Great Basin species occur in a few locations on the western slope of the Sierra and in the Great Western Divide of the southern Sierra Nevada. On the cismontane slopes these species are restricted to arid, exposed sites with very shallow soils. These plants may have occupied the cismontane sites at a time in the past when conditions in the Sierra Nevada were more arid than at present.

C. Northern Juniper Woodlands

Northern juniper woodlands (Fig. 19-7) occur from southeastern Washington and western Idaho to northeastern California. In California these communities occur as intermittent to extensive woodlands on the eastern slope of the Sierra-Cascade axis from Sierra County northward. *Juniperus occidentalis* var. *occidentalis* as the dominant if not the only tree species. This race of *Juniperus occidentalis* generally occurs between 700 and 2200 m (1600–6100 ft). Northern juniper woodlands are most extensive on the volcanic soils of the Modoc Plateau in Lassen and

Fig. 19-7. Northern juniper woodland near Yreka in Siskiyou County. *Juniperus occidentalis* var. *occidentalis* forms pure stands in these woodlands. Photo by V. L. Holland.

Modoc Counties and extend westward into eastern Siskiyou and Shasta counties. The understory vegetation may be composed of grasses but often is largely composed of shrubs and herbs of the Great Basin sagebrush scrub communities. Associated shrubs include the following:

Artemisia tridentata	Great Basin sagebrush
Artemisia arbuscula	sagebrush
Chrysothamnus nauseosus	rabbitbush
Purshia tridentata	bitter bush
Cercocarpus ledifolius	curl-leaf mountain-mahogany
Ribes velutinum	gooseberry

In mesic sites northern juniper woodlands grade into northern oak woodlands or montane forests dominated by *Pinus ponderosa* (yellow pine, ponderosa pine) or *Pinus jeffreyi* (Jeffrey pine). In these areas the junipers commonly occupy dry sites whereas the oaks or pines occur in more mesic situations.

D. Mountain Juniper Woodlands

Juniperus occidentalis var. *australis* (Mountain juniper) occurs at higher elevations than does the closely related western juniper, mostly above 2200 m (6100 ft). It has been reported to occur as far north as the Warner Mtns. of northeastern California. On transmontane slopes of the Sierra Nevada it commonly occupies a zone above the piñon pine woodlands and extends upward in elevation into subalpine forests. It is not limited to transmontane slopes but also extends westward to occur in dry upland sites in the subalpine zone on the cismontane slopes. Mountain juniper also occurs in localized populations in the San Bernardino and San Gabriel Mtns. of the Transverse Range and in the White, Inyo, and Panamint Mtns. of the western Basin and Range Province.

On transmontane slopes mountain juniper is commonly associated with open stands of *Pinus jeffreyi* (Jeffrey pine), and a shrubby understory dominated by *Artemisia tridentata* (Great Basin sagebrush), *Cercocarpus ledifolius* (curl-leafed mountain-mahogany), and *Purshia tridentata* (bitterbrush). At higher elevations it occurs in association with other subalpine species. At lower elevations it overlaps with the upper elevation stands of *Pinus monophylla* (piñon pine), sometimes forming high-elevation piñon-juniper woodlands.

E. Southern Piñon and Juniper Woodlands

In the Peninsular ranges of southwestern California and northern Baja California, desert woodlands are dominated by *Pinus quadrifolia*, *P. monophylla*, or *Juniperus californica*. These species are generally segregated along elevational or moisture gradients, with juniper woodlands occurring on the desert fringes bordering the creosote bush scrub communities and piñon pines at higher elevations in association with chaparral communities. *Pinus quadrifolia* usually occurs on slopes with western drainages and *Pinus monophylla* on eastern slopes. Associated species in lower elevation zones include such desert species as *Yucca schidigera* (Mojave yucca) and *Yucca baccata* (banana yucca). In higher elevation zones the piñon pines grow with desert chaparral or mixed chaparral communities. Common associates include *Adenostoma fasciculatum* (chamise) and *Ceanothus greggii* (desert ceanothus).

2. Joshua Tree Woodlands

Joshua tree woodlands (Fig. 19-8, 19-9) are located on sandy, well-drained desert mesas and slopes in transmontane California from the southern part of the Owens Valley to the Little San Bernardino Mountains. They extend a short way into cismontane California in the Walker Pass region. They are scattered in the Antelope Valley north of the San Gabriel and San Bernardino Mountains, but many stands in this region have been cleared for

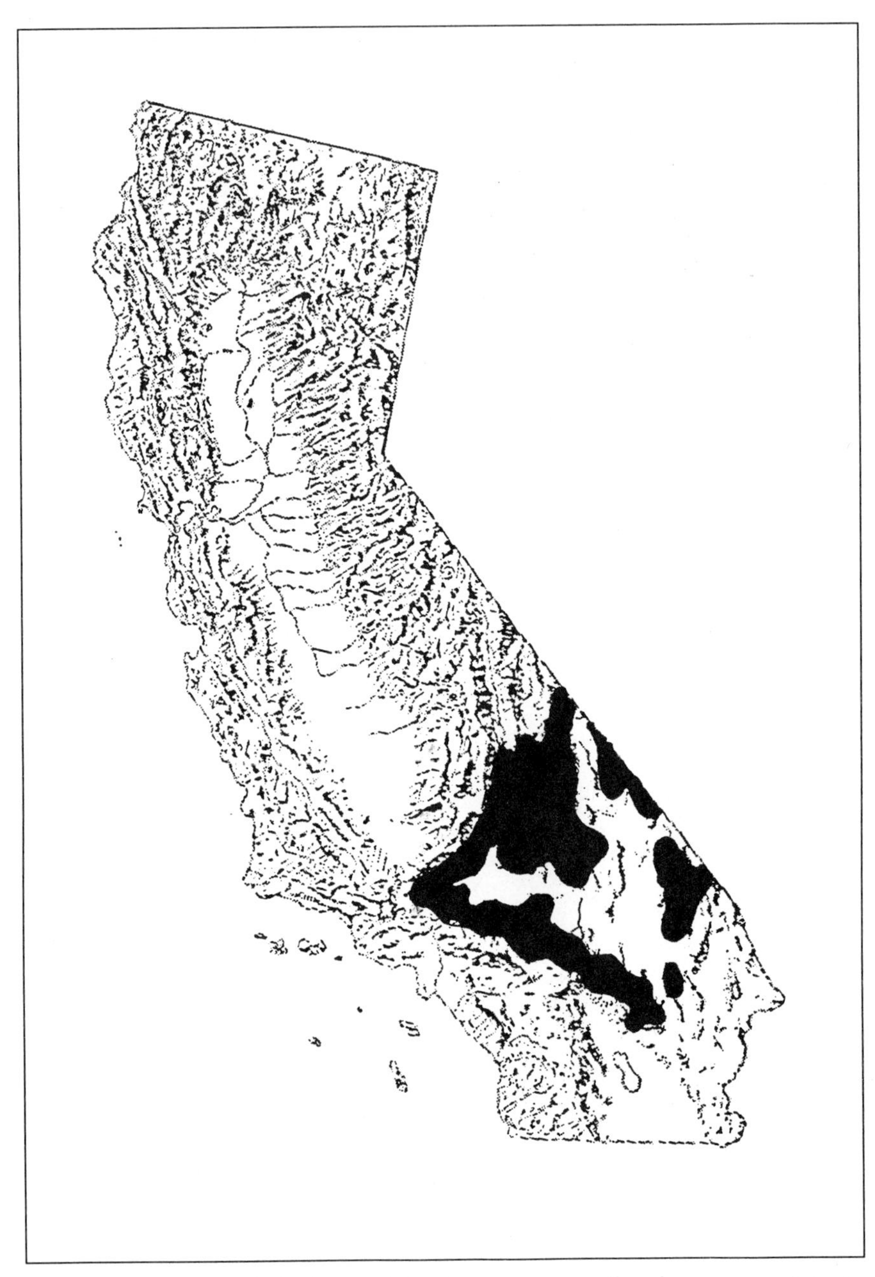

Fig. 19-8. Distribution of Joshua tree woodland communities in California.

Fig. 19-9. Joshua tree woodland east of Walker Pass, Kern County. The understory is a mixture of upland desert shrubs. Photo by David Keil.

agriculture or urban development. These communities extend eastward from California into southern Nevada and northwestern Arizona. They occur mostly between 600 and 1600 m elevation.

Joshua tree woodlands are a heterogeneous group of communities, united mainly by the presence of *Yucca brevifolia* (Joshua tree). This species occurs with a wide variety of associates. Several different desert scrub communities occur in essentially unmodified form as understory components of Joshua tree woodlands. The following types of desert scrub communities occur in association with Joshua trees in areas of southern California: Great Basin sagebrush scrub communities, blackbush scrub communities, saltbush scrub communities, and creosote bush scrub communities. In some areas Joshua tree woodlands intergrade with juniper woodlands or with piñon-juniper woodlands. In areas of the Transverse Ranges, Joshua trees even grow together with chaparral species.

Human impacts in Desert Woodlands

Until recently piñon pine and juniper woodlands have suffered less damage at the hands of man than have many of California's other plant communities. The lands on which desert woodlands occur are mostly not suitable for cultivation. During the late 1800's piñon pines and junipers were cut in some areas to provide fuel for smelting ores in mining camps. Piñons and junipers have been harvested for firewood or fence posts in some areas, but the trees in the past have not been considered large enough to produce commercial timber. Recently, however, juniper wood has become more valuable and has potential for commercial harvest. Piñon pine seeds have been harvested for human consumption since prehistoric times.

Large areas of piñon pine and juniper woodlands are used as grazing lands for both sheep and cattle. Prior to the onset of heavy grazing pressure, frequent fires tended to maintain open woodlands with scattered piñons and junipers. In many areas of western North America heavy grazing in the 19th and 20th centuries reduced the coverage of herbaceous species, especially grasses. With less herbaceous fuel to carry a fire, fire frequency decreased and junipers expanded their range and density, sometimes forming closed-canopy communities. Mature piñon pines and junipers have allelopathic effects on herbaceous understory species. In a dense piñon or juniper woodland the understory is often very sparse. Overgrazing in some areas has greatly reduced the amount of palatable grasses and enhanced the growth of *Bromus tectorum* (cheat grass), *Salsola tragus* [*S. kali*] (Russian thistle) and other undesirable species.

Range management practices have included clearing, poisoning, or burning of the trees to promote growth of grasses and other forage species, often with the intent of permanently converting the vegetation from woodland to grassland. This is sometimes accompanied by seeding of forage grasses. In the absence of seeding, communities of highly flammable introduced annuals have resulted that inhibit the natural recovery of both palatable forage plants and the dominant woody species.

Many areas of Joshua tree woodlands have been damaged or destroyed by human activities. In the Antelope Valley much of the area formerly occupied by Joshua trees has been cleared for agriculture or other development. These plants are also popular in desert landscaping, and many have been dug up and removed from the California deserts. Fortunately many stands of this picturesque species still survive and some are enclosed in parks or other areas where they are afforded legal protection.

Desert Woodlands—References

Aldon, E. F., and H. W. Springfield. 1973. The southwestern pinyon-juniper ecosystem: a bibliography. U.S.D.A. For. Serv. Gen. Tech. Rep. RM-4. Ft. Collins, Colorado.

Antevs, E. 1938. Rainfall and tree growth in the Great Basin. Carnegie Inst. Wash. Publ. 469. 97 pp.

Arno, R. S. 1971. Evaluation of pinyon-juniper conversion to grassland. J. Range Manage. 24:188–197.

Applegate, E. I. 1938. Plants of the Lava Beds National Monument. Amer. Midl. Natururalist 19:334–368.

Axelrod, D. I. 1950. Evolution of desert vegetation in western North America. Carnegie Inst. Wash. Publ. 590:215–260.

Bailey, D. K. 1970. Phytogeography and taxonomy of *Pinus* subsection *Balfourianae*. Ann. Missouri Bot. Gard. 57:210–249.

Billings, W. D. 1951. Vegetational zonation in the Great Basin in western North America. Pp. 101–122. *in* Comp. rend. du colloque sur les bases ecologiques de regeneration de la vegetation des zones arides, Union Internat. Soc. Biol., Paris.

Brown, D. E. 1982. Great Basin conifer woodland. Pp. 52–57 *in* D. E. Brown (ed.). Biotic communities of the American Southwest—United States and Mexico. Desert Plants 4:1–341.

Castagnoli, S. P., G. C. de Nevers, and R. D. Stone. 1983. Vegetation and flora. Pp. 43–104 *in* R. D. Stone and V. A. Sumida (eds.), The Kingston Range of California: A resource survey. Publ. 10, Environmental Field Program, Univ. California, Santa Cruz.

Critchfield, W. B., and G. L. Allenbaugh. 1969. The distribution of Pinaceae in or near northern Nevada. Madroño 20:12–26.

Devine, R. 1993. The cheatgrass problem. Atlantic 271(5):43–48.

Everett, R. L., and W. Clary. 1985. Fire effects and revegetation on Juniper-pinyon woodlands. Pp. 33–37 *in* K. Sanders and J. Durham (eds.). Rangeland fire effects. A symposium. U.S.D.I., Bureau of Land Management. Idaho State Office, Boise.

———, and S. D. Koniak. 1981. Understory vegetation in fully stocked pinyon-juniper stands. Great Basin Naturalist 41:467–476.

———, S. H. Sharrow, and R. O. Meeuwig. 1983. pinyon-Juniper woodland understory distribution patterns and species associations. Bull. Torrey Bot. Club 110:454–463.

Glock, W. S. 1937. Observations on the western juniper. Madroño 4:21–28.

Griffin, J. R., and W. B. Critchfield. 1972. The distribution of forest trees in California. Pacific S.W. Forest and Range Exp. Stn., Berkeley, Calif. 114 pp. (U.S.D.A. Serv. Res. Paper PSW 82).

Holmgren, N. H. 1972. Plant geography of the intermountain region. Pp. 77–161 *in* A. Cronquist, A. H. Holmgren, N. H. Holmgren and J. L. Reveal. Intermountain Flora. Vol. 1. New York Bot. Gard. and Hafner Publ. Co., N.Y.

Johnston, V. R. 1994. California forests and woodlands. A natural history. Univ. California Press, Berkeley & Los Angeles.

King, T. J. 1976. Late Pleistocene-early Holocene history of coniferous woodlands in the Lucerne Valley region, Mojave Desert, California. Great Basin Naturalist 36:227–238.

Koniak, S., and R. L. Everett. 1983. Seed reserves in soils of successional stages of pinyon woodlands. Amer. Midl. Naturalist 108:295–303.

Lanner, R. M. 1981. The pinyon pine: A natural and cultural history. Univ. Nevada Press, Reno. 208 pp.

Larson, P. 1970. Deserts of America. Prentice-Hall, Englewood Cliffs, N.J.

Ligon, J. D. 1978. Reproductive interdependence of pinyon jays and pinyon pines. Ecol. Monogr. 48:111–126.

Little, E. L. 1976. Atlas of United States trees. Vol. 3. Minor western hardwoods. U.S.D.A. Forest Service Misc. Publ. 1314. 13 pp. + 210 maps.

MacMahon, J. A. 1979. North American deserts: their floral and faunal components. Pp. 21–82 *in* D. W. Goodall, R. A Perry, and K. M. W. Howes (eds.). Arid-land ecosystems: structure, functioning and management. Vol. 1. Cambridge Univ. Press, Cambridge.

Meeuwig, R. O., and S. V. Cooper. 1981. Site quality and growth of pinyon-juniper stands in Nevada. For. Sci. 27:593–601.

Miller, R. F., and P. E. Wigand. 1994. Holocene changes in semiarid pinyon-juniper woodlands. Bioscience 44:465–474.

Minnich, R. A. 1976. Vegetation of the San Bernardino Mountains. Pp. 99–124 *in* J. Latting (ed.). Plant communities of southern California. Calif. Native Plant Soc. Spec. Publ. 2.

Mooney, H. A. 1973. Plant communities and vegetation. Pp. 7–17 *in* R. M. Lloyd and R. S. Mitchell. A flora of the White Mountains, California and Nevada. Univ. Calif. Press, Berkeley.

Nord, E. C. 1965. Autecology of bitterbrush in California. Ecol. Monogr. 35:307–334.

Nowak, C. L., R. S. Nowak, R. J. Tausch, and P. E. Wigand. 1994. Tree and shrub dynamics in northwestern Great Basin woodland and shrub steppe during the Late-Pleistocene and Holocene. Amer. J. Bot. 81:265–277.

Phillips, E. A., K. K. Page, and S. D. Knapp. 1980. Vegetational characteristics of two stands of Joshua tree woodland. Madroño 27:43–47.

Shreve, F. 1942. The desert vegetation of North America. Bot. Rev. 8:195–246.

St. Andre, G., H. A. Mooney, and R. C. Wright. 1965. The pinyon woodland zone in the White Mountains of California. Amer. Midl. Naturalist 73:225–239.

Tausch, R. J., N. E. West, and A. A. Nabi. 1981. Tree age and dominance patterns in Great Basin pinyon-juniper woodlands. J. Range Manage. 3:259–264.

Taylor, D. W. 1976. Disjunction of Great Basin plants in the northern Sierra Nevada. Madroño 23:301–310.

Thorne, R. F. 1982. The desert and other transmontane plant communities of southern California. Aliso 10:219–257.

Trombulak, S. C., and M. L. Cody. 1980. Elevational distributions of *Pinus edulis* and *P. monophylla* (Pinaceae) in the New York Mountains, eastern Mojave desert. Madroño 27:61–67.

Van Devender, T. R. 1977. Holocene woodlands in the southwestern deserts. Science 198:189–192.

Vasek, F. C. 1966. The distribution and taxonomy of three western junipers. Brittonia 18:350–372.

———, and M. G. Barbour. 1977. Mojave desert scrub vegetation. Pp. 835–867 *in* M. G. Barbour and J. Major (eds.). Terrestrial vegetation of California. John Wiley and Sons, N.Y.

———, and R. F. Thorne. 1977. Transmontane coniferous vegetation. Pp. 797–832 *in* M. G. Barbour and J. Major (eds.). Terrestrial vegetation of California. John Wiley and Sons, N.Y.

Vogl, R. J. 1976. An introduction to the plant communities of the Santa Ana and San Jacinto Mountains. Pp. 77–98 *in* J. Latting (ed.). Plant communities of southern California. Calif. Native Plant Soc. Spec. Publ. 2.

Wells, P. V. 1983. Paleobiogeography of montane islands in the Great Basin since the last glaciopluvial. Ecol. Monogr. 53:341–382.

———, and R. Berger. 1967. Late Pleistocene history of coniferous woodland in the Mojave desert. Science 155:1640–1647.

———, and C. D. Jorgensen. 1964. Pleistocene wood rat middens and climatic change in Mojave Desert: a record of juniper woodlands. Science 143:1171–1174.

———, and D. Woodcock. 1985. Full-glacial vegetation of Death Valley, California: juniper woodland opening to *Yucca* semidesert. Madroño 32:11–23.

West, N. E. 1988. Intermountain deserts, shrub steppes, and woodlands. Pp. 209–230 *in* M. G. Barbour and W. D. Billings (eds.)., North American Terrestrial Vegetation. Cambridge Univ. Press, Cambridge.

———, K. H. Rea, and R. J. Tausch. 1975. Basic synecological relationships in juniper-pinyon woodlands. Pp. 41–53 *in* G. F. Gifford and F. E. Busby (eds.). The pinyon-juniper ecosystem: a symposium. Utah Agric. Exp. Sta.

———, R. J. Tausch, K. H. Rea, and P. T. Tueller. 1978. Phytogeographical variation within juniper-pinyon woodlands of the Great Basin. Great Basin Naturalist Mem. 2:119–136.

———, R. J. Tausch, and A. A. Nabi. 1979. Pattern and rates of pinyon-juniper invasion and degree of suppression of understory vegetation in the Great Basin. Range Improve. Notes, U.S.D.A. Forest Serv., Intermountain Region, Ogden, Utah. 14 pp.

Young, J. A., R. A. Evans, and P. T. Tueller. 1975. Great Basin plant communities—pristine and grazed. Pp. 186–215, *in* R. Elston (ed.). Holocene climates in the Great Basin. Occas. Paper, Nevada Archeol. Survey, Reno.

Zamora, B., and P. T. Tueller. 1973. *Artemisia arbuscula, A. longiloba* and *A. nova* habitat types in northern Nevada. Great Basin Naturalist 33:225–242.

Chapter 20

Riparian Communities

Most streams, lakes and springs in California are bordered by riparian communities. These are communities that usually consist of one or more species of deciduous trees plus an assortment of shrubs and herbs, many of which are restricted to the banks and flood plains of these waterways. Sometimes the trees of these communities are tall enough and dense enough to form a forest, and at other times the trees are more scattered. The extent of the vegetation away from the watercourse depends on the size and nature of the banks and flood plains, the amount of water carried by the stream or present in the lake and very importantly on the depth and lateral extent of subterranean aquifers. Additionally historical patterns of land use often determine the actual extent of the riparian corridor. Along small stream channels a riparian community may form a very narrow band, whereas along larger streams or in broad valleys with meandering stream courses, the riparian woodland or forest areas may be quite extensive. Today less than one percent of California is occupied by riparian communities, but in pre-colonial times these communities occupied considerably larger areas.

Riparian communities are not restricted to particular climatic or edaphic conditions in the same manner as are many terrestrial communities. They are primarily dependent upon a permanent water supply. Other environmental factors, however, are very significant in determining community structure and composition. Riparian communities are extremely diverse and difficult to characterize. They vary on both north-south and east-west axes and along altitudinal gradients as well. In addition, waterways pass through almost every plant community in California, and as they do so, some of the species of these contiguous communities become part of the riparian community.

The microenvironments of a riparian community can be quite varied. There is seasonal fluctuation in light available to riparian understories because most of the dominant trees are deciduous (Fig. 20-1). When the trees are in their winter-dormant leafless condition, direct sunlight can reach the ground in most parts of a riparian community. Some herbaceous species and shrubs grow

Fig. 20-1. Riparian corridor with Fremont cottonwoods (*Populus fremontii*) and red willow (*Salix laevigata*) along Shell Creek in La Panza Mountains of San Luis Obispo County. The winter-deciduous trees have not yet leafed out in early spring and direct sunlight can reach the ground beneath the tree canopy. Photo by Robert F. Hoover.

actively and flower while the trees are leafless. The leafing out of the canopy results in a sharp change in conditions at the ground level. Broad-leaved deciduous trees cast rather dense shade when they are in full leaf. This shade results in a reduction in the quantity of light energy reaching the ground and a moderation of temperature fluctuation. Daytime temperatures beneath the tree canopy are often several degrees lower than temperatures in full sunlight. Wind velocity is usually decreased by the tree cover and relative humidity is increased. The moisture evaporated from the soil and released by transpiration can significantly raise the humidity in a riparian corridor. The evaporation also tends to decrease the temperature. Overall, the microclimate within a riparian woodland or forest is often much more mesic than that in adjacent areas. In regions where the non-riparian sites are also rather mesic, riparian and adjacent non-riparian communities may be very similar in structure and species composition.

In drier parts of California the riparian zones bordering small, often intermittent watercourses often consist of scattered trees, such as *Platanus racemosa* (sycamore) growing on the stream banks with open treeless areas in between (Fig. 20-2). This results in an alternation of groups of shade-tolerant and sun-tolerant plants. The band of vegetation that can actually be considered as a riparian zone may be extremely narrow in the treeless areas, with scant differentiation from the surrounding

Fig. 20-2. Riparian zone along small coastal stream in San Luis Obispo County. **A.** Sycamore (*Platanus racemosa*) grows in the foreground and *Carex senta* forms a clump in the stream bed. **B.** Willows (*Salix lasiolepis* and *S. laevigata*) form a major part of the canopy. The boulders are washed over each year by winter floods. The riparian corridor is narrow and gaps exist in the tree canopy. Photos by Elizabeth Bergen.

communities. Herbaceous riparian species may temporarily occupy a stream channel during the spring months but die back as the stream dries. The persistence of subterranean water supplies, however, is indicated by the scattered trees and/or shrubs.

Riparian environments are often considerably more mesic than adjacent, non-riparian environments. The presence of soil moisture and the localized mesic conditions along streams permits some plants to grow as riparian species in areas that are otherwise outside their limits of drought tolerance. For example, *Quercus lobata* (valley oak) is widely distributed in low-elevation foothill woodland sites with moderate precipitation. In the arid interior of the Central Valley it is almost exclusively a riparian species. Both *Acer macrophyllum* (big-leaf maple) and *Sequoia sempervirens* (coast redwood) become progressively restricted to riparian environments in the southern parts of their ranges. In the foothills of the Sierra Nevada, *Pinus ponderosa* (ponderosa pine) extends to considerably lower elevations in riparian than in non-riparian habitats.

The soils of a riparian area generally consist of interbedded layers of coarse and fine sediments ranging from large boulders and rounded river-rocks to gravel, sand, silt and clay (Fig. 20-2). The fine-grained particles tend to collect in backwater areas where water movement is slight and the coarser particles accumulate where water flows more swiftly. Meandering stream channels with broad flood-plains often deposit and redistribute sediments over time, creating both a vertical layering and a horizontal patchwork of sediments. The soils are often relatively coarse and the upper horizons are well-aerated during a portion of the year. Organic matter is often present in the form of decomposing plant litter from riparian plants or washed in from the surrounding watershed. Nutrient levels are comparatively high, both in the soil solution and as a reserve in unweathered rock particles. Soils closest to the stream are often relatively young, whereas those of seldom-flooded areas of the flood-plain may be deep and well developed.

Unlike the plants of many other communities of California, riparian dominants are summer-active and winter-dormant. Many of the subordinate plants are similarly summer growing species. The high water table of a riparian corridor allows the plants to remain metabolically active at times of the year when moisture stress is extreme in adjacent areas. Most of the riparian dominants, however, lose their leaves during the winter when active growth is taking place among the members of many lowland communities. Consequently the riparian plants often seem out of phase with the surrounding vegetation.

Riparian areas are very important as wildlife habitats. The multilayered canopy (Fig. 20-3) provided by the assorted trees, shrubs and herbs provides a diversity of nesting and feeding sites for birds and mammals. Riparian areas are productive habitats, especially at times when plants of other communities are dormant. The moisture of the stream is an important summer water source in the dry California landscape. The nutrients added to the stream and the alternating shaded and sunny zones of the patchy vegetation are important in stream ecology. The vegetation is an important component of the habitat for fish and other aquatic animals.

Riparian Dynamics

Riparian areas are dynamic and ever-changing environments. Stream-channels may be swept clean of vegetation during floods as sediments are shifted from place to place. Fast-flowing water may cut into the banks of a stream, undercutting existing vegetation. Flowing water, with its load of sediments, is abrasive, not only rounding the contours of rocks and boulders and at times carving into exposed bedrock, but also grinding organic debris into small, readily decomposed fragments. Swiftly flowing floodwaters may scour the bark from trees and shrubs or may uproot or deform plants in their path. In areas of slowly flowing water, sediments may be deposited as sand- or mud-bars, exposing new areas to colonization by plants. As the processes of channel-cutting and sediment deposition take place, streams may shift their channels, particularly in terrain with a gentle grade. Bordering either side of a stream is a flood-plain through which a stream meanders over periods ranging from a few years to centuries. Streams in steeper areas gradually cut downward, forming ravines and canyons.

Stream-flow is not uniform from season to season. The period between floods may be sufficient for plants to establish themselves on newly exposed or denuded sites. The low water-flow of summer may shrink the flow of a stream to a trickle or the flow may take place wholly underground within buried sediments. The exposed stream channel is often colonized by a mixture of seedlings of annual and perennial herbs and woody species. Some of these may be characteristic streamside plants, such as *Mimulus guttatus* (monkeyflower), *Rorippa nasturtium-aquaticum* [*Nasturtium officinale*] (watercress), *Polypogon monspeliensis* (rabbitfoot grass), *Veronica anagallis-aquatica* (speedwell) and *Juncus bufonius* (toad rush). Others may be waifs, seedlings of terrestrial species from surrounding areas or from higher elevations, grown from seeds that were swept into the channel during floods and germinated in the stream sediments. Still others are opportunistic weeds that grow in many different kinds of disturbed environments. Plants such as *Salix* spp. (willows) that are able to spread both by seeds and rhizomes may rapidly invade an exposed sand- or gravel-bar. All of these plants may be swept away in the next winter's floodwaters.

Stream flow is not uniform from year to year, either. Because precipitation in a watershed may be much higher in some years than in others, or may be concentrated at times of exceptionally heavy precipitation, a stream's volume varies greatly. Some areas of a stream's flood plain may be sufficiently elevated above the stream channel that they are flooded only during years with extremely high stream-flow. Areas closer to the stream and lower in elevation may be flooded on an annual basis. Various intermediate conditions exist as well. Areas that receive floodwaters on a fairly regular basis are often in a condition of perpetual natural succession. The portion of a riparian community closest to the stream may not advance beyond a pioneer stage. Areas further from the channel may undergo considerable successional change, only to be swept clean or de-

Fig. 20-3. Riparian forest along South Fork of the Kern River in northeastern Kern County. The dominant trees are *Salix gooddingii* (black willow) and *Populus fremontii* (Fremont cottonwood). The canopy is multilayered and continuous. The winter-deciduous trees are leafless for several months of the year. Photos by David Keil.

graded by periodic floods. Areas furthest from the stream may advance to seemingly mature forest or woodland communities.

Many of the woody dominants and some of the herbaceous species of a riparian community are adapted in one way or another to the effects of periodic flooding. Some have deep root systems that firmly anchor them against the force of flowing water. Some, such as various species of *Salix* (willows), *Baccharis salicifolia* [*B. glutinosa*, *B. viminea*] (mule-fat, seep-willow) and various kinds of sedges have flexible stems that are bent by the water but recover after the floodwaters have subsided. Many have rhizomes that are protected to some extent from flood-damage by layers of sediments. Others have no particular adaptations but are able to persist if they become established among large rocks or other sites that protect them from the full force of floodwaters.

Lakes and reservoirs, like streams, are subjected to changing water levels. Unlike rivers and streams, though, the waters are not swiftly flowing. Wave action may be sufficient to cause some zonation on the banks of a lake or reservoir, particularly if the body of water has a large surface area. Smaller bodies of water may be vegetated to the shoreline. The rising and falling of water levels associated with natural drought periods or with withdrawal of water from reservoirs creates an unstable zone within which successional changes occur on a periodic basis. High- and low-water conditions usually persist for much longer periods, however, than is the case for streams and rivers. If the difference between high and low water-levels is relatively small, marshes or swamps may develop in the zone affected by the fluctuation in water level. If the difference is extreme, however, as often is the case with reservoirs, a permanent band of riparian vegetation may not be able to become established. Plants that become established at times of low water-level are completely submerged at times that the reservoir is full. Plants that become established when the reservoir is full are left high and dry when the water-level drops.

Classification Of Riparian Vegetation

Classification of California's riparian communities is difficult because of the diversity and complexity of the communities and the great alterations imposed by human activities. As treated here, these communities are grouped into three main categories: (1) valley and foothill riparian communities, (2) montane riparian communities, and (3) desert riparian communities. Each of these is variable. Subgroupings based upon local habitat conditions or upon species composition might be recognized, but we have chosen to not do so here because of the almost unlimited combinations possible.

Fig. 20-4. Riparian corridor of Coon Creek in San Luis Obispo County. This coastal stream supports a riparian woodland corridor dominated by willows (*Salix* spp.) and coast live oak (*Quercus agrifolia*). Along its course the creek traverses areas of grassland, coastal scrub, chaparral, and other communities. A seasonally developed sand bar separates the mouth of the creek from the ocean. Photo by V. L. Holland.

1. Valley and Foothill Riparian Communities

These communities occur along waterways from near sea level to the lower elevation margins of the montane coniferous forest areas of cismontane California (Figs. 20-1–20-5). They range (or ranged) from broad valley flood plain forests to narrow steep canyon streams. Their elevational extent corresponds roughly with that of the grassland and woodland-chaparral zones. The climate is comparatively warm in the winter with precipitation falling mostly as rain. The summers are long and dry. Often surface water-flow ceases in the summer but subsurface moisture remains available. The dominant trees are almost all deciduous trees or large shrubs. These include:

Acer macrophyllum	big-leaf maple
Acer negundo	box-elder
Alnus rhombifolia	white alder
Alnus rubra [*A. oregona*]	red alder
Fraxinus latifolia	Oregon ash
Platanus racemosa	sycamore
Populus balsamifera ssp. *trichocarpa* [*P. trichocarpa*]	black cottonwood
Populus fremontii	Fremont cottonwood
Quercus lobata	valley oak
Salix laevigata	red willow
Salix gooddingii	black willow
Salix lasiolepis	arroyo willow
Salix lucida ssp. *lasiandra* [*S. lasiandra*]	yellow willow

These species are very unlikely to all occur in any particular riparian community. Each has its own range in the state. Some, such as *Populus fremontii* and *Salix gooddingii* tend to be most common in drier areas of the state. Others, such as *Populus balsamifera* ssp. trichocarpa and *Alnus rhombifolia* occur most commonly in moister areas. Some species occur mostly in foothill regions and others are most common in broad valleys. *Acer negundo* and *Fraxinus latifolia* occur most commonly in mountainous regions. *Quercus lobata* and *Salix gooddingii* are most common in broad valleys. In the southern two-thirds of its range, *Platanus racemosa* occurs both in mountains and in valleys, but in the north it is restricted to the Sacramento Valley.

Locally, evergreen species may be common or dominant. These include *Quercus agrifolia* (coast live oak), *Quercus wislizenii* (interior live oak), *Umbellularia californica* (California bay-laurel), and *Sequoia sempervirens* (coast redwood).

Shrubs are common and sometimes dominant (Fig. 20-5). It is not at all unusual to find dense thickets of shrubs occupying extensive areas within a riparian corridor. Common shrubs include:

Baccharis salicifolia	mule-fat, seep-willow
Cephalanthus occidentalis	button-willow
Cornus spp.	dogwoods
Rosa californica	wild rose
Rubus spp.	blackberries

Fig. 20-5. Riparian thickets of *Baccharis salicifolia* (mule fat) and *Pluchea sericea* (arrow weed) along Cuyama River, eastern Santa Barbara County. Hillsides in background are occupied by foothill a woodland community. Photo by David Keil.

Salix spp.	willows
Sambucus mexicana	elderberries
Symphoricarpos spp.	snowberries
Toxicodendron diversilobum [*Rhus d.*]	poison-oak

Many other shrubs sometimes occur in these riparian communities. Lianas (woody vines) that are sometimes present including:

Aristolochia californica	pipe-vine
Clematis ligusticifolia	virgin's bower
Lonicera spp.	honeysuckle
Toxicodendron diversilobum	poison oak
Vitis californica	grape

Numerous herbaceous plants occur in riparian zones, and in pioneer areas where trees are absent, these may be the most significant plant cover. Common herbs include:

Aralia californica	spikenard
Artemisia douglasiana	mugwort
Carex spp.	sedges
Cyperus spp.	flat-sedges
Datisca glomerata	durango-root
Eleocharis spp.	spike-rushes
Epilobium spp.	willow-herbs
Equisetum spp.	horsetail, scouring-rush
Juncus spp.	rushes
Mimulus spp.	monkeyflowers
Rorippa nasturtium-aquaticum	watercress
Scirpus spp.	bulrushes

Typha spp.	cattails
Urtica dioica	stinging nettle
Veronica anagallis-aquatica	speedwell

Many other herbs can also be found in riparian zones. Perennial herbs are often more common in riparian than non-riparian habitats as a consequence of the year-round moisture supply. A long list of exotic species that have become established in riparian zones could be added as well.

2. Montane Riparian Communities

Riparian communities in high mountain areas (Fig. 20-6) differ in several respects from those in lower elevation zones. Several factors combine to keep temperatures low in montane riparian areas. Increased elevation produces a characteristic temperature decrease. Valleys in mountainous areas often serve as drainages both for water and for cold air. Nocturnal cold-air drainage can result in temperatures several degrees cooler along a valley floor than on nearby slopes. Additionally, mountain streams are commonly very cold and swiftly flowing and they keep the atmosphere around the stream chilled. These combined cold temperature effects are limiting for several of the trees that are dominant in riparian communities at lower elevations. Such species as *Quercus lobata* and *Platanus racemosa* occur in lower foothill areas but drop out with increased elevation. At the same time, the cold temperatures permit such conifers as *Pinus ponderosa* (ponderosa pine) and *Pinus jeffreyi* (Jeffrey pine) to grow in riparian areas at elevations lower than they occupy on exposed slopes. The result is a localized inversion of the usual vegetation zones with bands of coniferous forest growing in canyons below slopes occupied by foothill woodland and mixed chaparral communities.

Montane and foothill riparian communities overlap considerably in species composition in lower montane elevation zones. Characteristic species of the lower parts of the montane riparian communities include such deciduous trees as:

Alnus rhombifolia	white alder
Alnus incana [*A. tenuifolia*]	mountain alder
Acer macrophyllum	big-leaf maple
Fraxinus latifolia	Oregon ash
Populus balsamifera ssp. *trichocarpa*	black cottonwood

In addition to the trees listed below, which are for the most part only found in riparian zones, members of the montane coniferous forests often extend into these areas and become components of the riparian community. In rocky canyon areas, montane riparian communities may also contain *Quercus chryso-*

Fig. 20-6. Montane riparian community along small, swiftly flowing stream. Mountain alder (*Alnus incana*) is the dominant shrub. Photo by V.L. Holland.

lepis (canyon-live oak) and *Quercus kelloggii* (black oak). Shrubby members of these communities include:

Cornus sericea [*C. stolonifera*]	creek dogwood
Rhododendron occidentale	western azalea, western
Ribes spp.	gooseberries, currants
Rubus parviflorus	thimbleberry
Physocarpus capitatus	nine-bark
Calycanthus occidentalis	spicebush
Salix spp.	willows
Symphoricarpos spp.	snowberry

Herbs include *Carex senta* and other *Carex* species, *Aralia californica* (spikenard), *Mimulus* spp., *Juncus* spp., and numerous others.

Canyon areas at mid-elevations often have a vegetation composed of a mixture of riparian and non-riparian species. Strictly riparian taxa tend to occupy the areas immediately bordering drainage channels. This riparian zone intergrades with a mixed hardwood-conifer forest that occupies the valley bottoms and slopes. Species such as *Quercus chrysolepis*, *Quercus kelloggii*, *Umbellularia californica* (California bay-laurel), *Pinus ponderosa*, *Calocedrus decurrens* (incense cedar), etc., may at times grow in the riparian zone. These species are not as tolerant

Fig. 20-7. Montane riparian communities along small streams in Sierra Nevada of Tulare County. **A.** Trees along the stream are quaking aspen (*Populus tremuloides*) and lodgepole pines (*Pinus contorta* ssp. *murrayana*). Willows form thickets along the stream banks. **B .** Lodgepole pines (*Pinus contorta* ssp. *murrayana*) grow together with thickets of willows along the stream banks. Photos by David Keil.

of saturated soils as are the more typical riparian taxa and consequently grow mostly on coarse rocky soils in riparian areas.

Montane riparian communities are often poorly developed in headwater areas where active erosion is taking place. An actively down cutting stream often has a channel that is situated on bedrock with few places available for riparian plants to become established. In such areas after the snow has melted the steep rocky channel retains little available moisture.

High elevation riparian communities are dominated by *Populus tremuloides* (quaking aspen; Fig. 20-7A), *Populus balsamifera* ssp. *trichocarpa*, and *Betula occidentalis* (water birch). A frequent associate is *Pinus contorta* ssp. *murrayana* (Sierran lodgepole pine) which is able to tolerate both flooding and waterlogged soils. Species of *Salix* and *Ribes* are among the most common shrubs (Fig. 20-7B). In upper-elevation cismontane areas where the slopes surrounding a stream channel are forested with conifers, the riparian zone is often rather narrow. It may contain merely a border of shrubs and small trees, or where the stream opens into a meadow a zone of willows, aspens and lodgepole pine may be readily apparent. These communities grade into meadow communities, and as discussed in chapters 16 and 21, the boundary between forest and meadow may be dynamic.

Montane riparian vegetation sometimes forms along the shores of mountain lakes (Fig. 20-8). However, the development of high-elevation riparian communities on the shores of mountain lakes is often inhibited by ice movement on the shorelines. Alternate freezing, cracking and thawing wedge ice blocks onto the shore that damage woody plants growing close to the shoreline.

On some transmontane slopes *Populus tremuloides* forms extensive stands surrounded by sparse forests or by Great Basin sagebrush desert scrub vegetation (Fig. 20-9). Such transmontane communities may include *Populus fremontii* [at lower elevations], *Quercus kelloggii*, *Q. chrysolepis*, *Salix lucida* ssp. *lasiandra* and *S. lasiolepis*. From the Owens Valley region southward on transmontane slopes, the montane riparian communities often contain *Fraxinus velutina* (velvet ash).

The highest elevation streamside communities border the rivulets that form when snow melts above timberline. These areas generally have a mixture of alpine shrubs and sedges. This vegetation is considered to be a part of the alpine vegetation.

3. Desert Riparian Communities

Desert riparian communities occur along waterways in the desert areas of California (Fig. 20-10). These communities occur along the margins of the Mojave and Colorado Rivers, along the

Fig. 20-8. Montane riparian zone around mountain lake in the Sierra Nevada. Willows (*Salix* sp.) form a shrubby ring around the lake. The adjacent forest is dominated by lodgepole pine (*Pinus contorta* ssp. *murrayana*). Photo by V. L. Holland.

Fig. 20-9. Stand of quaking aspens (*Populus tremuloides*) on eastern slope of Sierra Nevada in Mono County. Great Basin sagebrush scrub occupies the unforested areas. Photo by David Keil.

Fig. 20-10. Desert riparian community near Lone Pine in Inyo County. The vegetation of the riparian corridor is dominated by willows (*Salix* spp.) and water birch (*Betula occidentalis*). Conifers such as *Pinus jeffreyi* and *P. monophylla* descend to lower elevations along the desert stream than on open slopes. Surrounding desert slopes are vegetated by a Great Basin sagebrush scrub community. Photo by David Keil.

courses of some smaller streams that flow into desert regions, along irrigation canals, and around some desert springs. Streams that flow into the deserts often drain into interior drainage basins and are consequently associated with alkali sink communities. Not all, however, drain into interior drainage basins. Some dry out along the way, changing from free-flowing streams to intermittent streams and dry washes as they cross the desert. Desert riparian communities consequently grade into desert dry wash communities (Chapter 18).

Desert riparian communities are not entirely free from drought despite the stream flow that provides their water supplies. When drought occurs in a desert region, stream flow may diminish to a trickle or may disappear altogether. Subsurface flow may greatly diminish as well. The members of desert riparian communities are a mixture of shrubs and trees with varying degrees of drought adaptations. Most are phreatophytes. Common native species include:

Fig. 20-11. Fan palm oasis in Joshua Tree National Park, San Bernardino County. *Washingtonia filifera* (California fan palm) grows along a spring-fed stream. The surrounding desert slopes are vegetated by an open Mojave Desert scrub community with various shrubs. Photo by David Keil.

Baccharis salicifolia	seep-willow
Baccharis sergilloides	baccharis
Cercidium floridum	palo verde
Fraxinus velutina	velvet ash
Phragmites australis [*P. communis*]	common reed
Pluchea sericea	arrow-weed
Populus fremontii	Fremont cottonwood
Prosopis glandulosa	honey mesquite
Prosopis pubescens	screw-bean mesquite
Salix exigua	sand bar willow
Salix gooddingii	black willow
Salix laevigata	red willow

Desert streams that originate in the Transverse or Peninsular Ranges (e.g., the Mojave River, Borrego Palm Canyon) sometimes have species such as *Alnus rhombifolia* and *Platanus racemosa* that are mostly restricted to cismontane regions.

Several introduced species have become important members of these communities. The most important are various species of *Tamarix* (salt-cedar). These Eurasian phreatophytes now occupy extensive areas in desert riparian communities. Another common introduced species in some areas is *Arundo donax* (giant reed).

One interesting type of desert riparian community that deserves special mention is the fan palm oasis (Fig. 20-11). *Washingtonia filifera* (California fan-palm), the only palm native to the western United States, is a dominant in oasis communities. The palms occur around moist, often somewhat alkaline springs in the creosote bush scrub areas from Twentynine Palms south through the Colorado desert and into northern Baja California. Two fan palm oases also occur in western Arizona. A number of the California oases are located at springs that come to the surface along the San Andreas Fault. Various other riparian species commonly grow in association with *Washingtonia filifera*. These include *Alnus rhombifolia*, *Platanus racemosa*, *Populus fremontii*, *Prosopis glandulosa*, and species of *Salix*.

Human Impacts On Riparian Systems

Riparian communities have been greatly modified and reduced in areal coverage by human activities. There are several reasons for this. Formerly the most extensive riparian communities occurred along the San Joaquin and Sacramento Rivers and their tributaries. These communities formed densely forested bands across the otherwise treeless Central Valley. The alluvial soils that occur in broad valleys are ideal agricultural land, and the early American immigrants who settled the Central Valley soon began to place these lands under cultivation. Almost all the Central Valley's riparian forests were eventually cut and replaced by farmlands and orchards. Less than two percent of the original riparian vegetation in the Sacramento Valley still remains. On a lesser scale this transformation of the land has occurred in many other parts of California. Most of the once extensive desert riparian woodlands of the Colorado River Valley are now gone. Many areas that have not been actively cultivated have been modified by the grazing of domestic stock in riparian areas. This has resulted in the removal of palatable plants, eating and trampling of seedlings of riparian tree species, invasion of non-palatable weed species and the degradation of stream banks and water quality.

A second major impact has been the efforts of Californians to conserve water and prevent flooding through the construction of dams and reservoirs. Downstream from the dams, this has drastically reduced the flow of many streams. Upstream it has inundated valleys once occupied by riparian communities. Reservoirs are often a major source of water for irrigation and often exhibit great fluctuation of water level. Natural lakes are often bordered by a riparian zone, but because of the rise and fall of the water level in artificial reservoirs, extensive riparian communities seldom develop. Although some riparian species may colonize the borders of a reservoir, water levels are often too variable for tree species to become permanently established.

A third major source of alteration of riparian communities is the location of many towns and cities along watercourses. Wherever this occurs, flooding and its concomitant economic damage have been inevitable. Efforts to control this flooding have included removal of riparian vegetation to speed the movement of floodwaters, dredging, channelization, and enclosure of the stream in concrete. All sorts of solid and liquid wastes have been dumped into streams. Exotic species cultivated as ornamentals have found the ample moisture of the stream banks congenial and have invaded riparian areas and sometimes replaced the native vegetation. Other introduced species have been deliberately planted along the streams and now grow as a part of an altered riparian community. In urban areas a riparian community may contain far more introduced species than native taxa. Additional losses occur from urbanization projects, highway construction and recreational developments.

Large areas of desert riparian woodlands have been destroyed in the past century. Trees and shrubs have been removed from some desert watercourses to increase stream flow. Along the Colorado River construction of dams has flooded areas that once were densely wooded. The lowering of water tables by pumping from wells and the separation of streams from their watersheds by dam construction have changed the characteristics of many formerly active stream courses. Trees have been cut and the fertile flood plains plowed for agriculture. Often all that remains of a once flourishing riparian corridor is a sandy wash channel across the desert.

Riparian Communities—References

Argent, G., ed. 1991. Delta-Estuary. California's inland coast. A public trust report. Prepared for the California State Lands Commission. 208 pp.

Benedict, N. B. 1984. Classification and dynamics of subalpine ecosystems in the south Sierra Nevada. Pp. 92–96 *in* R. E. Warner and K. M. Hendrix (eds.). California riparian systems. Univ. Calif. Press, Berkeley.

Bowler, P. A.. 1990. Riparian woodland: an endangered habitat in southern California. Pp. 80–97 *in* A. A. Schoenherr (ed.), Endangered plant communities of southern California. Proceedings of the 15th Annual Symposium. Southern California Botanists spec. publ. 3.

Brothers, T. S. 1984. Historical vegetation changes in the Owens River riparian woodland. Pp. 75–84 *in* R. E. Warner and K. M. Hendrix (eds.). California riparian systems. Univ. Calif. Press, Berkeley.

Campbell, C. J., and W. N. Green. 1968. Perpetual succession of stream-channel vegetation in a semiarid region. J. Ariz. Acad. Sci. 5:86–98.

Conard, S. G., R. L. MacDonald, and R. F. Holland. 1980. Riparian vegetation and flora of the Sacramento Valley. Pp. 47–57 *in* A. Sands (ed.). Riparian forests of California. Their ecology and conservation. Univ. Calif Div. Agric. Sci., Davis.

Cowardin, L. M., V. Carter, F. C. Golet, and E. T. LaRoe. 1979. Classification of wetlands and deepwater habitats of the United States. U.S.D.I. Fish and Wildlife Service, Office of Biological Services Publ. FWS/OBS-79/31. 103 pp.

Dick-Peddie, W. A., and J. P. Hubbard. 1977. Classification of riparian vegetation, Pp. 85–90 *in* R. R. Johnson and D. A. Jones (tech. coord.) Importance, preservation and management of riparian habitat: a symposium. U.S.D.A. Forest Service Gen. Tech. Rep. RM-43. Rocky Mt. Forest and Range Exp. Stn., Ft. Collins, Colorado.

Ellison, J. P. 1984. A revised classification of native aquatic communities in California. Calif. State Dept. Fish and Game Planning Br. Admin. Rep. 84-1. 30 pp.

Ferren, W. R.. 1989. A preliminary and partial classification of wetlands in southern and central California with emphasis on the Santa Barbara Region. Prepared for Wetland plants and vegetation of coastal southern California, a workshop organized for the California Department of Fish and Game and the United States Fish and Wildlife Service. University of California , Santa Barbara. 54 pp.

———, and P. L. Fiedler. 1993. Rare and threatened wetlands in central and southern California. Pp. 119–131 *in* J. E. Keeley (ed.), Interface between ecology and land development in California. Southern California Academy of Sciences, Los Angeles.

Griffin, J. R., and W. B. Critchfield. 1972. The distribution of forest trees in California. Pacific S.W. Forest and Range Exp. Stn., Berkeley, Calif. 114 pp. (U.S.D.A. Serv. Res. Paper PSW 82).

Haslam, S. M. 1978. River plants. The macrophytic vegetation of watercourses. Cambridge Univ. Press, Cambridge. 396 pp.

Holstein, G. 1984. California riparian forests: deciduous island in an evergreen sea. Pp. 2–22 *in* R. E. Warner and K. M. Hendrix (eds.). California riparian systems. Univ. Calif. Press, Berkeley.

Keller, E. A. 1980. The fluvial system: selected observations. Pp. 39–46 *in* A. Sands (ed.). Riparian forests of California. Their ecology and conservation. Univ. Calif Div. Agric. Sci., Davis.

Johnson, R. R., and D. M. Jones (tech. coord.). 1977. Importance, preservation and management of riparian habitat: a symposium. U.S.D.A. Forest Service Gen. Tech. Rep. RM-43. Rocky Mt. Forest and Range Exp. Stn., Ft. Collins, Colorado. 217 pp.

———, and J. K. McCormick. 1978. Strategies for protection and management of floodplain wetlands and other riparian ecosystems. U.S.D.A. Forest Service Gen. Tech. Rep. WO-12. Washington, D.C. 410 pp.

McBride, J. R., and J. Strahen. 1984. Fluvial processes and woodland succession along Dry Creek, Sonoma County, California. Pp. 110–119 *in* R. E. Warner and K. M. Hendrix (eds.). California riparian systems. Univ. Calif. Press, Berkeley.

Minckley, W. L., and D. E. Brown. 1982a. Interior Californian riparian deciduous forests and woodlands. Pp. 250–254 *in* D. E. Brown (ed.). Biotic communities of the American Southwest—United States and Mexico. Desert Plants 4:1–341.

———, and ———. 1982b. Riparian scrublands. *in* D. E. Brown (ed.). Biotic communities of the American Southwest—United States and Mexico. Desert Plants 4:1–341.

———, and ———. 1982c. Sonoran oasis forest and woodlands. p. 274 *in* D. E. Brown (ed.). Biotic communities of the American Southwest—United States and Mexico. Desert Plants 4:1–341.

———, and ———. 1982d. Sonoran riparian deciduous forest and woodlands. Pp. 269–273 *in* D. E. Brown (ed.). Biotic communities of the American Southwest—United States and Mexico. Desert Plants 4:1–341.

———, and ———. 1982e. Sonoran riparian scrub. Pp. 278–279 *in* D. E. Brown (ed.). Biotic communities of the American Southwest—United States and Mexico. Desert Plants 4:1–341.

———, and ———. 1982f. Southwestern wetlands. *in* D. E. Brown (ed.). Biotic communities of the American—Southwest United States and Mexico. Desert Plants 4:1–341.

———, and ———. 1982g. Warm-temperate interior strands. Pp. 265–267 *in* D. E. Brown (ed.). Biotic communities of the American Southwest—United States and Mexico. Desert Plants 4:1–341.

———, and J. N. Rinne. 1985. Large woody debris in hot-desert streams: an historical review. Desert Plants 7:142–153.

Nilsen, E. T., M. R. Sharifi and P. W. Rundel. 1984. Comparative water relations of phreatophytes in the Sonoran Desert of California. Ecology 65:767–778.

Norton, B. E., J. S. Tuhy, and R. P. Young. 1980. A riparian community classification study. Dept. of Range Science, Utah State Univ., Logan, Utah. 77 pp.

Reichenbacher, F. W. 1984. Ecology and evolution of southwestern riparian plant communities. Desert Plants 6(1):15–22.

Roberts, R. C. 1984. The transitional nature of northwestern California riparian systems. Pp. 85–97 *in* R. E. Warner and K. M. Hendrix (eds.). California riparian systems. Univ. Calif. Press, Berkeley.

Roberts, W. G., J. G. Howe, and J. Major. 1980. A survey of riparian forest flora and fauna in California. Pp. 3–19 *in* A. Sands (ed.). Riparian forests of California. Their ecology and conservation. Univ. Calif Div. Agric. Sci., Davis.

Robichaux, R. 1980. Geologic history of the riparian forests of California. Pp. 21–34 *in* A. Sands (ed.). Riparian forests of California. Their ecology and conservation. Univ. Calif Div. Agric. Sci., Davis.

Robison, T. W. 1958. Phreatophytes. U.S. Geol. Surv. Water Supply Pap. 1423. 84 pp.

Sands, A. (ed.). 1980. Riparian forests of California. Their ecology and conservation. Univ. Calif Div. Agric. Sci., Davis. 122 pp.

———. 1982. The value of riparian habitat. Fremontia 10(1):3–7.

Scott, L. B., and S. K. Marquis. 1984. An historical overview of the Sacramento River. Pp. 51–57 *in* R. E. Warner and K. M. Hendrix (eds.). California riparian systems. Univ. Calif. Press, Berkeley.

Sculthorpe, C. D. 1967. The biology of aquatic vascular plants. St. Martin's Press, N.Y. 610 pp.

Smith, F. E.. 1980. A short review of the status of riparian forests in California. Pp. 1–2 *in* A. Sands (ed.). Riparian forests of California. Their ecology and conservation. Univ. Calif Div. Agric. Sci., Davis.

———. 1982. The changing face of the San Joaquin Valley. Fremontia 10(1):24–27.

Stegman, J. L. 1976. Overview of current wetland classification and inventories in the United States and Canada. Pp. 102–120 *in* J. H. Sather (ed.). National wetland classification and inventory workshop proceedings—1975. U.S.D.I., Fish and Wildlife Service. Washington, D.C.

Strahan, J. 1984. Regeneration of riparian forests of the Central Valley. Pp. 58–67 *in* R. E. Warner and K. M. Hendrix (eds.). California riparian systems. Univ. Calif. Press, Berkeley.

Stromberg, J. C., and D. T. Patten. 1992. Mortality and age of black cottonwood stands along diverted and undiverted streams in the eastern Sierra Nevada, California. Madroño 39:205–223.

———, and ———. 1992. Response of *Salix lasiolepis* to augmented stream flows in the upper Owens River. Madroño 39:224–235.

Thomas, J. W., C. Maser, and J. E. Roderick. 1979. Wildlife habitats in managed rangelands—the Great Basin of southeast Oregon. Riparian zones. U.S.D.A. Forest Service Gen. Tech. Rep. PNW-80. Corvallis, Oregon. 18 pp.

Thompson, K. 1961. Riparian forests of the Sacramento Valley, California. Ann. Assoc. Amer. Geogr. 51:294–315.

———. 1980. Riparian forests of the Sacramento Valley, California. Pp. 35–38 *in* A. Sands (ed.). Riparian forests of California. Their ecology and conservation. Univ. Calif Div. Agric. Sci., Davis.

Vogl, R. J., and L. T. McHargue. 1966. Vegetation of California fan palm oases on the San Andreas Fault. Ecology 47:532–540.

Walters, M. A., R. O. Teskey, and T. M. Hinckley. 1980. Impact of water level changes on woody riparian and wetland communities. Vol. VII. Mediterranean region. Western arid and semiarid region. U.S.D.I. Fish and Wildlife Service. Office of Biol. Services. Washington, D.C. 83 pp.

Warner, R. E., and K. M. Hendrix (eds.). 1984. California riparian systems. Ecology, conservation and productive management. Univ. Calif. Press, Berkeley. 1035 pp.

Whitney, S. 1979. A Sierra Club naturalist's guide to the Sierra Nevada. Sierra Club Books. San Francisco. 526 pp.

Zembal, R.. 1990. Riparian habitat and breeding birds along the Santa Margarita and Santa Ana Rivers of southern California.. Pp. 98–114 in A. A. Schoenherr (ed.), Endangered plant communities of southern California. Proceedings of the 15th Annual Symposium. Southern California Botanists spec. publ. 3.

Chapter 21

Freshwater Wetland Communities

Freshwater wetland plant communities occur in both still and flowing water. The bodies of water range from small pools a few meters wide to large reservoirs and natural lakes. Some areas that are only seasonally wet also support freshwater wetland communities. Together these various aquatic environments occupy about 2.5 percent of California's total area.

Aquatic plant communities are quite variable and are dependent on several interacting environmental factors. Some of the factors that influence the species composition and physiognomy of a freshwater wetland community are water depth, fluctuations in water level, rate of water flow, water and air temperatures, other climatic variables, pH and dissolved salts, organic content of the water, nature and depth of bottom sediments, and the history of the body of water.

Water depth is particularly important in determining the composition of an aquatic community. In deep open water only submersed or floating aquatic plants are able to become established. In shallow water, however, the dominant plants are generally emergents, rooted below the water level but having their leaves, stems and flowers elevated above the water. Fluctuations in water level can result in seasonal changes in community structure. Some bodies of water are temporary and dry up during California's long summer drought. Artificial reservoirs are often allowed to fluctuate, and their water level may be so variable that the establishment of a well-developed aquatic community may be inhibited.

The rate of water flow is significant in determining the nature of an aquatic community. Rapidly flowing water is usually well-aerated and the moving water tends to sweep the bottom clean of fine sediments. Rooted plants must be anchored firmly enough to avoid being swept away and must be flexible enough that they are not torn apart by water currents. Slowly flowing or standing water ranges from well-aerated to anaerobic. In slow-moving or standing water fine-grained sediments including organic debris may settle to the bottom. A greater diversity of plants can become established, including floating aquatics.

The interaction between bottom sediments, water chemistry and temperature conditions is complex and highly significant to plants. The major sources of inorganic nutrients in an aquatic system are bottom sediments and runoff. In non-flowing or slowly flowing water the decomposition of organic sediments may be the most direct source of such nutrients as nitrates, phosphates, potassium, etc. Nutrient cycling is dependent upon complex interactions among various groups of organisms and upon the sources and nature of the nutrients. Decomposition may be inhibited by low temperatures or acidic pH conditions, and the result may be low nutrient availability. The species composition often reflects these chemical interactions.

The amount of organic matter in a body of water and the nature of its decomposition affect the plant life. Organic matter can be added from decomposition of vegetation, animal wastes or sewage pollution. In nutrient-rich water, much algal growth may occur at the water surface, forming a floating scum that blocks light penetration. Under such conditions, submersed aquatic vascular plants are unable to get adequate light to carry on photosynthesis. Decomposition of organic matter may deplete the supply of oxygen. The sediments at the bottom of a nutrient-rich pond or lake may be completely anaerobic. Because oxygen is essential for root growth, many plants are unable to grow in such environments.

The history of a body of water has a great influence on the aquatic communities that occur there. The ultimate fate of most inland lakes and ponds is to be filled in by accumulated sediments. This gradual filling-in of a body of water results in a progressive change in the aquatic plant communities that occupy the site. A pond or lake often supports a series of concentric rings of vegetation in different stages of succession. The changes in community structure that accompany the filling-in of a body of water also change the various chemical components of the system.

Small ponds, lakes and reservoirs are often island-like in their distribution, isolated from each other by broad expanses of dry land. Species composition often reflects the results of chance introductions. Bodies of water with similar habitats may have very different species compositions solely because of the element of chance involved. Dispersal of seeds or spores from other aquatic communities is dependent upon both random and non-random dispersal agents. Some plants, such as *Typha* spp. (cattails) have wind-borne seeds that may be blown from one body

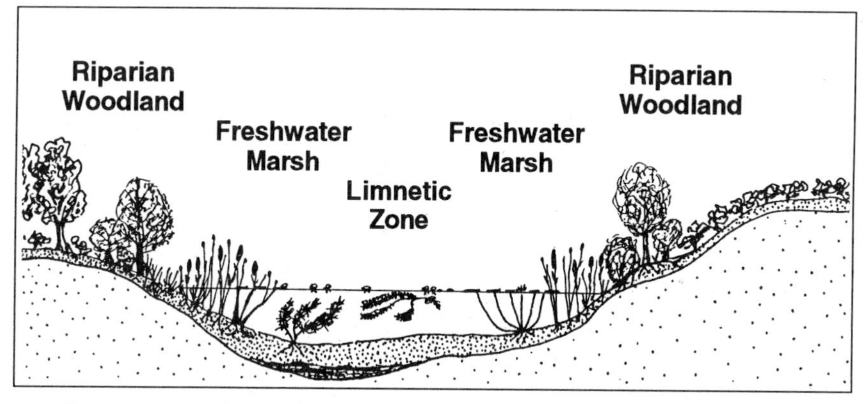

Fig. 21-1. Relationship among freshwater wetland and riparian communities. Limnetic communities composed of rooted and free-floating submersed aquatic plants occupy the deeper areas of open water in the center of a lake or pond. Rooted emergent herbaceous plants form a freshwater marsh (a littoral community) in shallow water. A riparian community dominated by shrubs and trees occurs around the margins of the body of water.

of water to another. Other propagules may be transported by waterfowl or other wildlife from one body of water to another. Human activities may involve deliberate or accidental introduction of plants or their seeds into a body of water.

Freshwater wetland communities in California form two main groups. Limnetic communities occur in open water and littoral communities occur in shallow water and along the shores of otherwise open bodies of water (Fig. 21-1). Most vascular aquatic plant communities are littoral. The littoral communities recognized in this treatment are freshwater marshes, bogs and fens, montane meadows, and vernal pools. In some cases the littoral zone is in turn surrounded by a riparian woodland or shrubland community (Chapter 20)

1. Limnetic Plant Communities

Limnetic plant communities have both an algal component and a higher plant component. The algal component may largely planktonic and consists of a mixture of various types of algae and cyanobacteria. If a body of water is sufficiently shallow (or is deeper and clear), algae that grow attached to bottom debris may be important as well. Vascular plants of the open-water environment are either rooted or planktonic. Rooted forms include:

Fig. 21-2. Lowland freshwater marsh bordering Laguna Lake, San Luis Obispo County. The marsh is dominated by tall tules (*Scirpus californicus* and *S, acutus*) and smartweed (*Polygonum amphibium*). The tules are evergreen in low elevation marshes. A limnetic community occupies the open water in the center of the lake. Photo by David Keil.

Ceratophyllum demersum	hornwort
Elodea canadensis	elodea
Isoetes spp.	quillwort
Myriophyllum spp.	water-milfoil
Najas spp.	water-nymph
Potamogeton spp.	pondweed
Sparganium spp.	bur-reed
Zannichellia palustris	horned pondweed

Plants that float at or near the water surface include:

Azolla filiculoides	mosquito fern
Ceratophyllum demersum	hornwort
Lemna spp.	duckweed
Ranunculus aquatilis	water buttercup
Utricularia spp.	bladderwort
Wolffia spp.	water-mote
Wolffiella spp.	mud-midget

The development of floating aquatics is usually restricted to small bodies of water or to sheltered embayments. In large open lakes floating aquatics are often swept ashore and stranded by wind action and waves.

Fig. 21-3. Freshwater marsh in montane lake in the Sierra Nevada. Many marshland species occur over a wide elevational range. The sedges that dominate this marsh are dormant during the winter. A limnetic community occupies the open water in the center of the lake. Photo by Robert Rodin.

2. Freshwater Marshes

Freshwater marshes occur throughout California in a diversity of climatic conditions (Figs. 21-2, 21-3, 21-5). Freshwater marsh communities, like riparian communities (Chapter 20), are not limited to particular elevational zones. Marshes occur in nutrient-rich mineral soils that are saturated through most or all of the year. These communities are best developed in locations with slow-moving or stagnant shallow water. Such sites commonly occur along the margins of ponds and lakes and in the flood plains of slow-moving streams and rivers. Often freshwater marshes entirely fill the basins of shallow ponds. In areas where freshwater marshes develop there is not always standing water. In some cases the water table is so close to the surface that it can be tapped by shallowly rooted marsh plants. Marshes are not always associated with ponds or other depressions. Small marsh communities often occur on hillsides where seepage from springs has dampened the slopes.

Marsh soils are often anaerobic or nearly so. The low oxygen supply results from two factors. The presence of water in the soil fills the spaces among the soil particles, thereby excluding atmospheric gasses. The presence of organic debris in various

Fig. 21-4. *Juncus* sp. in freshwater marsh. Emergent reed-like monocots are common components of marsh vegetation. Photo by David Keil.

stages of decomposition promotes the growth of bacteria and other microorganisms. Respiration of these decomposers greatly reduces the oxygen supply available to the roots and rhizomes of higher plants. Much of the decomposition that takes place in the soil of a marsh occurs under anaerobic conditions and results in the production of toxic by-products and such gasses as methane and hydrogen sulfide. Because of these soil conditions, relatively few kinds of higher plants are able to grow in marsh soils.

The dominant plants of freshwater marsh communities are mostly perennial monocots that can reproduce vegetatively by underground rhizomes. The stems of these plants are generally hollow or filled by aerenchyma, a porous tissue that allows gas exchange to occur between the exposed parts of the plants and their roots. Common marsh dominants are various species from the genera of monocots listed below:

Fig. 21-5. Freshwater marshes. **A.** Coreopsis Lake and Oso Flaco Lake on the Nipomo Dunes in San Luis Obispo County. The tall reeds are *Scirpus californicus* (tule). **B.** Marsh near New Cuyama in Santa Barbara County dominated by cattails (*Typha* sp.). Photos by David Keil.

Carex spp.	sedge
Eleocharis spp.	spike-rush
Juncus spp. (Fig. 21-4)	rush
Sagittaria spp.	arrowleaf
Scirpus spp. (Fig. 21-2, 21-5A)	bulrush, tule
Sparganium spp.	bur-reed
Typha spp. (Fig. 21-5B)	cattail

Some marshes are comparatively low-profile communities. In others, however, stands of tall reedy monocots are dense and nearly impenetrable. *Scirpus acutus* and *Scirpus californicus* often grow to several meters in height (Figs. 21-2, 21-5).

In addition to the dominant monocots there are also some common dicot components in marsh communities:

Bidens spp.	beggar-ticks
Cicuta spp.	water hemlock
Epilobium spp.	willow-herb
Mimulus guttatus	common monkeyflower
Polygonum spp.	smartweed
Rorippa nasturtium-aquaticum [*Nasturtium officinale*]	watercress
Rorippa spp.	yellow-cress
Rumex spp.	dock

Rooted aquatics with floating stems and leaves are important components of some marshes:

Brasenia schreberi	water-shield
Marsilea vestita	water-clover fern
Nuphar luteum ssp. *polysepalum* [*N. polysepalum*]	yellow pond-lily
Polygonum amphibium	water smartweed
Potamogeton spp.	pondweeds
Oenanthe sarmentosa	water-parsley
Hydrocotyle spp.	pennywort

Floating and submersed aquatics grow in marshes as well as in limnetic communities. *Azolla filiculoides*, *Lemna* spp., and other Lemnaceae are particularly common in marshy sites with some open water.

Some freshwater marshes are seasonal communities. During the winter and spring when ample moisture is available in the soil, communities dominated by *Juncus*, *Carex*, *Eleocharis*, etc., occur in some low areas. These sites may retain some soil moisture well into the summer, but the soil surface becomes dry and hard. Grassland species such as *Hemizonia* and *Madia* spp. (tarweeds) may predominate during the dry summer months. The perennial marsh species may die back to the ground level or may be grazed during the summer. Their rhizomes remain alive though, and in the following wet season these plants once again form a seasonal marsh.

In estuarine areas freshwater marshes grade into coastal salt marshes (Chapter 7) along a salinity gradient. In such areas ecotonal associations may blur the distinction between the communities. Some species of *Scirpus* and *Juncus* are relatively salt-tolerant. These often grow together with such species as *Potentilla anserina* [*P. egedii*] (silverweed), *Jaumea carnosa* (fleshy jaumea), and various other plants in a mixed community with considerable zonation. Similar gradations occur where streams or springs empty into saline desert basins (Fig. 18-18). Alkali sink communities and saline marshes may be bordered by freshwater marshes.

Marshes are bordered by many different communities. Often they occur in association with riparian woodland vegetation (Fig. 21-1). Willows often grow in the fringes of marshes and may successionally replace the marsh species. The ecotones with other communities may be very narrow, particularly if the transition from dry land to wet soil is abrupt (Fig. 21-5A).

Marshlands that occur along watercourses often are subjected to the effects of winter floods. A marsh may be eliminated by the flushing and scouring of stream flow. Generally, suitable environments for re-establishment of marshes are created by the same floods that destroy existing marshlands. Marsh species are often good colonizers and rapidly invade suitable habitats.

3. Bogs and Fens

In some wet areas decomposition is very slow and a deposit of peat forms from the partially decomposed remains of plants. When this occurs a large amount of the inorganic nutrients is tied up in undecomposed plant tissues and is unavailable for use in the community. Peaty soils are therefore nutritionally deficient. Mineral nutrients may be added to a peatland area from dust, rain, or stream flow. Peatland areas are often classified either as bogs or as fens depending on the source of the nutrients. Bogs in the restricted sense receive all of their moisture from precipitation and their nutrients from the air. These areas are therefore very mineral deficient. Fens receive a portion of their mineral supply from water that has percolated through or flowed across a mineral soil. The influx of mineral nutrients into a fen is proportionally greater than for a bog. Because bogs and fens are often confused with each other in California (for example, some areas like Mason Bog have been described as bogs but have the characteristics of fens), and because they have similar vegetational features, we are not differentiating them here. Some authors have adopted the term "mire" to include bogs, fens and other similar communities.

Fig. 21-6. Fen in Klamath Mountains of Siskiyou County. *Darlingtonia californica* (cobra plant, pitcher plant) is the aspect dominant along with various species of *Carex* (sedge). Photo by David Keil.

Bogs and fens commonly begin as open bodies of water that are gradually filled in by an accumulation of partially decayed vegetation. Some have a floating mat of vegetation and peat that extends from the margin toward the middle of the bog. Mosses, especially species of *Sphagnum*, along with the roots and rhizomes of sedges and other flowering plants form a tightly interwoven network that is sufficiently buoyant to support the weight of an adult human. The accumulated dead stems, leaves, etc., of these plants add a yearly layer to the mat, and the mat gradually thickens. All open water may be covered over as the mat expands and the entire basin may ultimately be filled in by the accumulation of organic sediments. The accumulated peat may reach a thickness of several meters.

Bogs and fens usually have a zonation that is correlated with successional events. The open water area in the center of a bog may be dominated by a mixture of floating and rooted aquatic plants. The bog mat is relatively thin next to the open water but becomes progressively thicker close to the shore. The mat often mounds up above the water table in its thicker areas. Where this happens the water content of the upper layers may drop low enough that shrubs and trees can become established. *Pinus contorta* ssp. *murrayana* (lodgepole pine) is a common invader of bog and fen margins at high elevations (Chapter 16). When a bog has completely filled in it may ultimately be replaced by a forest community, often dominated by lodgepole pine.

Plants that grow in bogs and fens must be tolerant of very poor soil conditions. Because a large proportion of the nutrients is tied up in peat, plants must be able to endure continuous nutrient deficiency. The partial decomposition of bog plants yields an excess of hydrogen ions, thereby making the bog soils highly acidic. Acid conditions inhibit most bacterial activities and this further reduces the rate of decomposition. Peatland soils are generally anaerobic a few centimeters below the surface and bog plants must therefore either grow in the uppermost soil layers or be adapted to anaerobic conditions. Most plants of bogs and fens grow very slowly as a result of these harsh conditions.

Peatland sediments preserve a natural record of past climatic conditions. Pollen grains and spores are rather resistant to decay under ordinary circumstances, and in the conditions of a bog these cells may be preserved in identifiable condition for many thousands of years. Windblown pollen grains that settle onto bog areas are subsequently incorporated into the ever-thickening layer of peat. By analyzing the changes in species of plants represented in different layers of peat and comparing them with modern patterns of pollen deposition, paleoecologists can reconstruct the climatic history of the area occupied by the bog. These peatland sediments often provide a continuous record of climatic events of the period since the end of the Pleistocene glaciation.

Bog and fen communities are not common in California. They occur both in coastal lowland areas and in high mountains. Bogs can occur within montane coniferous forests or at elevations above timberline. As is the case for other California plant communities, human activities have led to the destruction of some bogs.

California peatlands are dominated mostly by a mixture of mosses and sedges. Not all bogs contain *Sphagnum* and those that do not often have less acidic soils than those that do. *Carex* (sedge) is represented in bog communities by several species. Other sedges include *Eriophorum gracile* (cotton-grass) and *E. criniger* [*Scirpus criniger*]. Other species of bogs and fens include:

Darlingtonia californica (Fig. 21-6)	cobra plant, pitcher plant
Dodecatheon alpinum	shooting-star
Drosera rotundifolia	sundew
Hastingsia alba [*Schoenolirion album*]	hastingsia
Juncus spp.	rush
Menyanthes trifoliata	bogbean

Fig. 21-7. Montane meadow in the Sierra Nevada in Tulare County. Meadows are generally dominated by sedges and grasses. The taller plants in the foreground are *Veratrum californicum* (corn lily). The forest around the meadow includes lodgepole pine, Jeffrey pine, white fir, and quaking aspen. Photo by David Keil.

Parnassia palustris	grass-of-Parnassus
Platanthera spp.	bog orchid
Rudbeckia californica	California cone-flower
Saxifraga oregona	saxifrage
Tofieldia occidentalis [*T. glutinosa*]	tofieldia

The open water portion of a bog or fen may be occupied by a limnetic assemblage of floating or rooted aquatic species or by emergent littoral species.

4. Montane Meadows

Montane meadows occur as scattered open areas in montane coniferous forests (Chapter 16) and in level or gently sloping terrain above timberline in the alpine zone. These communities generally occur in areas where the water table is close to the surface. Most meadows occupy valleys with relatively slow drainage. Some are filled-in lakes and ponds. They are related to bog and fen communities, freshwater marshes and grasslands (Chapter 11). Usually the soils of meadows are somewhat more aerated than are those of marshes. Although organic content is often high, there is not as much peat formed as in bogs. The soil water is not as acidic in a meadow as in a bog and species diver-

Fig. 21-8. Invasion of mountain meadow by lodgepole pine (*Pinus contorta* ssp. *murrayana*. Lodgepole pine, the most moisture-tolerant of the montane conifers, can become established in meadow areas. The drainage provided by the shallow stream in the meadow removes enough water that the trees can become established. Photo by V. L. Holland.

sity is often greater. Moisture is supplied to meadow communities primarily by melting snow and generally to a much lesser extent by summer thunderstorms.

Montane meadows are dominated by a mixture of grasses, sedges and rushes, and often have a diversity of forbs (Fig. 21-7). Sometimes low-growing shrubs are present as well. The species composition is largely determined by the degree to which the soil remains moist during the growing season. Wet meadows have a higher proportion of sedges and rushes than do meadows that dry out during the summer. Many mountain meadows are moist only during spring and the early part of the summer and become completely dried out by mid-summer. Perennial grasses are often very common in such areas. Some meadows contain both moist and seasonally dry zones with species composition varying in relationship to the persistence of soil moisture. Within a meadow species distributions and vegetation composition depend to a large extent upon site-specific moisture conditions.

Montane meadows are rather productive environments and many of the species that grow there are palatable forage. These support native herbivores and during the 1800's and 1900's many meadows were grazed by domestic stock, particularly sheep.

Grazing still occurs in some meadows. The grazing had several effects that had long-lasting results. Trampling of meadow soils in some areas produced severe soil damage. As a result bare spots were created and species composition changed. Grazing pressure on palatable species caused them to decline and unpalatable plants to increase. For example, *Torreyochloa pallida* [*Puccinellia pauciflora, Glyceria pauciflora*] (weak manna grass), a very palatable species has been extirpated in some areas by heavy grazing. In areas where succession favors replacement of meadows by trees, grazing tended to prevent tree seedlings from becoming established.

Montane meadow communities that develop in forested regions generally occupy sites where the shallow water table inhibits tree growth. In some cases the ecotone between forest and meadow is very abrupt. In others there is a zone of moisture-tolerant trees bordering the meadow. Trees vary in their tolerance of wet soil conditions. Species such as *Pinus ponderosa* (ponderosa pine), *Abies concolor* (white fir) and *Calocedrus decurrens* (incense-cedar) are intolerant of saturated soils. However, such trees as *Pinus contorta* ssp. *murrayana* (lodgepole pine), *Populus tremuloides* (quaking aspen) and *Populus balsamifera* ssp. *trichocarpa* (black cottonwood) are relatively tolerant of moist soils (Chapter 16). In areas where the meadows are relatively dry or are seasonally dry, these trees may invade the meadow and form an open forest (Fig. 21-8). Ultimately tree density may increase to the extent that a forest successionally replaces a meadow. The reverse can happen too, with meadow species increasing their representation and the trees dying out. Shifts in drainage patterns can result in the water table being raised in a forest zone resulting in death of trees. Meadows can then develop in place of the forest.

Meadow communities commonly are composed of a mixture of vascular and non-vascular species. Characteristically meadows are dominated by sedges, rushes and grasses including:

Achnatherum lemmonii [*Stipa columbiana*]	needlegrass
Achnatherum occidentale [*Stipa occidentalis*]	needlegrass
Agrostis spp.	bent grass
Bromus carinatus	California brome
Calamagrostis breweri	reedgrass
Carex spp.	sedge
Deschampsia spp.	hair grass
Eleocharis bolanderi	spike-rush
Eleocharis pauciflora	spike-rush
Elymus glaucus	blue wild-rye
Elymus trachycaulis [*Agropyron subsecundus, A. t.*]	slender wheatgrass
Eriophorum criniger [*Scirpus c.*]	cotton-grass

Eriophorum gracile	cotton-grass
Festuca brachyphylla	fescue
Glyceria elata	manna grass
Juncus spp.	rushes
Poa spp.	bluegrass
Scirpus microcarpus	small-headed bulrush
Trisetum spicatum	trisetum

Other grass species may be locally abundant. In many meadows introduced European grasses have invaded and now grow with the native species.

Relatively few shrubs occur in meadows. Most are species of *Salix* (willows) or members of the Ericaceae, a family that is commonly represented in areas with acid soils. These include:

Cassiope mertensiana	white-heather
Kalmia polifolia	bog-laurel
Ledum glandulosum	Labrador-tea
Phyllodoce breweri	pink-heather
Potentilla fruticosa	bush cinquefoil
Salix spp.	willows
Vaccinium caespitosum [*V. nivictum*]	Sierra bilberry
Vaccinium uliginosum [*V. occidentale*]	western blueberry

Wildflowers are often conspicuous in meadow communities. During a growing season phenological changes in the floral displays of showy wildflowers may result in a series of seasonal color changes. A particular wildflower may assume aspect dominance for a short period of time when it comes into flower. Then its flowers fade and its display is replaced by that of another species as the season progresses. Species diversity of forbs is generally greater in dry, grass-dominated meadows than in moist, sedge-dominated areas.

The ground surface of a meadow often includes a layer of various mosses. *Sphagnum* and other mosses tolerant of very wet conditions are common in persistently wet meadows. Meadows that are seasonally dry support other mosses.

Some of the various forbs that occur in montane meadows vary with elevation. On this basis meadows can be grouped into elevational phases. The montane phase occurs mostly in red fir forests and montane mixed coniferous forests, usually below 2900 m (9500 ft) elevation. Conspicuous species include:

Aster spp.	aster
Camassia quamash	camas lily
Dodecatheon spp.	shooting stars
Epilobium angustifolium	fireweed
Erigeron spp.	fleabane daisy

Heracleum lanatum	cow-parsnip
Cirsium scariosum [*C. drummondii*, *C. foliosum*, *C. tioganum*]	elk thistle
Dugaldia hoopesii [*Helenium h.*]	mountain sneezeweed
Iris missouriensis	iris
Helenium bigelovii	sneezeweed
Lilium pardalinum	leopard lily
Lilium parvum	alpine lily
Platanthera leucostachys [*P. dilatata*]	white bog orchid
Oenothera elata	evening primrose
Barbarea orthoceras	winter cress
Sidalcea spp.	checkerbloom
Lupinus spp.	lupine
Pedicularis groenlandica	elephant heads
Potentilla spp.	cinquefoil
Senecio spp.	groundsel
Veratrum californicum	corn lily

Subalpine and alpine meadows occur above about 2900 m in lodgepole pine forests, subalpine forests and above the tree line. Common wildflowers in these communities include *Pedicularis groenlandica*, *Gentiana* spp. (gentian), *Dodecatheon alpinum* (shooting star), *Castilleja lemmonii* (paintbrush), and many other species. Dwarf shrubs include *Phyllodoce breweri* and *Cassiope mertensiana*. Numerous other species are common to both lower elevation and high elevation meadows. Herbs are often dwarfed in high-elevation meadows where the summer growing season is short and cool and high winds are very common.

5. Vernal Pools

Vernal pools are shallow ephemeral bodies of water that occupy depressions in grassland and woodland areas (Fig. 21-9). These pools are filled by water from winter storms and subsequently dry up during the spring or early summer. The pools are underlain at a shallow depth by an impervious layer that may be a calcareous hardpan, a claypan or bedrock. The floor of the pool is generally covered with a layer of clay or silt. Less commonly the pools have a sandy bottom. The pools are seldom more than a meter deep and often are no more than a few centimeters in depth. During the summer drought, vernal pools become completely dry with their basins occupied by dry cracked mud. As the pools evaporate the concentration of solutes in the remaining water increases. The centers of some vernal pools become rather saline and alkaline, though not as much so as in alkali sink areas. Not all vernal pools are saline and alkaline and consequently there is much variation in species composition from one extreme to the other. Some vernal pools dry up soon after they form and others

Fig. 21-9. Vernal pool on the Carrizo Plain in eastern San Luis Obispo County. The area where this and other vernal pools occur is a flat plain with an underlying claypan that prevents water from penetrating deeply into the soil. The soils are somewhat salty and alkaline. The pool is surrounded by grassland vegetation with scattered saltbushes. Partially or wholly submersed vernal pool plants plus mats of algae fill the pool. The showiest wildflowers in the pool are *Lasthenia fremontii* (water goldfields). On dry ground around the pool are *Lasthenia californica* (common goldfields). Within a month after this photograph was taken the pool was dry and all the plants were dead. Photo by David Keil.

persist into early summer, particularly in wet years. The length of time that water persists has a major effect on species composition.

Vernal pools are very widely distributed in California. During the long history of the state's vegetation a specialized group of ephemeral species evolved that occupy the drying pools. Most of the species that occur in vernal pools are endemic to California and many are restricted to small areas of the state. Vernal pools are much like islands, surrounded on all sides by drier areas. Plant species that disperse from one pool to another must do so across areas of unsuitable habitat. Evolution of vernal pool endemics is comparable to evolution of island endemics, but on a smaller scale. This is similar to the evolution of serpentine endemics on "island" outcrops of serpentine.

Vernal pool vegetation is characterized by herbaceous plants that begin their growth as aquatic or semiaquatic plants and make a transition to a dry-land environment as the pool dries. This generally results in the development of concentric rings of vegetation that develop around the margins of the drying pool. Most vernal pool plants are annual herbs. The relatively few perennial species grow from deeply seated rhizomes or rootstocks. Shrubs and trees are absent from vernal pool communities. Some species from vernal pool communities have very showy flowers and act as aspect dominants.

Species composition of vernal pool vegetation often varies markedly from pool to pool. As indicated above, a part of this variation is attributable to differences in salinity and alkalinity or to the duration of the pool. The island-like nature of pools is also a factor in that the flora of an individual pool is determined in part by chance dispersal of propagules from other pools. The following partial list of vernal pool species is therefore general in nature and unlikely to characterize any particular vernal pool.

Annual herbs of vernal pools include:

Alopecurus spp.	foxtail
Blennosperma nanum	blennosperma
Callitriche spp.	water starwort
Crassula aquatica [*Tillaea a.*]	water pygmyweed
Deschampsia danthonioides	hairgrass
Downingia spp.	downingia
Epilobium spp. [*Boisduvalia* spp.]	mud-sprite
Gratiola ebracteata	hedge-hyssop
Juncus bufonius	toad rush
Lasthenia spp. (Fig. 21-9)	goldfields
Lilaea scilloides	flowering-quillwort
Limnanthes spp.	meadowfoam
Mimulus tricolor	tricolored monkeyflower
Myosurus minimus	mousetail
Navarretia spp.	navarretia
Neostapfia colusana	neostapfia
Orcuttia spp.	orcuttia
Plagiobothrys spp.	miniature popcorn flower
Psilocarphus spp.	woolyheads
Veronica peregrina	speedwell

Perennials include the spore producers, *Isoetes orcuttii* (quillwort), *Pilularia americana* (pillwort), and *Marsilea mucronata* (water-clover fern), and such flowering plants as *Triteleia hyacinthina* [*Brodiaea h.*] (white brodiaea), *Eleocharis acicularis* (slender spikerush), *Eleocharis macrostachya* (common spikerush), and *Eryngium* spp. (coyote-thistle).

Human Impacts on Freshwater Wetland Communities

Humans have had many influences on the vascular aquatic plants of California. On one hand, many shallow lakes, ponds and marshes have been drained to make way for agriculture and urban development. On the other hand, rivers and streams have been dammed throughout California, creating bodies of water ranging in size from stock tanks and farm ponds to large multipurpose reservoirs. Extensive systems of irrigation canals have been dug in many areas. The result of all this disturbance has been the reduction of some types of aquatic communities, the increase in others, and modifications of many more. As in most other plant communities of California, the introduction of exotic species has altered the original vegetation. Additionally wetland areas are often managed to promote the populations of various game birds, mammals and fish.

Pollution of bodies of water by the addition of treated or untreated sewage, industrial wastes, fertilizers, pesticides and other substances, has greatly altered aquatic ecosystems. These chemicals may modify the environment by increasing the turbidity of the water, thereby decreasing the penetration of sunlight. Decay of organic matter from pollutants or dead vegetation may reduce the oxygen content of a body to an anaerobic condition. Wastes from mines or industries may contain toxic chemicals that poison all aquatic life.

When rainwater falls through polluted air, it dissolves such pollutants as nitrogen and sulfur oxides, resulting in formation of nitric and sulfuric acids. The resulting "acid rain" may greatly alter the aquatic environment. Lakes in California have not yet been as seriously affected by this phenomenon as lakes in other areas, but the danger is clearly present. Mountain lakes situated in granite basins in the Sierra Nevada and other high mountains of California are not buffered against changes in pH.

Aquatic ecosystems are often very complex. In addition to higher plants, numerous algae are important components of the vegetation. These include both anchored and planktonic species. The algae support a complicated food web. The details of such a system are outside the scope of the present discussion, but the effects of human disturbance on the system are not. The complex ecological relationships that are dependent upon the algal component of the vegetation are easily upset by water pollution. Pollution of largely oligotrophic bodies of water by sewage and fertilizers often results in the development of algal blooms and subsequent eutrophication. This can so alter the environment of vascular aquatic plants as to change the community structure. Submersed aquatics such as *Potamogeton* spp. and *Elodea canad-*

Fig. 21-10. Impacts of grazing in freshwater wetlands. The disturbances of grazing include consumption of plant matter, trampling, selective elimination of some species and promotion of unpalatable species. Photo by David Keil.

ensis arc particularly vulnerable to the shading effects created by algal blooms.

Extensive freshwater marshes in California formerly occupied large portions of the San Joaquin and Sacramento Valleys. Most of these areas have been drained for agriculture and only small remnant marshes still remain. Fairly extensive freshwater marshes still can be found in the Sacramento River delta area. Small patches of freshwater marsh occur in drainage ditches and irrigation ditches in the Central Valley and these are usually dominated by the same species that formerly dominated the much more extensive natural marsh communities.

Wetland areas are often used as pasture land for domesticated livestock (Fig. 21-10). These communities are often quite productive and they provide a source of water in California's dry landscape. Mountain meadows are commonly used as summer pasture in various areas of the state. The effects of livestock include the consumption of the plants, trampling of the soil and accompanying erosion, the elimination of species intolerant of disturbance, and the increased reproduction of the hardier species.

Many introduced plants have invaded marsh habitats. Cultivation of rice in the Central Valley and Sacramento Delta areas has resulted in the introduction of weedy plants that have developed in other rice-growing areas. Because marshes are a widespread type of vegetation, the species of marsh plants native to one area are pre-adapted to similar habitats in other areas. Man has provided both the disturbance necessary for establishment of these species and the transport of their seeds and spores. With the popularity of boating as a form of recreation has come the wide dispersal of many aquatic and semiaquatic plants.

Many vernal pools have been damaged or destroyed by human activities. Much of the area formerly occupied by vernal pools has been converted to agricultural uses. The majority of the vernal pools that once were located in the Central Valley occupied land that is now under cultivation. As a result, some vernal pool endemics, including two genera of vernal pool grasses *Orcuttia* spp. and *Neostapfia colusana*, are in serious danger of extinction. Urbanization has also destroyed many vernal pools and threatens still more. Urban expansion in the San Diego area in the 1970's and 1980's obliterated many vernal pools. Other vernal pools have been destroyed by mining or road building. Significantly, vernal pools have been minimally affected by the introduction of Eurasian weeds. Although weed species may grow close to the margins of vernal pools, the weedy plants mostly lack the ecological adaptations that would allow them to grow in the pools. The vegetation of surviving vernal pools is therefore still largely natural.

Freshwater Wetland Communities—References

Argent, G., ed. 1991. Delta-Estuary. California's inland coast. A public trust report. Prepared for the California State Lands Commission. 208 pp.

Barry, W. J. 1981. Jepson prairie—will it be preserved? Fremontia 9(1):7–11.

Baskin, Y. 1994. California's ephemeral vernal pools may be a good model for speciation. Bioscience 44:384–388 + cover picture.

Benedict, N. B. 1982. Mountain meadows: Stability and change. Madroño 29:148–153.

———, and J. Major. 1982. A physiographic classification of subalpine meadows of the Sierra Nevada, California. Madroño 29:1–12.

Brown, D. E. 1982. Montane meadow grassland. Pp. 113–114 *in* D. E. Brown (ed.). 1982. Biotic communities of the American Southwest—United States and Mexico. Desert Plants 4:1–341.

Broyles, P. 1987. A flora of Vina Plains Preserve, Tehama County, California. Madroño 34:209–227.

Clark, L. J., and J. G. S. Trelawny. 1974. Lewis Clark's field guide to wild flowers of marsh and waterway in the Pacific Northwest. Gray's Publ. Ltd., Sidney, British Columbia.

Correll, D. S., and H. B. Correll. 1975. Aquatic and wetland plants of the southwestern United States. Stanford Univ. Press, Stanford. 2 vol., 1777 pp.

Cowardin, L. M., V. Carter, F. C. Golet, and E. T. LaRoe. 1979. Classification of wetlands and deepwater habitats of the United States. U.S.D.I. Fish and Wildlife Service, Office of Biological Services Publ. FWS/OBS-79/31. 103 pp.

Ellison, J. P. 1984. A revised classification of native aquatic communities in California. Calif. State Dept. Fish and Game Planning Br. Admin. Rep. 84-1. 30 pp.

Ferren, W. R., Jr.. 1989. A preliminary and partial classification of wetlands in southern and central California with emphasis on the Santa Barbara Region. Prepared for Wetland plants and vegetation of coastal southern California, a workshop organized for the California Department of Fish and Game and the United States Fish and Wildlife Service. University of California , Santa Barbara. 54 pp.

———, and P. L. Fiedler. 1993. Rare and threatened wetlands in central and southern California. Pp. 119–131 *in* J. E. Keeley (ed.), Interface between ecology and land development in California. Southern California Academy of Sciences, Los Angeles.

———, and D. A. Prichett. 1988. Enhancement, restoration, and creation of vernal pools at Del Sol open space and vernal pool reserve, Santa Barbara County, California. Environmental Report 13, Univ. California, Santa Barbara Herbarium.

Good, R. E., D. F. Whigham, R. L. Simpson, and C. G. Jackson (eds.). 1978. Freshwater wetlands. Ecological processes and management potential. Academic Press, New York. 378 pp.

Griggs, T. 1981. Life histories of vernal pool annual grasses. Fremontia 9(1):14–17.

Halpern, C. B. 1986. Montane meadow plant associations of Sequoia National Park, California. Madroño 33:1–23.

Helms, J. A. 1987. Invasion of *Pinus contorta* var. *murrayana* (Pinaceae) into mountain meadows at Yosemite National Park, California. Madroño 34:91–97.

———, and R. D. Ratliff. 1987. Germination and establishment of *Pinus contorta* var. *murrayana* (Pinaceae) in mountain meadows of Yosemite National Park, California. Madroño 34:77–90.

Holland, R. F. 1978. The geographic and edaphic distribution of vernal pools in the Great Central Valley, California. Calif. Native Plant Soc. Special Publ. 4.

———, and S. K. Jain. 1977. Vernal pools. Pp. 515–533 *in* M. G. Barbour and J. Major (eds.). Terrestrial vegetation of California. John Wiley and Sons, N.Y.

Jain, S. 1976. Some biogeographic aspects of plant communities in vernal pools. Pp. 15–21 *in* S. Jain (ed.). Vernal pools, their ecology and conservation. Univ. Calif. Davis Inst. Ecol. Publ. 9.

——— (ed.). 1976. Vernal pools, their ecology and conservation. Institute of Ecology Publ. 9, Univ. California, Davis.

———, and P. Moyle (eds.). 1984. Vernal pools and intermittent streams. Institute of Ecology Publ. 28, Univ. California, Davis.

Kopecko, K. J. P., and E. W. Lathrop. 1975. Vegetation zonation in a vernal marsh on the Santa Rosa Plateau of Riverside County, California. Aliso 8:281–288.

Loneragan, W. A., and R. del Moral. 1984. The influence of microrelief on community structure of subalpine meadows. Bull. Torr. Bot. Club 111:209–216.

Luckenbach, R. 1973. *Pogogyne*, polliwogs, and puddles—the ecology of California's vernal pools. Fremontia 1:9–13.

Mason, H. L. 1957. A flora of the marshes of California. Univ. California Press, Berkeley. 878 pp.

Muenscher, W. C. 1944. Aquatic plants of the United States.

Minckley, W. L., and D. E. Brown. 1982a. Arctic-boreal wetlands. Pp. 237–238 *in* D. E. Brown (ed.). Biotic communities of the American Southwest—United States and Mexico. Desert Plants 4:1–341.

———, and———. 1982b. Sonoran and Sinaloan interior marshlands and submergent communities. Pp. 282–283 *in* D. E. Brown (ed.). Biotic communities of the American Southwest—United States and Mexico. Desert Plants 4:1–341.

———, and———. 1982c. Southwestern wetlands. Pp. 224–236 in D. E. Brown (ed.) Biotic communities of the American Southwest—United States and Mexico. Desert Plants 4:1–341. Cornell Univ. Press, Ithaca, N.Y. 374 pp.

Niering, W. A. 1966. The life of the marsh. McGraw-Hill Book Co., N.Y.

Purer, E. A. 1939. Ecological study of vernal pools, San Diego County, California. Ecology 20:217–229.

Reid, G. K. 1961. Ecology of inland waters and estuaries. Reinhold Publ. Co., N.Y.

Schlising, R. A., and E. L. Sanders. 1982. Quantitative analysis of vegetation at the Richvale vernal pools, California. Amer. J. Bot. 69:34–742.

———, and———. 1983. Vascular plants of Richvale vernal pools, Butte County, California. Madroño 30(4 - suppl):19–30.

Sculthorpe, C. D. 1967. The biology of aquatic vascular plants. St. Martin's Press, N.Y. 610 pp.

Stegman, J. L. 1976. Overview of current wetland classification and inventories in the United States and Canada. Pp. 102–120 *in* J. H. Sather (ed.). National wetland classification and inventory workshop proceedings—1975. U.S.D.I., Fish and Wildlife Service. Washington, D.C.

Steward, A. N., L. R. J. Dennis, and H. M. Gilkey. 1963. Aquatic plants of the Pacific Northwest. 2nd ed. Oregon State Univ. Press, Corvallis. 261 pp.

Thorne, R. F. 1982. The desert and other transmontane plant communities of southern California. Aliso 10:219–257.

———, and E. W. Lathrop. 1969. A vernal marsh on the Santa Rosa Plateau of Riverside County, California. Aliso 7:85–95.

Vale, T. R. 1981. Tree invasion of montane meadows in Oregon. Amer. Midl. Naturalist 105:61–69.

Weller, M. W. 1981. Freshwater marshes. Ecology and wildlife management. Univ. Minnesota Press, Minneapolis. 146 pp.

Whitney, S. 1979. Meadow formation, and montane meadows. Pp. 297–301, 363–374 *in* A Sierra Club naturalist's guide to the Sierra Nevada. Sierra Club Books, San Francisco.

Zedler, P. H. 1987. The ecology of southern California vernal pools: a community profile. U.S. Fish and Wildlife Serv. Biol. Rep. 85 (7.11).

Chapter 22

Anthropogenic Communities

In large areas of California the vegetation is very different from that present when the first Spanish missionaries came into the state. Much of California's landscape has been altered by human activities. In many rural areas the original vegetative cover has been stripped away and replaced by crop plants. Roads have been built throughout the state. The construction of towns and cities has replaced much natural vegetation with houses, factories, shops, gardens, lawns, roadsides, and waste lots.

The removal of the vegetative cover from an area opens it up for natural succession. Disturbance is not new to California. Landslides, floods, fires, volcanic eruptions, etc., all have influenced the plants of the state throughout its history. Various native species are able to grow in particular kinds of disturbed sites. Entire assemblages of successional species are capable of colonizing naturally disturbed areas such as dune blowouts (Chapter 7), chaparral burns (Chapter 9), burned or cleared forests (Chapters 11, 12, 15), riverine sand and gravel bars (Chapter 19), etc.

The disruptions of the past 200 years have been different from the natural disturbances of the past, though. Both the extent of the disturbances and their intensity have been unprecedented. Extensive areas of California have been fundamentally transformed. Additionally, numerous plants not native to California have been introduced into the state, some deliberately as crops or ornamentals, and many inadvertently as weeds.

Although intense and prolonged human-caused disturbance is comparatively new to California, it has affected areas of Eurasia for thousands of years. During the long history of agriculture in southern Europe and the Middle East numerous plant species became adapted to life in areas affected by human-induced disturbance. These plants became weeds of fields, gardens and waste areas. Other species were not particularly weedy in their native regions where they were thoroughly integrated members of

What is a Weed?

One would think that the meaning of such a common word would be clear—but such is not the case. The word weed is variously defined and quite controversial.

Definitions Based on Value Judgments

Some definitions place value judgments on a species' economic impact. From the farmer's perspective any plant growing wild on cultivated ground is a weed if it excludes or injures crop plants or interferes with farm operations. It is certainly true that weeds of agriculture cause much economic damage, but limiting the definition of weed to those plants that cause economic harm may be unnecessarily restrictive. Some species are officially listed as Noxious Weeds by the state or federal governments. These plants are generally exotic species known to cause economic harm and are subject to eradication efforts when found.

Plants are often considered to be weeds if they lack aesthetic appeal. Small, inconspicuous flowers, unattractive foliage, or displeasing odors may make plants undesirable. Showy flowers, attractive foliage, and attractive scents may give a plant a better reputation. However, one person may find redeeming qualities in a plant another considers without value. One person's weed may be another person's wildflower.

Many people consider any plant that grows where it is not wanted to be a weed. This is a highly subjective value judgment that can be applied to plants that are a part of a natural plant community. Ranchers have long attempted to eradicate native species of *Delphinium* (larkspur) because they are poisonous to cattle. A farmer might consider a stand of a showy native wildflower to be a weed if it grows in his barley field, even though it was a part of the grassland community that was present before the land was placed into agriculture. Housing developments are often constructed in previously undisturbed vegetation. Homeowners may consider plants to be weeds even though they are indigenous to the site.

A value-based decision that a plant is a weed is a subjective judgment that can have unforeseen consequences. A farmer who considered all thistles as weeds, for instance, was responsible for eradicating one of the few populations of *Cirsium loncholepis* (La Graciosa thistle), a rare plant now listed as Threatened by the California Department of Fish and Game. Anyone can be an expert and can make decisions as to a plant's desirability. Any plant can be treated as a weed if someone does not like it.

Definitions Based on Biological Attributes

Other definitions emphasize aspects of the plant's biology. Ecologists consider weeds to be colonizing species that occur primarily in areas that have been subjected to disturbance, especially human-induced disturbance. H. G. Baker (1964) defined a plant as a weed "if in any specified geographical area, its populations grow entirely or predominantly in situations markedly disturbed by man." Such a definition places no value judgment on a species economic impact or aesthetic qualities.

Our definition of weed is somewhat broader than that of Baker. We include species introduced by human activities to areas outside their natural range that invade stands of native vegetation, not just those that grow in human-disturbed sites. Some introduced species are aggressive invaders of California's native plant communities, sometimes occurring in areas of natural disturbance (e.g., riparian corridors) and sometimes colonizing undisturbed areas as well. Many of California's weeds are plants out of place, native to some region far from California. They are often free from the parasites that limited their populations in their native lands. Many evolved weedy tendencies in areas such as the Mediterranean region where human caused disturbances have long been present. They were preadapted to California's climate and some compete effectively with native species.

Some of California's weeds are home-grown. Some grew in naturally disturbed sites in pre-colonial California. Others have adapted to human-caused disturbances in modern times. Some species of *Amsinckia* (fiddleneck), for instance, are as invasive in agricultural areas of the San Joaquin Valley as many of the exotic weeds from other regions and have become aggressive weeds in areas far from California. The number of species with weedy tendencies is small, however, in proportion to the many species native to the state.

The Problem of Common Names

Plants that have the word weed as part of their common names often receive bad reputations. Many species of milkweeds, locoweeds, tarweeds, etc., are neither economically undesirable nor aggressively colonizing. The name alone may prejudice many people against them, however. As a reaction of this name-induced bias, the California Native Plant Society's *Inventory of Rare and Endangered Plants of California* does not use common names that imply weediness to rare native species. Rare species of *Madia* and *Hemizonia*, for instance are called tarplants rather than tarweeds.

communities that also contained specialized parasites and herbivores. However, plants introduced into California have often left behind the diseases and parasites that in their native habitats kept them in check.

California has gained its weed flora by various means. Most weed species have been introduced from other parts of the world, but some native species also have weedy tendencies. Some of the introduced weeds came to California during or after the Mission Period as contaminants in grain shipments or livestock feed, on ballast, or in other undetermined ways. Some were deliberately introduced as ornamentals, as crops, as forage plants or for other purposes. According to some accounts, species of *Brassica* (mustard) were deliberately sowed by the Spanish missionaries to mark trails through the wilderness. In the vicinity of the missions, the development of settlements and gardens and heavy grazing of the surrounding lands opened habitats to the invasion of weeds. Spain and other regions around the Mediterranean Sea are climatically very similar to California and many of the weeds of the Mediterranean region that were introduced into California were easily able to become established.

California's climatic diversity makes one or another part of the state suitable for weedy plants of many different habitats. Some plants are widespread in temperate regions as weeds of agriculture and have found the agricultural areas of California to be suitable environments. Others have come from one or another region of the world climatically similar to a part of California. Plants suitable to the desert habitats of the state have been introduced from other desert regions of the world (e.g., *Tamarix* spp., the salt-cedars). Pasture grasses of various regions now grow side-by-side with native species in coastal grasslands and mountain meadows. Dune plants from Africa and Europe now grow on California's dunes. Each year still new introductions occur. At present about one of every six species growing wild in the state is introduced. In much of California these introductions have been integrated into the natural communities of the state. In some areas, however, the introduced weeds and some native weedy species have formed communities of their own.

Communities dominated by plants introduced by man and established or maintained by human disturbance are anthropogenic communities. Some of these are entirely artificial communities such as cultivated row-crops, lawns, vineyards, etc. Others are assemblages of weedy species that have invaded disturbed areas, often in spite of human efforts to control them. These weed-dominated communities *often* represent the early stages of natural succession. In the absence of disturbance many weedy plants do not persist, but are replaced by native vegetation. Many of human activities, however, cause continual or repetitive disturbances. The cultivation of agricultural fields and gardens repeatedly disturbs the soil. Roadsides are disturbed by vehicles, oil, dust, etc. Urban areas have many sorts or repetitive disturbances of the environment of the plants living there. As a result of disturbance, many areas of California support weed-dominated communities.

Although many weed communities are temporary associations that are successional to other communities, some assemblages of introduced plants apparently do not represent seral stages. In some areas of California introduced weedy plants now dominate the vegetation, often in the absence of ongoing disturbance. For example, large areas of California's grasslands are occupied by introduced annual grasses that are apparently permanent and dominant components of the state's vegetation. Many of the same annual grasses also dominate the herbaceous component of many oak woodland communities.

Some anthropogenic communities are dominated by annuals and others by perennials. In some the dominants are grasses. In others weedy dicots predominate. Associations from agricultural and urban areas may be quite different. Roadside habitats are quite different from cultivated fields. On the other hand, lawns and pastures are often similar. Some weeds are well adapted to one type of disturbance and not to another.

Most successful weeds produce large quantities of seeds and readily invade disturbed sites. Many have features that allow their seeds to be widely dispersed. The seeds of some weedy plants may remain dormant in the soil for many years. When the soil is disturbed the seeds germinate and establish a new successional community. Many are self-fertile or reproduce apomictically (by means not involving sexual reproduction), and a population can become established from a single seed.

Communities of weeds are not always easy to characterize. Chance and site history both play major roles in determining what species comprise the weed flora of a particular site. A weed with a nearby seed source may be disproportionately represented. Ecologically similar communities may have wholly different species composition. Areas where the soil has been plowed, grazed, trampled or chemically treated may each support a different assemblage of weeds.

The effects that exotic plants have on the native vegetation vary. In some cases, such as a lone English walnut tree in an oak woodland, the impact is relatively small. In other cases, such as when pampas grass takes over entire coastal hillsides displacing almost all native vegetation, the impact is very significant. Often in urban areas the woody vegetation is composed of a diverse assemblage of native and introduced species that form unique local communities. Because of its climatic diversity and the vast number of cultivated species that are potential invaders of the California landscape, the variety of local "urban mix" communities is too great for any attempt to classify. When examining the vegetation of California, one must always be aware of the introduced vegetation and weigh its importance for each area being examined.

We have organized the anthropogenic communities into five groupings: (1) agrestal communities, (2) pastoral communities, (3) ruderal communities, (4) plantations, and (5) the “urban mix”. The first two represent weed-dominated communities of rural, agricultural areas. The third are the communities commonly encountered in urban waste areas, along roadsides, and in similarly disturbed areas. There is considerable overlap in species composition, but some plants tend to occur most commonly in one or another. Plantations are areas that have been planted in trees either for windbreaks, firewood, ornamentals or agricultural uses, and "the urban mix" is a grouping of non-native and native vegetation common in open areas around urban developments.

1. Agrestal Communities

Agrestal communities form in areas that have been disturbed by cultivation. Many species of weeds thrive in the same environments as crop plants. The weed species are able to grow to maturity and to reproduce side by side with the crop. Because culture practices vary depending on the crop being grown, the weed communities often vary as well. Weedy grasses often grow best alongside grain crops. Broadleafed weeds may grow best with broadleafed crop plants. Other agrestal species grow in old fields that have been removed from cultivation. Approximately eight percent of California is devoted to croplands at present. At any given time a small proportion of this land is occupied by agrestal communities. However, almost all of this land is at one time or another occupied by weed communities.

Some agrestal weeds are annuals that are able to complete their life cycles within the same time span as the crop plants. Most are self-fertile or set seeds apomictically. Many are genetically variable and have local races that have been selected by the human activities designed to enhance the growth of the crop plants or to control the weeds. Mechanical harvesting of crops promotes the growth of weeds with seeds that can be disseminated by the machinery. Many of the seeds are long-lived and are able to persist as a buried seed bank in the soil for many years. Buried seeds can survive drought periods and re-establish weed populations when favorable conditions return. Some weedy annuals such as *Avena fatua* (wild oats) produce allelopathic chemicals that inhibit the growth of nearby plants. Others, such as *Sonchus* spp. (sow-thistles) develop a basal rosette that covers a small patch of ground, shading out seedlings of nearby plants.

Other agrestal weeds are perennials with underground rhizomes. Although above-ground parts of the plant may be destroyed by cultivation, the rhizomes are merely fragmented and each clone may yield a new plant. For this reason, plants such as *Convolvulus arvensis* (field bindweed) and *Acroptilon repens* [*Centaurea r.*] (Russian knapweed) are particularly difficult to eradicate once established.

Special kinds of agrestal communities develop in areas of the Central Valley where rice is grown. Because of the aquatic conditions in a rice field, various aquatic and semiaquatic herbs grow together with the rice plants, some of them capable of causing much economic harm through competition with rice plants. Some of these plants also invade canals and choke off water flow. Some of the ricefield weeds are natives such as species of *Potamogeton* (pondweed), *Typha* (cattail), *Scirpus* (bulrush) and *Leptochloa* (sprangletop) and others are exotics such as *Echinochloa crusgalli* (barnyard grass).

Agrestal weeds are economically very important. California's farmers spend many millions of dollars each year to combat the growth of weeds. Weed species harm crops in several ways. They can stunt the growth of crops by shading them, by allelopathic interactions, and by competitively absorbing nutrients and water from the soil. Weed seeds or other plant parts may contaminate grain or other crops. Combating weeds is time-consuming, expensive and sometimes potentially harmful to the environment as a whole. Cultivation is necessary to reduce weed growth in many crops. Chemical herbicides are often sprayed onto fields to reduce the growth of weeds and increase the yield of crop plants. These chemicals sometimes drift onto other nearby areas damaging natural communities or other crop plants. The long-term effects of their residues in the soil are not fully known.

The impacts of past cultivation of the land may persist long after the disturbances have ceased. A few years of cultivation are generally sufficient to destroy all but the most hardy of the native species. Perennial herbaceous natives are particularly susceptible to such disturbance. Once eliminated from an area they may not readily reinvade after the cultivation has ceased. Not all of the aggressive introduced species that thrive under cultivation persist when a field has been fallow for many years. Successional changes usually result in a community dominated at least in part by introduced species. A fallow field that is initially dominated by broadleafed weedy species may change to a

Fig. 22-1. Pastoral weeds. Unpalatable plants such as these coarse milk thistles (*Silybum marianum*) have invaded many pasture areas of California. as more palatable grasses and forbs are consumed by livestock. Photo by David Keil.

grassland dominated by introduced Eurasian grasses. The persistence of non-native weeds makes re-establishment of the native species very difficult. Additionally a seed source must be available, and suitable means of dispersal must be present if native species are to return to once-cultivated land. The absence of native species in an area of apparently undisturbed ground is often an indication of long-ago cultivation of the land.

2. Pastoral Communities

Pastoral communities are dominated by species well adapted to the grazing of livestock. At present about twenty-five percent of California is used as rangeland or pasture. Most of this area is considered in other chapters (e.g., grasslands in Chapter 10 and oak woodlands in Chapter 14). The intensity of grazing varies greatly in California and only those communities in which heavy grazing has brought about marked changes in dominance are considered here to be pastoral communities. The dominants of pastoral communities have various characteristics. Some have a low-growing habit that allows them to survive grazing and reproduce. Others are distasteful or bear spines that ward off herbivores. Some have rhizomes and spread vegetatively. Many are highly palatable but are able to reproduce despite being grazed.

Many of California's native species are poorly adapted to heavy grazing. There were comparatively few large herbivores in California until cattle, sheep and goats were introduced. Between the end of the Pleistocene Epoch, when several species of large herbivores became extinct, and the Mission Period, when domesticated livestock were introduced, grazing pressure was rather low. The introduction of heavy grazing and of weedy plants well adapted to grazing overwhelmed many of the native plants, forever changing the California landscape. Valley and southern coastal grassland communities (Chapter 11) are a type of pastoral community. Human activity and the grazing pressure of domesticated livestock were important in the establishment of the annual grasses that now dominate these communities, but the grasses now perpetuate themselves, whether grazing pressure is present or not.

Grazing does not affect all plants equally, however. Areas that have been grazed but never cultivated may retain numerous herbaceous annual and perennial native taxa. Those which grow prostrate on the ground or that reproduce in very large numbers may persist in spite of heavy grazing. Native species that are highly palatable, that are fragile and intolerant of trampling or that require a lengthy period for inflorescence development, flower and seed production are at a disadvantage under grazing pressure.

If introduced or native weedy species that have attributes that deter grazing are present, their proportion in the community will rise while that of the more palatable natives falls. Several species of introduced thistles often increase under such conditions. In heavily grazed areas such plants as *Carduus pycnocephalus* (Italian thistle), *Centaurea solstitialis* (yellow star thistle), *Cirsium vulgare* (bull thistle), and *Silybum marianum* (milk thistle; Fig. 22-1) have become dominant, forming spiny, impenetrable masses. These often grow together with such distasteful or poisonous herbs as *Brassica* spp. (mustard), *Rumex crispus* (curly dock), and *Conium maculatum* (poison hemlock).

3. Ruderal Communities

Ruderal communities are assemblages of plants that thrive in waste areas, roadsides and similar disturbed sites in towns and cities and along rural roadways. Some grow in heavily compacted soils with little available oxygen. Hard-packed soils of roadsides, parking lots, footpaths, etc. support communities of ruderals. Such species as *Chamomilla suaveolens* [*Matricaria matricari-*

Fig. 22-2. Ruderal community in coastal San Luis Obispo County. Native disturbance followers such as *Eremocarpus setigerus* (turkey-mullein) and *Heterotheca grandiflora* (telegraph weed) grow together with various exotic weed species. Photo by V. L. Holland.

oides] (pineapple-weed) and *Polygonum arenastrum* [*P. aviculare*] (common knotweed) thrive under such conditions. Small pockets of soil that accumulate on road shoulders, in cracks of pavement, on rooftops and other sites often are invaded by weeds such as *Sonchus oleraceus* (sow-thistle), *Conyza canadensis* (horseweed), or *Chenopodium* spp. (goosefoot). Some urban weeds are ornamentals, escaped from cultivation. Ruderal communities are difficult to characterize and often are temporary assemblages. In urban settings the majority of wild plants are often introduced weeds rather than native species.

Approximately 90 percent of California's populace lives in cities or towns. Some people see ruderal communities far more often than they see natural communities. For many city-dwellers, almost all of the familiar wild plants are weeds. Ruderal communities unfortunately are an ever-expanding part of the vegetation of California. The urban development of California continues to change both the landscape and the plants. In many urban areas so few of the native species are left that only ruderal communities remain.

The weed flora of California is large and it continues to grow as additional species are inadvertently introduced. The list that follows summarizes some of the more common and familiar weeds, but it is very far from complete. Over 1000 species of exotic plants have been introduced into California's wild flora. Some of these are only localized in their distribution and cannot be considered weeds. Others, such as the following, are widespread and often very aggressive invaders:

Amaranthus spp.	amaranth
Avena spp.	wild oats
Brassica spp.	mustards
Bromus spp.	brome grasses
Carduus pycnocephalus	Italian thistle
Centaurea spp.	star thistles, knapweeds
Chamomilla suaveolens	pineapple weed
Chenopodium spp.	goosefoots
Chrysanthemum coronarium	garland daisy
Cirsium vulgare	bull thistle
Conium maculatum	poison hemlock
Convolvulus arvensis	field bindweed
Conyza spp.	horseweeds
Cytisus spp.	brooms
Foeniculum vulgare	fennel
Hirschfeldia incana [*Brassica geniculata*]	perennial mustard
Hordeum murinum [*H. leporinum*]	foxtail barley
Lolium multiflorum	wild-rye
Malva spp.	mallows
Medicago polymorpha [*M. hispida*]	bur-clover
Melilotus spp.	sweet-clovers
Oxalis spp.	wood-sorrel, Bermuda-buttercup
Plantago spp.	plantains
Polygonum arenastrum	knotweed
Portulaca oleracea	purslane
Raphanus sativus	wild radish
Rumex spp.	docks
Salsola tragus [*S. kali*]	Russian thistle, tumbleweed
Silybum marianum (Fig. 22-1)	milk-thistle
Sonchus spp.	sow-thistles

Stellaria media	chickweed
Taraxacum officinale	dandelion
Xanthium spp.	cocklebur

These species may, at times, be integrated into various other communities. Few of California's plant communities are free from the changes brought about by human activities. Most contain a mixture of introduced and native species. In most the native plants are still dominant but the newcomers have made their presence known. As more and more of the state is permanently changed, more and more habitats suitable for weeds become available.

4. Plantations

In many areas of California plantations of trees have been established for various purposes. Some trees were planted in rows as windbreaks; some were planted over large areas for agricultural purposes (e.g., fruit trees); some were planted for ornamental purposes (e.g., Monterey pines and cypresses); and some were planted for firewood or lumber (e.g., blue gum). Trees in some of these plantations simply live out their life, die and are replaced by other forms of vegetation (e.g., abandoned walnut orchards). Others, such as *Eucalyptus globulus* (blue gum), reproduce, and in some cases spread into and replace native plant communities.

Plantation trees all have some of the same impacts as any other trees in forest or woodland communities. Collectively they shade the ground, draw nutrients and water from the soil, contribute litter to the soil beneath them, and influence the interactions and food webs of microorganisms, vertebrate and invertebrate animals, and other smaller plants with which they grow. As they grow in stature their influences increase and the associated organisms may undergo successional changes.

Although there are numerous examples of plantations in California that have influenced natural vegetation, we will restrict our discussion here to a few of the most commonly planted trees, especially in coastal regions of California. Communities dominated by these species serve here as examples of plantation communities that can be compared with plantations dominated by other species elsewhere in the state.

Blue gum (*Eucalyptus globulus*), Monterey pine (*Pinus radiata*), and Monterey cypress (*Cupressus macrocarpa*) have been planted widely in California. In some areas they occur as windrows, whereas in other areas they form extensive man-made forest communities. This is especially true of the blue gum. All of these trees are well adapted to the environmental conditions of

California, especially in coastal areas, and they grow fast and tall. In some areas the trees appear to have become naturalized, and they are reproducing and maintaining themselves. In some areas the introduced trees are spreading into surrounding communities and replacing native vegetation.

The most extensive of these man-made forests are large plantings of blue gum. Many of these plantations are characterized by having pure, dense stands of blue gum trees that grow tall and straight (although those in windy areas near the immediate coast are often wind-pruned). The blue gums tower over all of the species native to the area. The trees shade the ground and litter the soil surface with fallen branches, leaves, fruits, and bark. Fog-drip and rainwater passing through the leaves and branches carry dissolved chemicals that add to the substances leached from the fallen litter, producing a significant allelopathic (growth inhibiting) effect on understory vegetation. The net result is that very few other plant species are able to grow in the blue gum forests. Consequently, the understory is often sparse if present at all. Thus in areas where blue gum trees have been planted, they have almost completely replaced the native vegetation.

The California State Department of Parks and Recreation has removed some blue gum trees from state park land to save the integrity of the native vegetation of the parks and to restore the native vegetation that was prevalent before the introduction of blue gums. Additional removal of *Eucalyptus* has been proposed for areas such as Montaña de Oro State Park in San Luis Obispo County, but these proposals have met with much opposition from people who like the appearance of the forested areas even if they are composed of trees native to Australia.

Monterey pine and Monterey cypress are both native species and are endemic to small portions of the California coast (Chapter 12). However, both have been planted widely over the state for their natural beauty and because they grow so well in California's coastal climate. Like the blue gums, these trees grow tall and straight and also have a significant impact on understory vegetation though the allelopathic effect does not seem as pronounced. These trees have also been planted in pure stands as well as in mixtures. Sometimes they have been planted together with blue gum. The native vegetation has been modified significantly in areas where these trees grow outside their natural range. Usually the understory consists of a sparse scattering of weedy plants such as those discussed previously.

Plantations of a different sort are orchard crops. Nut and fruit crops such as walnuts (*Juglans* spp.), plums, almonds and peaches (all species of *Prunus*), and apples (*Malus sylvestris*) are widely planted in California. These trees tend to dominate their environment much in the same way as do natives such as oaks. Management of the land on which these trees are growing ranges from regular cultivation of the soil or mowing of the herbaceous understory to little or no secondary disturbance. Depending upon the nature and intensity of the disturbance associated with management of the plantation, variable assemblages of exotic or native species may grow under and between the trees. The overall vegetation may have the appearance and species composition of a cultivated field, a pasture, or a foothill woodland community.

5. The "Urban Mix"

In addition to trees being planted in plantations, there are many places where plants not native to a specific area have either escaped or been planted in areas around urban or residential developments. In urban areas, it is not uncommon to find mixtures of non-native and native vegetation in open areas. Common examples of non-native plants found in urban and native mixes of vegetation are trees such as Monterey pine, Monterey cypress, *Eucalyptus* spp. (eucalyptus), *Ailanthus altissima* (tree of heaven), *Acacia* spp. (acacias), *Prunus* spp. (cherries, almonds, etc.), and *Schinus molle* (pepper tree). In addition to trees there are many shrubs and perennials like *Agave* spp. (century plants) and *Cortaderia jubata* (pampas grass), *Cytisus* spp. (brooms), *Genista monspessulana* [*Cytisus monspessulanus*] (French broom) and *Spartium junceum* (Spanish broom) that are also common. Ornamental vines such as *Hedera helix* (English ivy), *Lonicera japonica* (Japanese honeysuckle), and *Vinca major* (periwinkle) often spread from developed lots into adjacent undeveloped areas.

A good example of an urban mix can be found in the Oakland-Berkeley Hills area. These hills are vegetated by a combination of native and introduced trees, shrubs, and herbs. Trees native to the area include *Sequoia sempervirens* (coast redwood), *Umbellularia californica* (California bay-laurel), *Quercus agrifolia* (coast live oak), and *Arbutus menziesii* (madrone). Non-native trees include several species each of *Eucalyptus* and *Acacia*. In addition, *Pinus radiata* (Monterey pine), a California native, is well-established although it is not indigenous the Oakland-Berkeley hills. Shrub cover includes native coastal scrub and chaparral species as well as an assortment of introduced species. Scattered through much of the vegetation are houses, each with an assortment of ornamentals, some of which escape on a local basis into the forest or brush. When a disastrous fire swept through the area in the fall of 1991, many of the trees and shrubs that fueled the fire were exotics that had become a part of the urban mix.

Anthropogenic Communities—References

Baker, H. G. 1962. Weeds—native and introduced. J. Calif. Hort. Soc. 23:97–104.

———. 1974. The evolution of weeds. Ann. Rev. Ecol. Syst. 5:1–24.

———. 1985. What is a weed? Fremontia 12(4):7–11.

———. Characteristics and modes of origin of weeds. Pp. 147–172 *in* H. G. Baker and G. L. Stebbins (eds.). 1965. The genetics of colonizing species. Academic Press, N.Y. 588 pp.

Beatty, S. W., and D. I. Licari. 1992. Invasion of fennel (*Foeniculum vulgare*) into shrub communities on Santa Cruz Island, California. Madroño 39:54–66.

Bowler, P. A. 1992. Shrublands. In defense of disturbed land. Restoration & Management Notes 10(2):144–149.

California Weed Conference. 1985. Principles of weed control in California. Thomson Publications, Fresno, Ca. 473 pp.

Devine, R. 1993. The cheatgrass problem. Atlantic 271(5):43–48.

Frenkel, R. E. 1970. Ruderal vegetation along some California roadsides. Univ. Calif. Publ. Geogr. 20:1–163.

Guillerm, J. L., and J. Maillet. 1982. [The agrestal weed flora and vegetation of] Western Mediterranean countries of Europe. Pp. 227–243 *in* W. Holzner and M. Numata (eds.), Biology and ecology of weeds. Dr. W. Junk, Publ., The Hague, Boston and London.

Harper, J. L. (ed.). 1960. The biology of weeds. Blackwell, Oxford. 256 pp.

Holzner, W. 1982. Concepts, categories and characteristics of weeds. Pp. 3–20 *in* W. Holzner and M. Numata (eds.), Biology and ecology of weeds. Dr. W. Junk, Publ., The Hague, Boston and London.

Holzner, W., and M. Numata (eds.). 1982. Biology and ecology of weeds. Dr. W. Junk, Publ., The Hague, Boston and London.

Keeley, J. E. 1993. Interface between ecology and land development in California. Southern California Academy of Sciences, Los Angeles.

Klingman, G. C., and F. M. Ashton. 1975. Weed science. Principles and practices. Wiley-Interscience, N.Y. 431 pp.

McBride, J. R. 1985. Natives that behave like weeds. Fremontia 12(4):12–15.

McClintock, E. 1985. Escaped exotic weeds in California. Fremontia 12(4): 3–6.

———. 1985. Some weeds called escaped exotics in California. Crossosoma 10(1):1–6.

Parish, S. B. 1920. The immigrant plants of southern California. So. Calif. Acad. Sci. Bull. 14(4):3–30.

Radosevich, S. R., and J. S. Holt. 1984. Weed ecology. Implications for vegetation management. John Wiley & Sons, New York, et al. x + 265 pp.

Rejmánek, M., and J. M. Randall. 1994. Invasive alien plants in California: 1993 summary and comparison with other areas of North America. Madroño 41:161–177.

Robbins, W. W., M. K. Bellue, and W. S. Ball. 1951. Weeds of California. State of Calif. Dept. Agric. 547 pp.

Rossiter, R. C. 1966. Ecology of Mediterranean annual-type pasture. Adv. Agronomy 18:1-50.

Stebbins, G. L. 1965. Colonizing species of the native California flora. Pp. 173–195 –172 *in* H. G. Baker and G. L. Stebbins (eds.). 1965. The genetics of colonizing species. Academic Press, N.Y. 588 pp.

Stuckey, R. L., and T. M. Barkley. 1993. Weeds. Pp. 193–198 *in* Flora of North America Editorial Committee, Flora of North America, Vol. 1. Introduction. Oxford University Pres, New York.

Sukopp, H. and P. Werner. 1983. Urban environments and vegetation. Pp. 247–260 *in* W. Holzner, M. J. A. Werger, and I. Ikusima (eds.). Man's impact on vegetation. Dr. W. Junk Publ., The Hague, Boston London.

Thurston, J. M. 1982. Wild oats as successful weeds. Pp. 191–199 *in* W. Holzner and M. Numata (eds.), Biology and ecology of weeds. Dr. W. Junk, Publ., The Hague, Boston and London.

University of California. 1985. Growers weed identification handbook. Division of Agricultural Sciences, Univ. California, Davis. Looseleaf, unpaginated.

Appendix

Comparison of Community Classification Systems

The primary purpose of community classification systems is to organize diverse vegetation into units with similarities so we can communicate about them. Because the vegetation in California is so diverse and because it is impossible to categorize every local aggregation of plants, community classification systems will always have some problems. To be useful, classification systems must simplify the vegetation into recognizable assemblages, often overlooking or ignoring some inconsistencies. Ours is no exception.

Several classification systems have been developed for California vegetation over the years, and the number of recognized plant communities in these systems varies from about 29 to 300. All have similarities and also some significant differences depending on the main purpose of the classification system. For example, the system used in *A Guide to Wildlife Habitats* (Mayer and Laudenslayer, 1988) was designed to serve as ecosystem-oriented resources for wildlife biologists, whereas Eyre (1980) classified vegetation as to forest cover type. In contrast, The California Department of Fish and Game's California Natural Diversity Data Base (CNDDB) community classification system was designed to inventory plant communities with special emphasis on rare communities so that they could be put into a data base. Cheatham and Haller (1975), designed the first system used by CNDDB and recognized about 250 communities. Their system was modified by Holland (1986) and now includes about 280 communities. Of these, 135 are considered rare and need more study and protection.

Currently, the California Native Plant Society (CNPS), in conjunction with CNDDB, is preparing a multi-faceted effort to inventory, define and protect California's natural communities and ecosystems. Once this inventory is completed and published (Sawyer and Keeler-Wolf, 1995), it will be used as the CNDDB classification system of natural communities. It will also be used by The Nature Conservancy in the California section of their national and western regional vegetation classification that is currently under development

We felt it would be useful to include a comparison of the most widely used systems with ours for cross referencing (Table A-1). This appendix compares our system to [the 1994 draft version of] the new CNPS System (Sawyer, 1994), the CNDDB System (Holland, 1986), and the California Wildlife Habitat Relationships System (Mayer and Laudenslayer, 1988). The comparison table allows the reader to go from one system to the other and to refer to specific chapters in our book for additional information. Refer to Chapter 5 for more detailed information about our classification system. The new CNPS classification system is a hierarchical arrangement using **series** as the organizational level. A series is a community that "represents areas of the landscape dominated by similar plants". A series is composed of stands with similar overstory composition. Stands are local examples of plant communities defined by floristic composition and structure. The CNPS classification system also has dichotomous keys to assist the reader in identifying the various series.

The CNPS system differs from ours in describing a much larger number of series than we do in our communities classification system, and yet there are many local assemblages that cannot be assigned to any of the CNPS series. California's plant communities are so diverse that attempting to describe all of the variations will be a difficult if not impossible task. In our system, communities are grouped in a more inclusive fashion with recognition of the existence of many local assemblages that may be described on an individual basis. Thus, our system is somewhat broader, allowing a detailed description of the local variation found within the community being studied.

We suggest using the CNPS system and our book together for a better understanding of the California vegetation. For example, within our Pioneer Coastal Dunes Communities CNPS lists four series: Sand-verbena-beach bursage, Native dunegrass, European beach grass, and ice plant. However, within a given area of Pioneer Coastal Dunes, one can find significantly more variation in species composition from site to site. Some areas may be dominated by beach-bursage and sea-rocket, others by beach-bursage, beach morning glory, and dundelion, others by salt bush and sea rocket, others by beach-bursage and ice plant and some by European beach grass, all within a short distance of one another. Thus, we prefer describing an area like this as Pioneer Coastal Dunes dominated by various mixtures of sand verbena, beach bursage, beach morning glory, ice plant, dundelion, sea rocket and European beach grass. By using the CNPS system in conjunction with our book, one can use the CNPS series when they fit and also describe the variations of their series when appropriate. We hope this combination will allow the user to better describe and understand the diversity of plant communities in California.

References

Cheatham, N. H., and J. R. Haller. 1975. An annotated list of California habitat types. University of California Natural Land and Water Reserve System, unpubl.

Eyre, F. H. (ed.). Forest cover types of the United States. Soc. Amer. Foresters, Washington, D.C.

Holland, R. F. 1986. Preliminary descriptions of the terrestrial natural communities of California. Nongame Heritage Program, Calif. Dept. Fish and Game, Sacramento. 156 pp.

Mayer, K. E., and W. F. Laudenslayer Jr. (eds.). 1988. A guide to wildlife habitats of California. Calif. Dept. Forestry and Fire Protection, Sacramento. 166 pp.

Sawyer, J., 1994. Series descriptions of California vegetation. Unpublished draft. 320 pp.

——— and T. Keeler-Wolf. 1995. Series level descriptions of California vegetation. California Native Plant Society, Sacramento (in press).

Table A-1. Comparison of the community classification system in ***California Vegetation*** with three other commonly used systems.

CALIFORNIA VEGETATION (HOLLAND AND KEIL, 1995)	***SERIES LEVEL DESCRIPTIONS OF CALIFORNIA VEGETATION,*** CALIFORNIA NATIVE PLANT SOCIETY (SAWYER, 1994 DRAFT)	***TERRESTRIAL NATURAL COMMUNITIES OF CALIFORNIA*** CALIFORNIA NATURAL DIVERSITY DATA BASE (R. HOLLAND, 1986)	***WILDLIFE HABITATS OF CALIFORNIA*** (MAYER AND LAUDENSLAYER, 1988)
Formation **Community Types** **Phases (Series)**	**Series, Stands, and Habitats**	**Plant Communities (Element Code)**	**Wildlife Habitat Relationships Type**
Marine Aquatic Communities			
Subtidal and Intertidal Communities			Marine
Coastal Estuarine Communities	Ditch-grass	Coastal Brackish Marsh (52200*)	Estuarine
Coastal Salt Marsh Communities	Pickleweed Ditch-grass Salt grass Cordgrass	Coastal Salt Marsh (52100) Coastal Brackish Marsh (52200) Cismontane Alkali Marsh (52310*)	Saline Emergent Wetland
Coastal Sand Dune and Beach Communities		Coastal Dunes (21000)	
Pioneer Dune Communities	Sand-verbena-beach bursage Native dunegrass European beach grass Iceplant	Active Coastal Dunes (21100) Foredunes (21200)	
Dune Scrub Communities	Dune lupine-goldenbush Yellow bush lupine (in part) Stands of Antioch dunes (interior) Coyote brush (in part)	Backdune Scrub (21300)	Coastal Scrub

Dune Wetland Communities	Sedge		
Coastal Scrub Communities		Coastal Scrub (32000)	
Northern Coastal Scrub Communities	Coyote brush (in part) Salal-black huckleberry (in part) Yellow bush lupine (in part)	Northern (Franciscan) Coastal Scrub (32100)	Coastal Scrub
Southern Coastal Scrub Communities	Black sage Brittlebush (in part) California buckwheat California encelia California sagebrush Coast prickly-pear (in part) Coyote brush (in part) Mixed sage Purple sage White sage California sagebrush-California buckwheat Yellow bush lupine (in part) California buckwheat-white sage California sagebrush-black sage Dune lupine-goldenbush	Coastal Sage--Chaparral Scrub (37G00) Central (Lucian) Coastal Scrub (32200) Venturan Coastal Sage Scrub (32300) Central (Lucian) Coastal Scrub (32200) Venturan Coastal Sage Scrub (32300) Diegan Coastal Sage Scrub (32500*) Diablan Sage Scrub (32500)	Coastal Scrub
Southern Semidesert Coastal Scrub Communities	Brittlebush Bladderpod-California ephedra-narrowleaf goldenbush	Riversidian Sage Scrub (32700*)	Coastal Scrub

* NDDB Rare Communities.

Table A-1 (continued).

CALIFORNIA VEGETATION	*SERIES LEVEL DESCRIPTIONS OF CALIFORNIA VEGETATION,*	*TERRESTRIAL NATURAL COMMUNITIES OF CALIFORNIA*	*WILDLIFE HABITATS OF CALIFORNIA*
Sea-Bluff Coastal Scrub Communities	Coast prickly-pear (in part) Yellow bush lupine Coyote brush (in part) Salal-black huckleberry (in part)	Coastal Bluff Scrub (31000)	Coastal Scrub
Chaparral Communities		Chaparral (37000)	
Mixed Chaparral Communities	Scrub oak-birchleaf mountain-mahogany Scrub oak-chamise Scrub oak-chaparral whitethorn Chamise-bigberry manzanita Chamise-black sage Chamise-Eastwood manzanita Chamise-wedgeleaf ceanothus Chamise-hoaryleaf ceanothus Chamise-mission-manzanita-woolyleaf ceanothus Chamise-woolyleaf ceanothus Chamise-white sage Chamise-cupleaf ceanothus Chamise-Eastwood manzanita-bigberry manzanita Bigpod ceanothus-hollyleaf redberry Bigpod ceanothus birchleaf mountain-mahogany Birchleaf mountain-mahogany-California buckwheat (in part)	Upper Sonoran Mixed Chaparral (37100) Mesic North Slope Chaparral (37E00) Coastal Sage--Chaparral Scrub (37G00) Ione Chaparral (37D00*) Poison-Oak Chaparral (37F00) Coastal Sage-Chaparral Scrub (37G00)[in par] Flannel Bush Chaparral (37J00) [in part] Birchleaf mountain-mahogany-California buckwheat (in part)	Mixed Chaparral

	Birchleaf mountain-mahogany-California buckwheat (in part) Birchleaf mountain-mahogany-California buckwheat (in part)Birchleaf mountain-mahogany-California buckwheat (in part) Shrub interior live oak-chaparral whitethorn		
Chamisal Chaparral Communities	Chamise	Chamise Chaparral (37200)	Chamise-Redshank Chaparral
Red-Shanks or Ribbon-Bush Chaparral Communities	Red shank Red shank-chamise (in part) Red shank-birchleaf mountain-mahogany (in part)	Red Shank Chaparral (37300)	Chamise-Redshank Chaparral
Manzanita Chaparral Communities	Eastwood manzanita Greenleaf manzanita Ione manzanita Woolyleaf manzanita (in part) Whiteleaf manzanita Bigberry manzanita	Upper Sonoran Manzanita Chaparral (37B00)	Mixed Chaparral
Ceanothus Chaparral Communities	Bigpod ceanothus Blue blossom (in part) Wedgeleaf ceanothus Chaparral whitethorn Hoaryleaf ceanothus Hairyleaf ceanothus Deer brush (in part) Tobacco brush (in part) Mountain whitethorn (in part)	Upper Sonoran Ceanothus Chaparral (37800)	Mixed Chaparral

Table A-1 (continued).

CALIFORNIA VEGETATION	*SERIES LEVEL DESCRIPTIONS OF CALIFORNIA VEGETATION,*	*TERRESTRIAL NATURAL COMMUNITIES OF CALIFORNIA*	*WILDLIFE HABITATS OF CALIFORNIA*
Scrub Oak Chaparral	Scrub oak Mixed scrub oak Shrub interior live oak Shrub interior live oak-shrub canyon live oak Shrub interior live oak-scrub oak	Scrub Oak Chaparral (37900) Interior Live Oak Chaparral (37A00)	
Maritime Chaparral Communities	Blue blossom (in part) Woolyleaf manzanita (in part)	Maritime Chaparral (37C00)	Mixed Chaparral
Island Chaparral Communities		Island Chaparral (37700*) Island Ironwood Forest (81700) [in part]	Mixed Chaparral
Serpentine Chaparral Communities	Leather oak Woolyleaf manzanita (in part)	Serpentine Chaparral (37600) Digger Pine--Chaparral Woodland (71320) [in part]	Mixed Chaparral
Montane Chaparral Communities	Whiteleaf manzanita Mountain whitethorn Tobacco brush Bush chinquapin Huckleberry oak Deer oak Holodiscus Brewer oak Deer brush (in part) Curlleaf mountain-mahogany (desert)	Montane Chaparral (37500) Montane Dwarf Scrub (38000)	Montane Chaparral
Semidesert Chaparral Communities	Cupleaf ceanothus-fremontia-oak Birchleaf mountain-mahogany-California buckwheat (in part) Birchleaf mountain-mahogany	Alluvial Fan Chaparral (37H00) Semi-Desert Chaparral (37400) Flannel Bush Chaparral (37J00) [in part]	Mixed Chaparral

Grassland Communities			
Native Bunchgrass Grasslands	Purple needlegrass Nodding needlegrass Foothill needlegrass One-sided bluegrass (in part) California oatgrass (in part)	Native Grassland (42100) Coastal Prairie (41000) Coastal Terrace Prairie (41100) Bald Hills Prairie (41200)	Perennial Grassland
Valley and Southern Coastal Grasslands	California annual grassland Kentucky bluegrass (in part) Creeping ryegrass California oatgrass (in part) Introduced perennial grassland	Valley and Foothill Grassland (42000) [in part] Non-Native Grassland (42200) Wildflower Field (42300*)	Annual grassland
Northern Coastal Grasslands	Pacific reedgrass Idaho fescue Tufted hairgrass (in part) California oatgrass	Coastal Prairie (41000)	Perennial grassland
Desert Grasslands	Needle-and-thread Indian ricegrass Crested wheatgrass Cheatgrass Bluebunch wheatgrass One-sided bluegrass Kentucky bluegrass (in part) Ashy ryegrass	Mojave Mixed Scrub and Steppe (34200) [in part] Sagebrush Steppe (35300) [in part] Great Basin Desert Grassland (43000*)	Perennial grassland
Closed-cone Coniferous Forest Communities		Closed-cone Coniferous Forest (83000)	
Coastal Closed Cone Conifer Forests		Coastal Closed-cone Coniferous Forest	Closed-Cone-Pine-Cypress

Table A-1 (continued).

CALIFORNIA VEGETATION	***SERIES LEVEL DESCRIPTIONS OF CALIFORNIA VEGETATION,***	***TERRESTRIAL NATURAL COMMUNITIES OF CALIFORNIA***	***WILDLIFE HABITATS OF CALIFORNIA***
Coastal Closed-cone Cypress Forests	Gowen cypress McNab cypress Monterey cypress stands Santa Cruz cypress	Monterey Cypress Forest (83150*)	Closed-Cone-Pine-Cypress
Coastal Closed Cone Pine Forests	Beach pine Bishop pine Monterey pine Torrey pine	Beach Pine Forest (83110*) Bishop Pine Forest (83120) Monterey Pine Forest (83130*) Torrey Pine Forest (83140*)	Closed-Cone-Pine-Cypress
Pygmy Forests	Pygmy cypress	Pygmy Cypress Forest (83160*)	Closed-Cone-Pine-Cypress
Interior Closed Cone Conifer Forest Communities		Closed-cone Coniferous Forest (83000)	Closed-Cone-Pine-Cypress
Knobcone Pine Forests	Knobcone pine	Knobcone Pine Forest (83210*)	Closed-Cone-Pine-Cypress
Interior Closed-cone Cypress Forests	Arizona cypress Baker cypress Piute cypress Sargent cypress Tecate cypress	Northern Interior Cypress Forest (83220*) Southern Interior Cypress Forest (83330*)	Closed-Cone-Pine-Cypress
Coastal Coniferous Forest Communities		North Coast Coniferous Forest (82000)	Redwood
North Coast Coniferous Forests	Grand fir Port Orford-cedar Sitka spruce Western hemlock	Sitka Spruce--Grand Fir Forest (82100) Western Hemlock Forest (82200) Port-Orford-cedar Forest (82500*)	

Coastal Redwood Forests	Redwood	Redwood Forest (82300) North Coast Alluvial Redwood Forest (61120) Upland Redwood Forest (82320)	Redwood
Mixed Evergreen Forest Communities		Broadleaved Upland Forest (81000) [in part]	
Northern Mixed Evergreen Forests	Douglas-fir-tanoak Canyon live oak California bay Tanoak Douglas-fir (in part) Mixed oak (in part)	Mixed Evergreen Forest (81100) California Bay Forest (81200) Canyon Live Oak Forest (81320) [in part] Tanoak Forest (81400) Mixed North Slope Forest (81500) Douglas-fir Forest (82400) Interior Live Oak Forest (81330) [in part]	Montane Hardwood Conifer Montane Hardwood Douglas Fir
Central and Southern Mixed Evergreen Forests	Bigcone Douglas-fir-canyon live oak (in part) Coulter pine-canyon live oak (in part) Mixed oak (in part) California bay Santa Lucia fir Tanoak Santa Lucia fir (in part)	Coulter Pine Forest (84140) [in part] Bigcone Spruce--Canyon Oak Forest (84150 Interior Live Oak Forest (81330) [in part] Santa Lucia Fir Forest (84120) Mixed North Slope Cismontane Woodland (71420) Canyon Live Oak Forest (81320) [in part] Interior Live Oak Forest (81330) Mixed Evergreen Frost (81100) California Bay Forest (81200) Santa Lucia Fir Forest (84120) Tanoak Forest (81400)	Montane Hardwood Conifer Montane Hardwood

Table A-1 (continued).

CALIFORNIA VEGETATION	*SERIES LEVEL DESCRIPTIONS OF CALIFORNIA VEGETATION*	*TERRESTRIAL NATURAL COMMUNITIES OF CALIFORNIA*	*WILDLIFE HABITATS OF CALIFORNIA*
Sierran Mixed Hardwood Forests	Mixed oak (in part) Canyon live oak Interior live oak (in part) California buckeye Black oak	Black Oak Woodland (71120) [in part] Interior Live Oak Woodland (71150) Mixed North Slope Cismontane Woodland (71420) Canyon Live Oak Forest (81320) [in part] Interior Live Oak Forest (81330) [in part] Mixed Evergreen Forest (81100) California Bay Forest (81200) Black Oak Forest (81340) [in part]	Montane Hardwood Conifer Montane Hardwood
Oak Woodland Communities		Oak Woodland (71000)	
Coastal Live Oak Woodlands	Coast live oak	Coast Live Oak Woodland (71160) Coast Live Oak Forest (81310)	Coastal Oak Woodland
Valley Oak Woodlands	Valley oak	Great Valley Valley Oak Riparian Forest (61430) [in part] Valley Oak Woodland (71130)	Valley Oak Woodland
Foothill Woodlands	Blue oak Interior live oak Foothill pine Mixed oak (in part) California buckeye (in part)	Digger Pine Woodland (71300) Digger Pine--Oak Woodland (71410) Blue Oak Woodland (71140) Interior Live Oak Woodland (71150) Interior Live Oak Forest (81330) [in part] Alvord Oak Woodland (71170) Interior Live Oak Forest (81330) Juniper--Oak Cismontane Woodland (71430) [in part]	Blue Oak-Digger Pine Blue Oak Woodland

Northern Oak Woodlands	Mixed oak Oregon white oak California buckeye	Oregon Oak Woodland (71110) Mixed North Slope Woodland (71420)	Montane Hardwood
Southern Oak Woodlands	California walnut Engelmann oak Hinds walnut (in part) Hollyleaf cherry (in part) Island oak	Englemann Oak Woodland (71180) Walnut Forest (81600*) Walnut Woodland (71200) Interior Live Oak Forest (81330) [in part]	Coastal oak woodland
Island Oak Woodlands	Island Oak Catalina Ironwood (in part) Hollyleaf cherry (in part)	Island Ironwood Forest (81700) Island Cherry Forest (81810) Island Oak Woodland (71190)	
Montane Coniferous Forest Communities			
Montane Mixed Coniferous Forest Communities		Lower Montane Coniferous Forest (84000)	
Coulter Pine Forests	Coulter pine-canyon live oak (in part) Coulter pine [in part]	Coulter Pine Forest (84140) [in part]	
Ponderosa Pine Forests	Ponderosa pine	Coast Range Ponderosa Pine Forest (84130) Westside Ponderosa Pine Forest (84210) Eastside Ponderosa Pine Forest (84220)	Ponderosa pine Eastside pine
Jeffrey Pine Forests	Jeffrey pine	Ultramafic Jeffrey Pine Forest (84170) Southern Ultramafic Jeffrey Pine Forest (84172) Jeffrey Pine Forest (85100) Jeffrey Pine--Fir Forest (85210) [in part] Jeffrey Pine Forest (85100)	Jeffrey pine Eastside pine

Table A-1 (continued).

CALIFORNIA VEGETATION	SERIES LEVEL DESCRIPTIONS OF CALIFORNIA VEGETATION,	TERRESTRIAL NATURAL COMMUNITIES OF CALIFORNIA	WILDLIFE HABITATS OF CALIFORNIA
Mixed Conifer Forests	Enriched stands in the Klamath Mountains Mixed conifer Black oak Alaska yellow-cedar stands (in part) Bigcone Douglas-fir Douglas-fir (in part) Engelmann spruce (in part) Incense-cedar Pacific silver fir stands (in part) Douglas-fir-ponderosa pine Washoe pine Western white pine (in part) Jeffrey pine-ponderosa pine Stands on San Benito Mountain Bigcone Douglas-fir-canyon live oak (in part) Subalpine fir stands (in part)	Ultramafic White Pine Forest (84160) Ultramafic Mixed Coniferous Forest (84180) Sierran Mixed Coniferous Forest (84230) Douglas-fir Forest (82400) Washoe Pine-Fir Forest (85220*) Coast Range Mixed Coniferous Forest (84110) Aspen Forest (81B00) Black Oak Woodland (71120) [in part] Black Oak Forest (81340) [in part] Sierran Mixed Coniferous Forest (84230) Upper Montane Mixed Coniferous Forest (85300) Jeffrey Pine-Fir Forest (85210) [in part] Siskiyou Enriched Coniferous Forest (85410*) Salmon-Scott Enriched Coniferous Forest (85420*)	Sierran Mixed Conifer Klamath Mixed Conifer Eastside Pine Montane Hardwood Aspen
Giant Sequoia Forests	Giant Sequoia	Big Tree Forest (84250)	Sierran mixed conifer forest
White Fir Forests	White fir	Sierran White Fir Forest (84240) Southern California White Fir Forest (84320) Desert Mountain White Fir Forest (85330)	White fir

Red Fir Forest Communities	Red fir	Red Fir Forest (85310)	Red fir
Lodgepole Pine Forest Communities	Lodgepole pine	Lodgepole Pine Forest (86100) Whitebark Pine--Lodgepole Pine Forest (86220) [in part]	Lodgepole pine
Subalpine Forest Communities	Mixed subalpine forest Mountain hemlock Subalpine fir stands (in part) Mountain juniper (in part) Foxtail pine Whitebark pine Western white pine (in part) Limber pine Bristlecone pine Low sagebrush (in part) Rothrock sagebrush (in part) The subalpine upland shrub habitat	Sierran Mixed Subalpine Coniferous Forest (86200) Foxtail Pine Forest (86300*) Bristlecone Pine Forest (86400*) Southern California Subalpine Forest (86500) Whitebark Pine Forest (86600) Limber Pine Forest (86700) Ultramafic White Pine Forest (84160) [in part] Whitebark Pine--Lodgepole Pine Forest (86220) [in part]	Sub-alpine conifer Low sagebrush
Alpine Communities			
Alpine Meadows	The alpine habitat (in part) Tufted hairgrass Sedge (in part) Shorthair sedge (in part) Shorthair (in part) The subalpine meadow habitat (in part) The subalpine wetland shrub habitat (in part)	Subalpine and Alpine Meadow (45200*) [in part] Alpine Snowbank Margin (91300)	

Table A-1 (continued).

CALIFORNIA VEGETATION	*SERIES LEVEL DESCRIPTIONS OF CALIFORNIA VEGETATION,*	*TERRESTRIAL NATURAL COMMUNITIES OF CALIFORNIA*	*WILDLIFE HABITATS OF CALIFORNIA*
Rocky Alpine Communities	The alpine habitat (in part)	Alpine Dwarf Scrub (94000) [in part] Alpine Fell-field (91100) [in part] Alpine Talus and Scree Slope (91200)	Alpine Dwarf-shrub
Desert Alpine Communities	The alpine habitat (in part)	Alpine Fell-field (91100) [in part] Subalpine Sagebrush Scrub (35220) [in part]	Low sagebrush
Desert Woodland Communities			
Piñon Pine and Juniper Woodland Communities		Piñon-Juniper Woodlands	
Piñon-Juniper Woodlands	Singleleaf piñon (in part) California juniper Singleleaf piñon-Utah juniper Utah juniper Mountain juniper (in part)	Great Basin Piñon-Juniper Woodland (72121) Great Basin Piñon Woodland (72122) Mojavean Piñon Woodland (72210) [in part] Peninsular Piñon Woodland (72310) Juniper-Oak Cismontane Woodland (71430) [in part] Mojavean Juniper Woodland and Scrub (72220) Cismontane Juniper Woodland and Scrub (72400) [in part]	Piñon Juniper
Sierran Piñon Pine Woodlands	Singleleaf piñon (in part)	Mojavean Piñon Woodland (72210) [in part]	
Northern Juniper Woodlands	Western juniper	Northern Juniper Woodland (72110) Great Basin Juniper Woodland and Scrub (72123)	Juniper

Mountain Juniper Woodlands	Mountain juniper	Great Basin Juniper Woodland and Scrub (72123)	Juniper
Southern Piñon and Juniper Woodlands	Singleleaf piñon (in part) Parry piñon Twoleaf piñon stands	Peninsular Juniper Woodland and Scrub (72320)	
Joshua Tree Woodland Communities	Joshua tree	Joshua Tree Woodland (73000)	Joshua tree
Desert Scrub Communities			
Great Basin Sagebrush Scrub Communities	Big sagebrush Black sagebrush Low sagebrush (in part) Parry rabbitbrush Rothrock sagebrush (in part) Rubber rabbitbrush	Great Basin Mixed Scrub (35100) Sagebrush Scrub (35200) Sagebrush Steppe (35300) [in part] Rabbitbrush Scrub (35400)	Sagebrush Bitterbrush
Saltbush Scrub Communities	Fourwing saltbush Allscale Desert-holly (in part) Hopsage Shadscale series Spinescale Winter fat Mixed saltbush	Desert Saltbush Scrub (36110) Shadscale Scrub (36140) Alkali Playa (46000) Valley Saltbush Scrub (36210*) [in part] Foothill Chenopod Scrub (36300) [in part]	Alkali Desert Scrub
Blackbush Scrub Communities	Black bush	Blackbush Scrub (34300)	Desert Scrub
Creosote Bush Scrub Communities	Brittlebush (in part) Creosote bush Brittlebush-white bursage Creosote bush-white bursage Elephant tree Ocotillo series	Sonoran Creosote Bush Scrub (33100) Sonoran Desert Mixed Scrub (24200) Mojave Creosote Bush Scrub (34100) Mojave Mixed Scrub and Steppe (34200) [in part] Sonoran Thorn Woodland (75000)	Desert succulent shrub Desert scrub

Table A-1 (continued).

CALIFORNIA VEGETATION	*SERIES LEVEL DESCRIPTIONS OF CALIFORNIA VEGETATION,*	*TERRESTRIAL NATURAL COMMUNITIES OF CALIFORNIA*	*WILDLIFE HABITATS OF CALIFORNIA*
Creosote Bush Scrub Communities (continued)	Teddy-bear cholla White bursage All-thorn stands Catclaw acacia (in part) Crucifixion-thorn Nolina Foothill palo verde-saguaro Mojave yucca Desert-holly Big galleta (in part)		
Desert Sand Dune Communities	Desert sand-verbena Mesquite (in part) Creosote bush (in part)	Desert Dunes (22000)	
Desert Dry Wash Communities	Scalebroom (in part) Mesquite Blue palo verde-ironwood-smoke tree Tamarisk (in part) Arrow weed (in part) Catclaw acacia (in part)	Desert Dry Wash Woodland (62200) Modoc-Great Basin Riparian Scrub (63600) Mojave Desert Wash Scrub (63700) [in part]	Desert wash
Alkali Sink Communities	Greasewood Iodine bush Bush seepweed Alkali sacaton (in part) Ditch-grass Salt grass	Desert Sink Scrub (36120) Desert Greasewood Scrub (36130) Valley Sink Scrub (36210*) [in part] Alkali Meadows and Seeps (45300) Alkali Playa (46000*) Transmontane Alkali Marsh (52320*)	Wet Meadow (in part) Desert Scrub
West Central Valley Desert Scrub Communities	Fourwing saltbush Allscale	Valley Sink Scrub (36210*) [in part]	

	Hopsage Shadscale Spinescale Winterfat Mixed saltbush Bladderpod-California ephedra-narrow leaf goldenbush	Valley Saltbush Scrub (36210*) [in part] Foothill Chenopod Scrub (36300) [in part] Upper Sonoran Subshrub Scrub (39000)	
Riparian Communities		Riparian Forests (61000)	Lacustrine (in part) Riverine (in part)
Valley and Foothill Riparian Communities	Black cottonwood California sycamore Fremont cottonwood Red alder White alder Arroyo willow Black willow (in part) Hooker willow Narrowleaf willow Red willow Sandbar willow Pacific willow Mule fat Buttonbush Mexican elderberry Mixed willow Giant reed (introduced) Scalebroom (in part) Giant reed (in part) Scalebroom (in part) Hinds walnut (in part)	Freshwater Swamp (52600) [in part] North Cost Riparian Forest (61100) Central Coast Riparian Forest (61200) Southern Riparian Forest (61300) Great Valley Riparian Forest (61400) Sycamore Alluvial Woodland (62100) Southern Sycamore-Alder Riparian Woodland (62400) North Coast Riparian Scrub (63100*) Central Coast Riparian Scrub (63200*) Southern Riparian Scrub (63300*) Great Valley Riparian Scrub (63400) Central Coast Live Oak Riparian Forest (61220) Great Valley Valley Oak Riparian Forest (61430) [in part] Red Alder Forest (81A00)	Valley foothill riparian

Table A-1 (continued).

CALIFORNIA VEGETATION	*SERIES LEVEL DESCRIPTIONS OF CALIFORNIA VEGETATION,*	*TERRESTRIAL NATURAL COMMUNITIES OF CALIFORNIA*	*WILDLIFE HABITATS OF CALIFORNIA*
Montane Riparian Communities	Narrowleaf willow White alder Aspen Black cottonwood Mountain heather-bilberry (in part) Mountain alder Sitka alder Mixed willow The Montane wetland shrub habitat Canyon live oak (in part) The Montane wetland shrub habitat (in part)	Montane Riparian Forest (61500) Montane Riparian Scrub (63500) Freshwater Swamp (52600*) [in part] Aspen Forest (81B00)	Aspen Montane Riparian
Desert Riparian Communities	Scalebroom (in part) Fremont cottonwood Fan palm Mesquite Tamarisk Black willow (in part) Narrow leaf willow Mule fat Mixed willow Giant reed Arrow weed (in part)	Modoc--Great Basin Riparian Forest (61600) Mojave Riparian Forest (61700) Colorado Riparian Forest (61800) Desert Dry Wash Woodland (62200) [in part] Modoc Great Basin Riparian Scrub (63600) Mojave Desert Wash Scrub (63700) [in part] Colorado Riparian Scrub (63800) Desert Fan Palm Oasis Woodland (62300)	Desert Riparian Palm Oasis
Freshwater Wetland Communities		Freshwater Marsh (52400)	Lacustrine (in part) Riverine (in part)

Limnetic Plant Communities	Duckweed Mosquito fern Pondweeds with floating leaves (in part) Yellow pond lily (in part)	Freshwater Marsh (52400)	
Freshwater Marsh Communities	Sedge Nebraska sedge Beaked sedge Rocky Mountain sedge Spikerush Bulrush-cattail Bulrush Cattail Bur-reed Pondweeds with floating leaves Pondweeds with submerged leaves Yellow pond lily Quillwort	Freshwater Marsh (52400) Freshwater Seep (45400*) Vernal Marsh (52500) Freshwater Swamp (52600*) [in part]	Freshwater Emergent Wetland
Bog and Fen Communities	Darlingtonia The fen habitat	Bog and Fen (51000)	Freshwater Emergent Wetland
Montane Meadow Communities	Mountain heather-bilberry Shorthair sedge Shorthair The Montane Meadow habitat The Montane wetland shrub habitat The subalpine meadow habitat The subalpine shrub habitat Tufted hairgrass	Montane Meadow (45100) Subalpine and Alpine Meadow (45200*) [in part]	Wet Meadow

Table A-1 (continued).

CALIFORNIA VEGETATION	*SERIES LEVEL DESCRIPTIONS OF CALIFORNIA VEGETATION,*	*TERRESTRIAL NATURAL COMMUNITIES OF CALIFORNIA*	*WILDLIFE HABITATS OF CALIFORNIA*
Vernal Pool Communities	The northern hardpan vernal pool habitat The northern claypan vernal pool habitat The northern basalt flow vernal pool habitat The northern volcanic mudflow vernal pool habitat The Santa Rosa Plateau vernal pool habitat The San Diego mesa vernal pool habitat	Vernal Pool (44000)	Annual grassland
Anthropogenic Communities			
Agrestal Communities	no corresponding vegetation type		**Cropland**
Pastoral Communities	no corresponding vegetation type	no corresponding vegetation type	Pasture
Ruderal Communities	Pampas grass Giant reed (riparian) Broom Iceplant		
Plantations	Eucalyptus	no corresponding vegetation type	Urban Eucalyptus Orchard and Vineyard
The Urban Mix	Eucalyptus Giant reed (in part) Iceplant Pampas grass Broom	no corresponding vegetation type	Urban Eucalyptus

Index